PROCESS ANALYSIS
AND DESIGN FOR
CHEMICAL ENGINEERS

McGraw-Hill Chemical Engineering Series

BUILDING THE LITERATURE OF A PROFESSION

Fifteen prominent chemical engineers first met in New York more than 50 years ago to plan a continuing literature for their rapidly growing profession. From industry came such pioneer practitioners as Leo H. Baekeland, Arthur D. Little, Charles L. Reese, John V. N. Dorr, M. C. Whitaker, and R. S. McBride. From the universities came such eminent educators as William H. Walker, Alfred H. White, D. D. Jackson, J. H. James, Warren K. Lewis, and Harry A. Curtis. H. C. Parmelee, then editor of *Chemical and Metallurgical Engineering,* served as chairman and was joined subsequently by S. D. Kirkpatrick as consulting editor.

After several meetings, this committee submitted its report to the McGraw-Hill Book Company in September 1925. In the report were detailed specifications for a correlated series of more than a dozen texts and reference books which have since become the McGraw-Hill Series in Chemical Engineering and which became the cornerstone of the chemical engineering curriculum.

From this beginning there has evolved a series of texts surpassing by far the scope and longevity envisioned by the founding Editorial Board. The McGraw-Hill Series in Chemical Engineering stands as a unique historical record of the development of chemical engineering education and practice. In the series one finds the milestones of the subject's evolution: industrial chemistry, stoichiometry, unit operations and processes, thermodynamics, kinetics, and transfer operations.

Chemical engineering is a dynamic profession, and its literature continues to evolve. McGraw-Hill and its consulting editors remain committed to a publishing policy that will serve, and indeed lead, the needs of the chemical engineering profession during the years to come.

THE SERIES

Bailey and Ollis: *Biochemical Engineering Fundamentals*
Bennett and Myers: *Momentum, Heat, and Mass Transfer*
Beveridge and Schechter: *Optimization: Theory and Practice*
Carberry: *Chemical and Catalytic Reaction Engineering*
Churchill: *The Interpretation and Use of Rate Data—The Rate Concept*
Clarke and Davidson: *Manual for Process Engineering Calculations*
Coughanowr and Koppel: *Process Systems Analysis and Control*
Danckwerts: *Gas Liquid Reactions*
Gates, Katzer, and Schuit: *Chemistry of Catalytic Processes*
Harriott: *Process Control*
Johnson: *Automatic Process Control*
Johnstone and Thring: *Pilot Plants, Models, and Scale-up Methods in Chemical Engineering*
Katz, Cornell, Kobayashi, Poettmann, Vary, Elenbaas, and Weinaug: *Handbook of Natural Gas Engineering*
King: *Separation Processes*
Knudsen and Katz: *Fluid Dynamics and Heat Transfer*
Lapidus: *Digital Computation for Chemical Engineers*
Luyben: *Process Modeling, Simulation, and Control for Chemical Engineers*
McCabe and Smith, J. C.: *Unit Operations of Chemical Engineering*
Mickley, Sherwood, and Reed: *Applied Mathematics in Chemical Engineering*
Nelson: *Petroleum Refinery Engineering*
Perry and Chilton (Editors): *Chemical Engineers' Handbook*
Peters: *Elementary Chemical Engineering*
Peters and Timmerhaus: *Plant Design and Economics for Chemical Engineers*
Ray: *Advanced Process Control*
Reed and Gubbins: *Applied Statistical Mechanics*
Reid, Prausnitz, and Sherwood: *The Properties of Gases and Liquids*
Resnick: *Process Analysis and Design for Chemical Engineers*
Satterfield: *Heterogeneous Catalysis in Practice*
Sherwood, Pigford, and Wilke: *Mass Transfer*
Slattery: *Momentum, Energy, and Mass Transfer in Continua*
Smith, B. D.: *Design of Equilibrium Stage Processes*
Smith, J. M.: *Chemical Engineering Kinetics*
Smith, J. M., and Van Ness: *Introduction to Chemical Engineering Thermodynamics*
Thompson and Ceckler: *Introduction to Chemical Engineering*
Treybal: *Mass Transfer Operations*
Van Winkle: *Distillation*
Volk: *Applied Statistics for Engineers*
Walas: *Reaction Kinetics for Chemical Engineers*
Wei, Russell, and Swartzlander: *The Structure of the Chemical Processing Industries*
Whitwell and Toner: *Conservation of Mass and Energy*

PROCESS ANALYSIS AND DESIGN FOR CHEMICAL ENGINEERS

William Resnick

Wolfson Professor of Chemical Engineering
Israel Institute of Technology

McGraw-Hill Book Company

New York   St. Louis   San Francisco   Auckland   Bogotá   Hamburg
Johannesburg   London   Madrid   Mexico   Montreal   New Delhi
Panama   Paris   São Paulo   Singapore   Sydney   Tokyo   Toronto

This book was set in Times Roman.
The editors were Julienne V. Brown and Susan Hazlett;
the production supervisor was Leroy A. Young.
The drawings were done by ANCO/Boston.
Fairfield Graphics was printer and binder.

PROCESS ANALYSIS AND DESIGN FOR CHEMICAL ENGINEERS

1234567890 FGFG 89876543210

Library of Congress Cataloging in Publication Data

Resnick, William, date
    Process analysis and design for chemical engineers.

    (McGraw-Hill series in chemical engineering)
    Bibliography: p.
    Includes index.
    1. Chemical processes. I. Title.
TP155.R47   660.2   80-10678
ISBN 0-07-051887-4

To Marcia

CONTENTS

Chapter 7   Economic Analysis and Evaluation

Chapter 8   Forecasting the Future

Chapter 9   Dealing with Uncertainty

PREFACE

Significant changes have taken place in chemical engineering education during the past two decades. The dominant factor in most university chemical engineering departments is the study of engineering sciences, and only limited time is available for the application of this theoretical and fundamental material to the practical problems of chemical processing. In their classroom work students of chemical engineering are usually presented with reasonably well-defined situations and problems. In addition, they usually are expected to deal with only one aspect of a problem at a time,, e.g., the calculation of a distillation column for a given separation, the volume of a reactor required for a reaction of specified kinetics, or the power required to pump a fluid through a defined piping system. The limited time available in the usual curriculum has resulted in a decrease in the study of process design and analysis and of economics and financial evaluations.

As a result of these factors, the young graduate frequently finds it difficult to acclimatize to the industrial environment. In practice, the engineer will deal with poorly defined situations in which the problems to be solved will not be precisely stated, in which the data available will be incomplete and probably inaccurate, in which the decisions made with regard to one factor can reflect through an entire processing system, and in which the economic considerations will have an importance that may not have been sufficiently stressed in the student's classroom exposure.

This book attempts to help the chemical engineering student and the practicing engineer sharpen their skills and apply them to the problem of analyzing a number of the factors that can affect the design, operation, and commercial success of chemical processing systems.

In the first two chapters some of the characteristics of the chemical industry are detailed, and the role of the chemical engineer in this dynamic industry is briefly presented. Some of the factors involved in the generation and screening of alternatives are considered, along with the roles of analysis and synthesis in design.

One of the shocks young chemical engineers receive as they enter upon their industrial careers comes from the realization that for practically any process or compound with which they must contend either there are very few or no physical and thermodynamic data available or there is an abundance of conflicting data. Chapter 3 is therefore devoted to methods of estimating data. These methods can also be used to interpolate and extrapolate with existing data and to serve as a guide in the evaluation of their reliability.

Several chapters are devoted to the analysis of reaction equilibrium, to nonreacting process analysis, to chemical kinetics, and to reactor design and analysis. Although no attempt is made to cover chemical engineering thermodynamics and chemical reactor design exhaustively, the presentation is sufficiently broad to show, with little or no need for recourse to other sources, how these fundamental tools can be and are used in the analysis of processing systems.

Chapters 7 to 9 are devoted to economic evaluation, forecasting and prediction, and dealing with uncertainty. No process analysis can be complete or well founded without a well-based economic evaluation, and to be realistic this evaluation requires that projections into the future be made. The chemical engineer should therefore be aware of some of the principles and procedures involved in forecasting and predicting such factors as markets, prices, costs, and productivity. The effects of inflation and how to account for erosion in the purchasing power of currency are also considered. The practicing chemical engineer is always confronted with uncertainty, and some analytical methods and procedures for quantifying and dealing with uncertainty are presented.

In Chapter 10 several of the more formal and systematic procedures for the synthesis of chemical processes are discussed. Some elements of optimization and chemical process simulation are presented, and a computer-aided process synthesizer is described. In Chapter 11 five case studies are presented, and detailed analyses are made so that the reader can see how the various principles are applied and how they work out in practice. Numerous examples are solved throughout the book, and a large number of problems of varying complexity are included. Many of them are open-ended and can provide stimulating challenges to the reader. For the reader's convenience all the references are concentrated at the end of the book.

The book is suitable for a one- or two-semester course for advanced undergraduate or graduate chemical engineers. The recommended sequence of material is that presented in the book although there are a number of appropriate sequences. Another suitable sequence would be Chapters 1, 7, 8, 9, 2, 10, 3, 4, 5, 6, and 11. Depending upon the particular department and curriculum, Chapters 4, 5, and 6 may be treated very lightly if the instructor feels that the students have had sufficient and suitable exposure to this material in their undergraduate studies. Although in our department the course based on this book is accompanied by a separate plant design course, this book can be used in the traditional senior design course with lectures being given as needed in accordance with the progress of the design project.

It is hoped that this book, which is based on notes used in teaching a course in process analysis and design for chemical engineers for a number of years, will be useful in easing the young chemical engineer's transition from the academic world to the world of industrial chemical engineering practice where design decisions must be made. Practicing chemical engineers should find the integrative presentation of a number of chemical engineering tools and their application to process analysis and process design of value in their day-to-day confrontation with practice.

Thanks and acknowledgments are due to many: To my students, from whom I have learned much, but especially to the class of '79, who not only taught me much but also scrutinized the final manuscript for typographical errors and for errors of fact. To P. V. Danckwerts, J. F. Davidson, and K. Østergaard, for their hospitality: major parts of the manuscript were written when I was a visitor at the Shell Department of Chemical Engineering and the Danmarks Tekniske Højskole Instituttet for Kemiteknik. To the Master and Fellows of Churchill College, for the hospitality of their friendly and stimulating college during a sabbatical leave. To my departmental colleagues, and especially to D. Hasson for our many hours of discussion and for his encouragement. To my learned colleague from the Department of Chemistry, A. Halevi for the quotations from our Sages and from even earlier sources that he suggested for the chapters and that are as appropriate and relevant now as they were then.

The typing has been the work of Norma Jacob with an able assist by Talma Shavit. Some of the earliest parts were typed in Cambridge by Miss Margaret Sansom and in Lyngby by my wife.

The debt I owe to my wife cannot be measured. Suffice it to say that I have relied on her for advice, support, and encouragement through all our years of marriage.

William Resnick

LIST OF SYMBOLS

a = van der Waals constant

A = availability of component; course of action; heat-transfer coefficient

b = van der Waals constant

B = availability

c_p = heat capacity at constant pressure

c_p^0 = ideal-gas constant-pressure heat capacity

c_v = heat capacity at constant volume

C = cost

C_i = concentration of component i

D = annual depreciation; product demand

E = activation energy; economic measure of performance; internal energy

f = fraction of market; fugacity; inflation rate

f_f = failure probability density function

f_r = repair-time probability density function

F = annual fixed cost; molar flow rate

$\tilde{F}$ = function

$\bar{F}$ = annual maintenance cost per unit of investment; base case

g = acceleration of gravity; general economy growth rate

G = Gibbs free energy; mass flow rate per unit cross-sectional area

H = enthalpy

i = interest rate

I = investment

k = ratio of heat capacities; reaction rate constant; thermal conductivity

k_f = market-share rate factor

k_F = price-floor decay-rate factor

k_g = economy growth-rate factor

k_i = capacity of ith plant addition

k_L = learning-rate factor

k_M = price-margin decay-rate factor

k_0 = frequency factor

K = capitalized cost; reaction equilibrium constant

L = length

m = exponent in cost-capacity equation

M = molecular weight

n = number; number of interest periods; number of years; moles

N = number of moles

N_i = rate of change of number of moles of component i

NPW = net present worth

O = objective function

p = probability

P = parachor; pressure; present value; principal

$\bar{P}$ = production capacity

P_i = vapor pressure of component i

$\bar{P}_i$ = partial pressure of component i

P_F = price floor

P_M = price margin

P_v = vapor pressure

Q = heat; production rate

Q_d = design production rate

r = annular thickness

r_i = rate of formation of i per unit reactant volume

r_i' = rate of formation of i per unit reactor volume

r_i'' = rate of formation of i per unit interfacial area

R = gas constant; reflux ratio; region of available technology; reliability; uniform cash flow

$\dot{R}$ = uniform-cash-flow rate

$\mathscr{R}$ = reaction rate as moles per unit time per unit mass of catalyst

s = sensitivity

S = cross-sectional area; entropy; future value; interfacial area; state of nature

t = time

$\bar{t}$ = average residence time

t_r = reaction time

T = absolute temperature; number of trials; tax rate

T_c = temperature of coolant stream

u = velocity

U = unavailability; utility

v = specific volume; unit variable cost; volumetric feed rate

$$V = \text{annual sales volume; variable; volume}$$

V = annual sales volume; variable; volume
W = catalyst weight; work
W_l = lost work
W_{rev} = reversible work
W_u = useful work
x = mole fraction (general or in liquids)
X = fractional conversion; task constraints
$X_i \cup X_j$ = union of sets X_i and X_j
$X_i \subset X_j$ = set X_i is contained in set X_j
$X_i \cap X_j$ = intersection of sets X_i and X_j
y = mole fraction in gases
z = elevation above a datum
Z = compressibility factor
α = pessimism-optimism index; stoichiometric coefficient
γ = fugacity coefficient
λ = mean failure rate
μ = dynamic viscosity; mean repair rate
ρ = density
ρ_b = bulk density
σ = standard deviation; surface tension
ϕ = instantaneous fractional yield
Φ = overall fractional yield

Subscripts

b = boiling
c = critical condition
f = final condition; formation
l = liquid
0 = initial condition
o = overhead product
r = reduced conditions
v = vapor, vaporization

Superscripts

b = bottom product
L = liquid
$\circ$ = standard conditions
V = vapor phase

CHAPTER ONE

THE CHEMICAL INDUSTRY

Behold, as the clay is in the potter's hand, so are
you in my hand.

Jeremiah, 18: 6

The products of the chemical industry, the raw materials it uses, and the markets
it serves cover a range so diverse that the chemical industry defies a straight-
forward description. In this chapter we shall attempt to characterize the chemical industry, to indicate its diversity, and to point out not only its economic
significance but also its importance in our day-to-day lives. The role played by
the chemical engineer in this industry will also be described.

1-1 WHAT IS THE CHEMICAL INDUSTRY?

Because of the diversity of its products, raw materials, and markets, the chemical
industry is one of the most difficult industries in the world to define. Even the
definition of the chemical industry or of its products differs somewhat from
country to country and in some cases even between different groups within the
same country. A descriptive definition is possible, however, by defining
the chemical process industries as those industries in which processes of a chemical nature predominate or which are closely related to such industries.

Production of chemicals as well as the production of certain raw materials
used by the chemical industry are thus considered to be part of the chemical
process industries. This latter category would include, for example, the production and purification of refinery gases and preparation of naphtha feedstocks for

the petrochemical industry. Products that reach the ultimate consumer, such as plastics, pharmaceuticals, glass, rubber, and paper, are also products of the chemical process industry.

The Pervasiveness of Products of the Chemical Industry

The products of the chemical process industries are requisites for all sectors of the economy—agricultural, industrial, and consumer. A few examples of these products can be mentioned. Fertilizers, insecticides, fungicides, feed additives, and veterinarian drugs are necessary in modern agriculture. In the industrial sector, products of the chemical process industries are essential: the pulp and paper industries rely heavily on bleaches and adhesives; alkalies, detergents, bleaches, dyes, and resins are vital to the textile industry; paints, wood preservatives, resins, and adhesives are indispensable to the wood-products industries; leaching agents, flotation agents, and pickling acids are used in metallurgical industries. All these industries are parts of any industrialized economy. Ultimately, the consumer uses chemical products either directly for consumption or indirectly in the form of consumer durable goods. Some consumable items include cosmetics and pharmaceuticals, soaps and detergents, and photographic chemicals.

All of us who live in an industrialized economy are surrounded by products of the chemical industry. Our very standard of living would be impossible without the contributions of this industry. Thus it is that the chemical industry and its products are fundamental in all aspects of life: health, food, clothing, shelter, transportation, communication, recreation, and defense.

Classifications for Statistical Purposes

Government agencies which assemble and classify industrial data for statistical purposes generally use the classifications established in the Standard Industrial Classification (SIC) Manual published by the U.S. Office of Management and Budget. The chemical and allied products industry is major group 28 in the SIC Manual and includes establishments producing basic chemicals as well as establishments manufacturing products by predominantly chemical processes. Establishments classified in this group manufacture three general classes of products: basic chemicals, e.g., acids, alkalies, salts and organic chemicals; chemical products to be used in further manufacture, e.g., synthetic fibers, plastic materials, and pigments; and, finally, chemical products to be used for ultimate consumption, e.g., drugs, cosmetics, and soaps, or to be used as materials or supplies in other industries, e.g., paints, fertilizers, and explosives.

Petroleum refining and related industries (SIC group 29) are also included in the process industries and represent major employers of chemical engineers. Other industries, parts of which are commonly considered to be among the chemical process industries, include, for example, the wood, pulp, paper and board (SIC group 26); rubber products (SIC group 30); and stone, clay, glass and ceramics (SIC group 32) industries.

Other codes and classifications in addition to SIC are in common use. They include enterprise-classification, tariff-classification, and commodity-classification codes. The classification picture becomes even more complicated at the international level because codes and classifications differ from country to country. Analysis and comparison of data from different sources, for different periods, and for different purposes require care and discrimination to allow for duplications, omissions, and differences of scope and definition.

1-2 THE CHEMICAL INDUSTRY AS A GROWTH INDUSTRY

One of the most dynamic sectors of the industrialized economy is the chemical industry. The production and consumption of chemicals has been growing at a considerably faster pace than that of the overall economy or of the manufacturing sector. In addition, the rate of investment in the chemical industry is high and among the highest in the manufacturing sector. In industrialized or industrializing countries investment in this industry represents from 10 to 15 percent of all investments made in the manufacturing industries.

Growth Elasticity

The growth rate of the chemical industries and its comparison with overall economic development can be expressed quantitatively by its growth elasticity. The *growth elasticity* of chemical production can be defined as the ratio of the growth rate of chemical production to the growth rate of the gross national product. It therefore expresses the dynamics of chemical development. The growth elasticity of the chemical industry is usually greater than 2; that is, the chemical industry grows at a rate more than twice the growth rate of the gross national product. Chemical-industry growth elasticities are remarkably similar in Western and Eastern Europe, the United States, and Japan, a fact that illustrates the important role of the chemical industries in overall economic and industrial development.

Reasons for the dynamics of the chemical industries are not difficult to find. One of the most important reasons is the continual replacement of conventional materials, often of agricultural origin, by materials of synthetic origin, such as synthetic fibers, plastics, and synthetic rubber. Another strong factor is the high demand for fertilizers and pesticides, required to raise and maintain high agricultural productivity. Rising standards of health care have resulted in increased demand for pharmaceuticals. These factors are also fed by a spirit of innovation, characteristic of this industry, that has resulted in a constant stream of new or improved products.

The Chemical Process Industries in the National Economy

Some relevant statistics of employment, capital spending, sales, and exports are presented in Table 1-1 for the chemical process industries and all manufacturing industries. The importance of the chemical process industries in the national

Table 1-1 Chemical process industry compared with all manufacturing, 1977

SIC code	Industry	Employment, thousands	Capital spending, billions	Sales, billions	Exports,† billions
	All manufacturing	19,554	$60.16	$1335.7	$120.1
26	Paper and allied products	698	3.36	41.9	
28	Chemicals and allied products	1,057	6.83	113.4	10.83
29	Petroleum and coal products	209	13.87	161.7	
30	Rubber and plastic products not elsewhere classified	675	1.45	33.7	

† Standard international tariff classification.

Source: *Chem. Eng. News,* Facts and Figures issue, June 12, 1978.

economy becomes apparent from an examination of the figures. During 1977, the year considered in Table 1-1, the chemicals and allied products industry accounted for more than 5 percent of the employment, 11.3 percent of the investment, and 8.5 percent of the sales of all manufacturing industries. The SIC groups 26, 28, 29, and 30 together accounted for 13.5 percent of the employment, 42.4 of the investment, and 26.3 percent of the sales of the manufacturing sector of the economy.

The chemical industries are highly capital-intensive. The investment per employee in petroleum companies is more than $350,000, and in chemical companies the investment is about $85,000, compared with the industrial average of about $20,000 per employee in manufacturing companies. The trend is toward even higher capital intensity. The industry also continues to invest at a high rate. The annual investment per employee in the chemical industry rose in the United States from $3330 in 1970 to $6500 in 1977.

Chemical-Industry Linkages

Products of the chemical industry must have a linkage with the ultimate consumer, and the consumer does, of course, use some of them directly. An overwhelming fraction of chemical-industry production, however, goes to other industries, including the chemical industry itself. The chemicals and selected chemical products industry provides material for industries such as pulp and paper, wood products, textiles, and metallurgy, some of which were mentioned earlier. In view of the high proportion of interindustry demand to total demand of the chemical industry, it is apparent that chemical demand depends on the productive structure of the economy as a whole as well as on the per capita level of income and consumption. In addition, because the products of the chemical industry are so pervasive throughout the industrial economy, its interindustry sales are not sharply concentrated. The chemical industry is therefore characterized by moderately strong forward linkages to many industries rather than by overwhelmingly strong ones to a few industries.

These linkage characteristics are illustrated in Table 1-2, in which the sales pattern of the chemical and related industries to other industries, which bought $1 million or more annually, is shown. Chemicals and related products were purchased by 78 other industries, and the largest buyer industry purchased only 19 percent of the total sales. Petroleum refining and related industries sold to 80 other industries, but only 7 percent of their sales went to their largest buyer industry.

Chemical Industry as Its Own Buyer Industry

Although they are not shown in Table 1-2, it is important to note that the chemical process industries are the largest buyers of products of the chemical process industry. This is so because the product of one part of the chemical process industry becomes the raw material for another part. The route to the ultimate consumer can be a long one. This characteristic of the chemical industry is illustrated in Fig. 1-1, which is a partial "tree" of products from

Table 1-2 Sales distribution patterns of the chemical and related industries, 1967

Industry	Number of buyer industries	Percent of total sales sold to largest buyer industry
Crude petroleum and natural gas	7	76
Chemical and fertilizer mineral mining	19	59
Paper and allied products except containers	75	20
Chemicals and selected chemical products	78	19
Plastics and synthetic materials	42	27
Drugs; cleaning and toilet preparations	42	9
Paints and allied products	63	30
Petroleum refining and related industries	80	7
Rubber and miscellaneous plastics products	80	7
Glass and glass products	49	26
Primary iron and steel manufacturing	62	19
Primary nonferrous-metal manufacturing	19	32

Source: Wei et al. (1979), by permission.†

† Dates in parentheses indicate references, listed before the index.

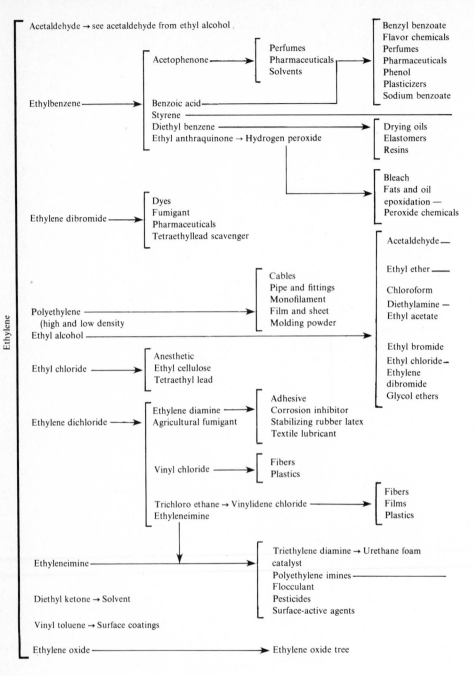

Figure 1-1 Partial tree of products from ethylene.

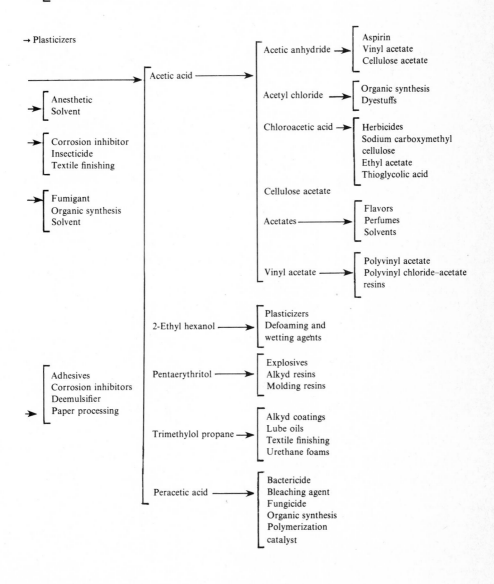

the petrochemical raw material ethylene. Some of the many chemical products from ethylene and the various stages to the ultimate consumer or until a product enters another industry are shown.

1-3 PRODUCT CLASSIFICATION

In general, products of the chemical industries can be roughly classified into three categories:

1. Long-established products, e.g., basic inorganic chemicals, dyestuffs, and paints
2. Products with lives of some 20 years which have recently moved into the bulk-commodity class, e.g., some of the petrochemicals, a number of major plastics, and synthetic fibers
3. A host of newer products that are still in the innovative stage insofar as applications and markets are concerned

Growth elasticities of these three categories are likely to approximate 1, 2, and 3, respectively. In other words, long-established products would grow at approximately the same rate as total industrial production whereas newer products that are still finding new applications and developing their markets would grow at a much faster rate than total industrial production.

This broad classification can be verified with the help of some production indexes, as shown in Table 1-3. The base year for the indexes is 1967, for which a production index of 100 was assigned. Between 1966 and the period covered by this table, although the growth in industrial production averaged almost 2.8 percent annually, the production of chemicals and products grew at an average rate of 5.9 percent annually, representing a growth elasticity relative to industrial production slightly greater than 2. This was slightly larger than the growth rate for basic chemicals, a category that includes alkalies and chlorine, industrial

Table 1-3 Selected production indexes (1967 = 100)

	Year				
Index	1966	1973	1974	1975	1976
Manufacturing	97.9	129.8	129.4	116.3	129.4
Chemicals and products	95.7	154.5	159.4	147.2	169.4
Basic chemicals	99.6	147.7	153.3	135.9	158.5
Basic organic chemicals	100.8	170	177.9	155.3	183.5
Plastics	97.3	243.3	258.5	213.1	283.1
Paints	100.3	114.7	119.2	114.2	124.2

Source: Chem. Eng. News, **55** (23): 1977 (Facts and Figures).

gases, coal tar, basic organic chemicals, and some inorganic chemicals. The basic organic chemicals within that classification grew at an average annual rate of 6.2 percent, a rate more than double that of industrial production. Plastics production, however, expanded at a rate of 11.3 percent annually, representing a growth elasticity greater than 3 with respect to industrial production. This high rate of growth would indicate that the plastics industry is continuing to develop new applications and markets for its products. It should also be stated that production figures for paint, varnish, and lacquer show that their growth rate averaged about 2 percent annually, slightly less than the industrial production growth rate.

Differentiated and Undifferentiated Products

Individual products can be classified as being undifferentiated or differentiated. *Undifferentiated products* are those which have specific chemical formulas and physical specifications regardless of who produces them. Such products as benzene, ammonia, ethylene, and sulfuric acid are undifferentiated. A differentiated product, on the other hand, is or purports to be different from a similar product made by a competitor. Polymers made by different producers, for example, may have different processing characteristics. A manufacturer may claim that his paint has a longer service life, higher gloss, and easier application characteristics than a competitor's product. A drug product may be sold as a differentiated product because the size distribution of its active ingredients, its excipients, etc., result in a superior physiological response.

1-4 ROLE OF RESEARCH AND DEVELOPMENT

A characteristic of the chemical industry is that it is science-oriented and depends on continued research and development efforts to create new industry and to stabilize the existing industry. It is characterized by innovation and improvement geared to the technological, sociological, and economic requirements of society.

Some of the benefits and effects of research and development should be mentioned. New and improved processes for existing products, as well as the development of new and previously unknown products and services, can result from research. Continuing research results in lower costs and lower prices for products and broadens the raw-material base. Materials which formerly were by-products or waste materials can be converted into useful products, and uses can be found for them. Only by research will problems of waste disposal and pollution be solved. And a continuing research and development program is essential to stabilize industrial employment and business.

Efforts aimed at improving existing processes can result in savings through more efficient use of energy, raw materials, and equipment, while successful product research can result in products of higher quality that are more effective and efficient in their applications. Research aimed at new processes has resulted in a multiplicity of manufacturing procedures for almost every product of the chemical process industry. It is as a result of these factors that the chemical industry has been characterized by rapid technological changes.

Expenditure for research and development in the chemical and allied industries averages about 3 percent of annual sales. The risks of research are high, and it is estimated that nine out of ten research programs are abandoned for technological or economic reasons. In addition, large sums of money may be required before profits can be realized from achievements in research and development. Du Pont invested $27 million in the 1930s and 12 years of effort before getting into satisfactory commercial production of what is now a universally accepted synthetic fiber, nylon.

1-5 ROLE OF THE CHEMICAL ENGINEER

The chemical engineer is charged with the responsibility for designing, constructing, and operating chemical processing systems. In this work the approach and attitude of the engineer are and must be different from that of the scientist. Some of the differences can be indicated in a few words. Whereas science is concerned with the observation and classification of facts and the discovery of general laws related to the physical world, in engineering the properties of matter and sources of energy are made useful in structures, machines, and products. Thus, the engineer is, in general, problem-oriented and is motivated by a need rather than by curiosity. Engineers must work within the constraints of existing science, and they must always consider the technological and economic restrictions. Scientists are generally motivated by curiosity and are not necessarily constrained by considerations of immediate practicality or usefulness. Scientists are sometimes fortunate enough to be able to pursue their activities within relatively loose economic limitations.

This distinction does not mean that the chemical engineer is not required to have or use scientific and mathematical skills. Quite the contrary. The tools the engineer must use are many and varied. They include knowledge of physics, chemistry, and thermodynamics, to mention several scientific disciplines, and the chemical engineer must be skilled in the use of mathematics. These skills and knowledge are required to handle and solve the problems that arise in the engineer's work of transforming raw materials and energy into useful finished products.

Process Analysis and Design

Several stages must be passed through to translate the work of the chemist or of the development engineer into a profitable, operating plant. The first and essential stage is that an economic opportunity must be recognized. Then come the stages of design conception, analysis, improvement, and evaluation. Only when the design is completed and implemented and the plant comes on stream does it become possible to exploit the economic opportunity whose recognition sparked off the process analysis and design.

A successful design requires more than the determination of the equipment requirements and the design parameters, frequently with the assistance of simulation and optimization techniques. A successful design implies not only an efficiently operating plant but also a plant that is economically viable. This, in turn, requires economic evaluations, as well as comprehension, appreciation, and evaluation of market conditions.

Thus, the role of the engineer is to analyze and synthesize chemical processing systems on the basis of equipment, thermodynamic, kinetic, economic, and marketing considerations and constraints. In this task the chemical engineer must often work beyond the boundaries of full scientific information and support, and many decisions must be made in the face of uncertainty.

The chemical engineer who designed it can bask in the reflected glory of a successful, safe, and reliable chemical plant that serves a useful need and can take pride in it. The chemical engineer who was forced to make decisions in the face of scientific, economic, equipment, and marketing uncertainties and who was perceptive and fortunate enough to have made the correct decisions will be the successful chemical engineer. In the chapters that follow some of the technical and economic tools necessary for the successful chemical engineer are presented.

PROBLEMS

Information and data for the solution of these problems can be found in sources like the following:

Chemical and Engineering News, Facts and Figures issue, published annually in June
"Chemical Statistics Handbook," Manufacturing Chemists Association, published at 5-year intervals
U.S. Department of Commerce Bureau of the Census Reports
U.S. Tariff Commission reports
Company annual reports to stockholders

1-1 Obtain annual production figures for sulfuric acid over a 10-year period. Calculate the annual percentage change in production and compare with the annual change in the industrial production index. Do your results support the claim that sulfuric acid production can be used as a measure of industrial production?

1-2 Calculate the growth elasticity for noncellulosic fibers such as acrylic, olefinic, and polyester fibers. Compare with values obtained for cellulosic fibers such as acetate and rayon fibers. Discuss.

1-3 Obtain annual production figures for detergents and for soaps over a 10-year period. Discuss.

1-4 Obtain annual production figures for aspirin over a 10-year period. Calculate the annual percentage change and compare with the annual change in the industrial production index. Discuss.

1-5 Obtain information for several chemical companies on:

    (*a*) Annual net sales

    (*b*) Annual net income

    (*c*) Net value of plant and equipment

    (*d*) Annual capital expenditure

    (*e*) Annual net sales per employee

    (*f*) Annual capital expenditure per employee

    (*g*) Net value of plant and equipment per employee

    (*h*) Annual expenditure on research and development (if available)

TWO

THE GENERATION AND SCREENING OF ALTERNATIVES

> Butter and honey shall he eat, that he may know to refuse the evil and choose the good.
>
> *Isaiah, 7: 15*

One of the functions of the chemical engineer is to synthesize processing systems, i.e., systems that transform raw materials, energy, and know-how into a more useful product or products. In all probability this process synthesis will require the solution of a multitude of problems, some apparently trivial in nature and some extremely complex.

To find the optimal solution to problems the engineer must generate as many alternative solutions as possible and then analyze and screen them to eliminate the poorer ones. While generating alternative solutions, thinking should not be restricted, and imagination should be given a loose rein. As a result of technological advances, a solution that may not have been possible in the past may be possible at the present time. Or an alternative solution should not be discarded if, for example, a reasonable research effort could provide the technological advance needed to make it a feasible solution. In this chapter we shall examine some of the factors involved in the generation and screening of alternatives.

2-1 ANALYSIS AND SYNTHESIS

If we examine dictionary definitions of analysis and synthesis, we can conclude that they are opposites. Analysis is a detailed examination of anything complex made in order to understand its nature. It involves separating or breaking up a whole into its fundamental elements or component parts. Synthesis is a composition or combination of parts or elements to form a whole. Analysis implies taking a process which already exists and breaking it apart for study. Synthesis implies creation of the process. As we shall see, analysis and synthesis are practically inseparable. They interplay and interact, supplement and complement, like yang and yin.

Polya (1945) presents a free rendering of a text dealing with analysis and synthesis in the solution of mathematical problems written by Pappus, a Greek mathematician who lived around A.D. 300.†

In analysis, we start from what is required, we take it for granted, and we draw consequences from it, and consequences from the consequences, till we reach a point that we can use as starting point in the synthesis. For in analysis we assume what is required to be done as already done (what is sought as already found, what we have to prove as true). We inquire from what antecedent the desired result could be derived; then we inquire again what could be the antecedent of that antecedent, and so on, until passing from antecedent to antecedent, we come eventually upon something already known or admittedly true. This procedure we call analysis, or solution backwards, or regressive reasoning.

But in synthesis, reversing the process, we start from the point which we reached last of all in the analysis, from the thing already known or admittedly true. We derive from it what preceded it in the analysis, and go on making derivations until, retracing our steps, we finally succeed in arriving at what is required. This procedure we call synthesis, or constructive solution, or progressive reasoning.

If we have a "problem" we are required to find a certain unknown x satisfying a clearly stated condition. We do not know yet whether a thing satisfying such a condition is possible or not; but assuming that there is an x satisfying the condition imposed, we derive from it another unknown y which has to satisfy a related condition; then we link y to still another unknown, and so on, until we come upon a last unknown z which we can find by some method. If there is actually a z satisfying the condition imposed upon it, there will be also an x satisfying the original condition, provided that all our derivations are convertible. We first find z; then, knowing z, we find the unknown that preceded z in the analysis; proceeding in the same way, we retrace our steps, and finally, knowing y, we obtain x, and so we attain our aim. If, however, there is nothing that would satisfy the condition imposed upon z, the problem concerning x has no solution.

Analysis and Synthesis in Action

Polya then presents a nonmathematical illustration:‡

A primitive man wishes to cross a creek; but he cannot do so in the usual way because the water has risen overnight. Thus, the crossing becomes the object of a problem; "crossing the creek" is the x of this primitive problem. The man may recall that he has crossed some other creek by walking along a fallen tree. He looks around for a suitable fallen tree which becomes

† From G. Polya, "How To Solve It: A New Aspect of Mathematical Method," copyright 1945 © 1973 by Princeton University Press, © 1957 by Polya; Princeton Paperback, 1971, pp. 142–143; reprinted by permission of Princeton University Press.

‡ From G. Polya, "How To Solve It: A New Aspect of Mathematical Method," copyright 1945 © 1973 by Princeton University Press, © 1957 by Polya; Princeton Paperback, 1971, pp. 145–146; reprinted by permission of Princeton University Press.

his new unknown, his y. He cannot find any suitable tree but there are plenty of trees standing along the creek; he wishes that one of them would fall. Could he make a tree fall across the creek? There is a great idea and there is a new unknown; by what means could he tilt the tree over the creek?

This train of ideas ought to be called analysis if we accept the terminology of Pappus. If the primitive man succeeds in finishing his analysis he may become the inventor of the bridge and of the axe. What will be the synthesis? Translation of ideas into actions. The finishing act of the synthesis is walking along a tree across the creek.

The same objects fill the analysis and the synthesis; they exercise the mind of the man in the analysis and his muscles in the synthesis; the analysis consists in thoughts, the synthesis in acts. There is another difference; the order is reversed. Walking across the creek is the first desire from which the analysis starts and it is the last act with which the synthesis ends. Analysis comes naturally first, synthesis afterwards; analysis is invention, execution; analysis is devising a plan, synthesis carrying through the plan.

Rudd and Watson (1968) outline the steps in the creation of plausible alternatives:

1. Define the primitive problem
2. Gather the cognate facts
3. Create specific problems
4. Screen specific solutions
5. Engineer a solution

Step 1 is analogous to the Pappus x and step 3 to y and z. Steps 4 and 5 will determine whether we can, indeed, "satisfy the condition imposed upon z."

2-2 SCREENING OF ALTERNATIVES

The screening operation comes into play to eliminate alternative specific solutions until only one alternative survives as the solution to be engineered. This screening operation requires that a number of questions be asked and answered. Some of the major questions are: What will be space, energy, and material requirements? Is the present state of technology sufficient? What capital costs, operating costs, and revenues are involved? Is the process or product compatible with the potential or capabilities of the company? Is there an alternative activity that could be more profitably pursued with the talent or money available in the company? What are the health, safety, legal, and environmental considerations?

In some cases, a very preliminary and almost trivial screening can result in the elimination of an alternative.

Example 2-1 A patent has recently appeared for an ion-exchange resin that is specific for gold. Should our company consider purchasing rights to this patent with a view to using the resin to recover gold profitably from seawater?

Screening The fact that the concentration of gold in seawater is extremely low, of the order of 5×10^{-6} ppm, would mean that large quantities of seawater would have to be pumped per unit of gold product. This, in turn, would suggest that energy requirements and material handling should be looked at in the early screening stages. Let us begin the screening by making an estimate of the energy cost for seawater pumping on the basis of the following set of reasonable assumptions:

1. Electricity costs approximately 5 cents per kilowatthour.
2. Overall pressure drop through the equipment is 10 m of fluid head
3. The specific gravity of seawater is 1.0.

Then

$$\frac{\text{Pumping cost}}{\text{Grams Au}} = \frac{10^6 \text{ g seawater}}{5 \times 10^{-6} \text{ ppm Au}} (10 \text{ m}) \frac{1 \text{ kg}}{1000 \text{ g}}$$

$$\times \frac{\text{kWh}}{3.671 \times 10^5 \text{ kg} \cdot \text{m}} \frac{\$0.05}{\text{kWh}} = \$272 \text{ pumping cost/g Au}$$

Since the price of gold is of the order of \$20 per gram, it is obvious from this simple screening operation that the recovery of gold from seawater would not be a profitable venture. The seawater pumping costs alone would exceed the value of the gold that could be recovered.

In the following example, a study of comparative energy requirements will be used to make the preliminary selection of the correct distillation sequence for a multicomponent separation.

Example 2-2 At one point in the production of chloromethanes, the preliminary product purification has produced a scrubbed, dried mixture comprising 35 mol % methylene chloride (A), 45 mol % chloroform (B), and 20 mol % carbon tetrachloride (C). This mixture is to be fractionated into essentially pure A, B, and C. The operation requires two fractionation columns operating in series, the second column receiving as its feed either the overhead or bottom product of the first column. What distillation sequence would you recommend?

Screening Two separate distillation operations will be necessary to meet the requirements of producing three "essentially pure" products, and two distillation sequences are possible. For our system, A (bp 40.7°C) is more volatile than B (bp 61.2°C), which in turn is more volatile than C (bp 76.8°C). The two possible sequences (Fig. 2-1) are shown in Table 2-1.

A preliminary, semiquantitative comparison of energy requirements for the two competing alternative sequences can be made if we make two reasonable simplifying assumptions:

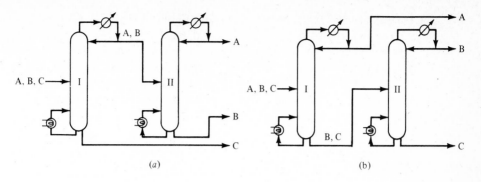

Figure 2-1 Distillation sequence: (a) alternative 1 and (b) alternative 2.

1. The separations are all of approximately equal difficulty, so that the reflux ratio requirements will be equal for both alternatives. The overhead vapor load and reboiler boil-up requirements for each column will therefore be $1 + R$ times the number of moles of overhead product, where R is the reflux ratio.
2. The molar heats of vaporization ΔH_v are equal for the three components A, B, and C.

We shall take as the basis of our calculations 100 mol of the mixture comprising 35 mol of A, 45 mol of B, and 20 mol of C.

Alternative 1 For column I

Moles of bottom product = 20 mol C

Moles of overhead product = 35 mol A + 45 mol B

Overhead vapor to condenser = $(1 + R)(35 + 45) = 80 + 80R$

Approximate heat input to reboiler = $(80 + 80R)\,\Delta H_v$

Table 2-1 Two possible distillation sequences

	Alternative 1		Alternative 2	
	Column I	Column II	Column I	Column II
Feed	...	Overhead product from column I (A + B)	...	Bottom product from column I (B + C)
Product:				
Overhead	A + B	Pure A	Pure A	Pure B
Bottom	Pure C	Pure B	B + C	Pure C

For column II

Moles of bottom product = 45 mol B

Moles of overhead product = 35 mol A

Overhead vapor to condenser = $(1 + R)(35)$

Approximate heat input to reboiler = $(35 + 35R)\,\Delta H_v$

Total reboiler heat requirement = $(115 + 115R)\,\Delta H_v$

Alternative 2 For column I

Moles of bottom product = 45 mol B + 20 mol C

Moles of overhead product = 35 mol A

Overhead vapor to condenser = $(1 + R)(35)$

Approximate heat input to reboiler = $(35 + 35R)\,\Delta H_v$

For column II

Moles of bottom product = 20 mol C

Moles of overhead product = 45 mol B

Overhead vapor to condenser = $(1 + R)(45)$

Approximate heat input to reboiler = $(45 + 45R)\,\Delta H_v$

Total reboiler heat requirement = $(80 + 80R)\,\Delta H_v$

The reboiler energy requirements for alternative 2 are substantially smaller than those for alternative 1. In addition, practically all the energy supplied in the reboiler must be removed in the condenser. Thus, the cooling-water requirements for the condenser will also be correspondingly lower for alternative 2 compared with alternative 1. Alternative 2 therefore provides the preferable sequence. This calculation also illustrates the reason for one of the multicomponent-distillation rules of thumb, or heuristics: remove the components one by one in the distillate.

The screening done in Examples 2-1 and 2-2 was based upon intuitive judgments of what the crucial factors in the screening process would be. In both the examples the question answered, however, was one of the major ones posed earlier: What will the energy requirements be? Wells (1973) proposed that the screening and examination use a standard format to ask what, when, where, how, who, and why. Table 2-2 shows his critical-examination sheet.

Table 2-2 Critical-examination sheet

What is system? (present facts)	Ask†	Proposed alternatives	Selection for development
What is the objective function? (usually economic)	Why?	What else influences system?	What should economic criteria be?
What is achieved? (sentence in minimum detail)	Why? (ask each part of sentence)	What else could be achieved? (eliminate, modify, avoid, other concepts)	What should be achieved? (short-term, long-term, economic criteria)
How achieved? (outline method; sequence and activities)	Why that way?	How else could it be achieved? (materials can be changed, extra or different method, change equipment specification, optimize)	How should it be achieved? (select items best fulfilling objective function)

† When, where, and who may also be asked.
Source: Wells (1973), by permission.

2-3 SEPARATION SEQUENCES

In the simple case examined in Example 2-2 two distillation columns were required to effect the separation of a ternary mixture into three products, and two arrangements, or sequences, of these columns were possible. The selection of the optimum sequence was based on energy requirements, which were much larger for the incorrect sequence than for the correct sequence. The number of possible sequences increases rapidly as the number of components to be separated increases.

King (1971) discusses the general case in which a mixture of R components is to be separated into R products. This operation can be performed in $R - 1$ columns, each column receiving a single feed and producing two products. The first column in the sequence is assumed to take j components overhead, leaving $R - j$ components in the bottom product. There will be S_j sequences by which the j overhead components can be separated into j products in subsequent distillations and S_{R-j} sequences by which the bottoms components can be separated in subsequent distillations. For any given value of j the number of possible sequences will therefore be $S_j S_{R-j}$. Since the number of components

Table 2-3 Number of possible column sequences for separating R-component feed into R products

Number of components R	Number of column sequences S_R
2	1
3	2
4	5
5	14
6	42
7	132
8	429
9	1430
10	4862

taken overhead in the first column can take on any value between 1 and $R - 1$, the number of different column sequences will be

$$S_R = \sum_{j=1}^{R-1} S_j S_{R-j}$$

The values for S_R can now be generated for any number R by taking $S_1 = S_2 = 1$ and solving the equation recursively. The number of possible column sequences is shown in Table 2-3. When more than only a few components are to be separated, it may be impractical to consider all possible sequences in detail to determine the optimum sequence. The number of sequences to be examined can be reduced by eliminating those which are not favored by any rules of thumb.

Four rules of thumb are listed by King (1971) for sequence selection. The development of the rules was based primarily on energy considerations, and particular attention was paid to difficult separations:

1. Direct sequences which remove the components one by one in the distillate are generally favored unless one of the subsequent heuristics applies.
2. Sequences which result in a more equimolal division of the feed between distillate and bottoms products should be favored.
3. Separations where the relative volatility of two adjacent components is close to unity should be performed in the absence of other components; i.e., reserve such a separation until the last column in the sequence.
4. Separations involving high specified recovery fractions should be reserved until last in the sequence.

2-4 ALTERNATIVE PROCESSING ROUTES

For almost any chemical product, alternative manufacturing and processing routes are available. These alternative methods are the result not only of research and development but also of requirements stemming from different raw material availability, by-product considerations, and considerations involving such factors as utilities, energy, and labor requirements, and capital investment. The relative importance of these factors can vary from company to company and from location to location.

As an example of alternative processing routes, five different manufacturing routes from the many processes for producing acetic acid, a relatively simple organic chemical, can be mentioned:

1. *Aerobic fermentation of ethyl alcohol with bacteria*

$$C_2H_5OH + O_2 \rightarrow CH_3COOH + H_2O$$

2. Destructive *distillation of wood* yields pyroligneous liquor containing approximately 7% acetic acid, 4% methanol and acetone, and a remainder of tars and oils. Recovery of the acetic acid from the pyroligneous liquor can be effected by one of several routes, namely, distillation and extraction, distillation and adsorption, or azeotropic distillation.
3. *Oxidation of acetaldehyde*

$$CH_3CHO + \tfrac{1}{2}O_2 \rightarrow CH_3COOH$$

Several alternative routes are available for producing the acetaldehyde.
4. *Carbonylation of methanol*

$$CH_3OH + CO \rightarrow CH_3COOH$$

5. *Direct oxidation of ethanol*

$$C_2H_5OH + O_2 \rightarrow CH_3COOH + H_2O$$

All these processes for the production of acetic acid are carried out commercially, and subalternatives within each process are available. The choice of the process will require that the chemical engineer use the full range of knowledge and expertise available. Some of the screening may be trivial. For example, if wood is not available as a raw material, obviously this factor would rule out wood distillation as a manufacturing route. As the screening progresses, it becomes more involved, more comprehensive, and more detailed. The selection and screening process continues through the evaluation of bids and proposals by potential licensors, equipment venders, contractors, and construction companies.

2-5 ALTERNATIVE PROCESSING TECHNIQUES

Not only are alternative processing routes available for almost any chemical product, but alternative processing techniques are usually available for every operation. A number of physical separation techniques are shown in Fig. 2-2. The bands between the wedges show phase states, and in each wedge between the phases the typical mechanisms used to effect a separation are shown. The physical separation techniques shown are based on the application of some forces or combination of forces that will result in differential motion between dissimilar elements of a heterogeneous system. The differential motion permits a physical separation between phases to take place. None of the separations shown in Fig. 2-2 involves a change of phase or a mass-transfer operation.

A similar figure could be produced for separation techniques other than those based on physical separation in heterogeneous systems. Some of these techniques would include, as examples, distillation, absorption, adsorption,

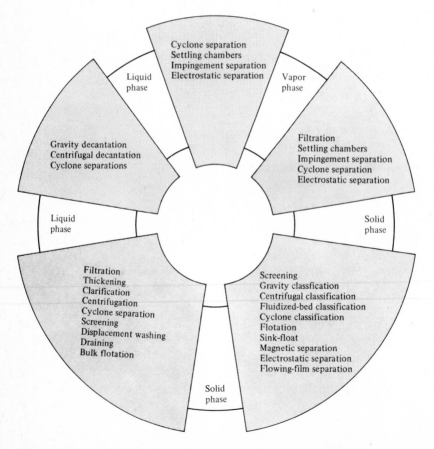

Figure 2-2 Alternative physical-separation techniques. [*From Melnechuk (1963), by permission.*]

crystallization, and extraction. King (1971) summarizes separation operations on the basis of whether the separation is a result of phase equilibria, a rate difference, or a mechanical process. He lists 42 different processes, points out the property that is exploited to effect the separation, and presents a practical example of the use of each process.

The reader is referred to Perry and Chilton (1973) for a number of alternative techniques for almost every processing operation that would be carried out in a chemical plant. Good engineering demands that the engineer select the best available technique. This requires that the alternatives be recognized and then evaluated in terms of the specific problem or duty. This evaluation of alternative techniques will involve an optimum matching of the process requirements on the one hand with the processing technique on the other. It may even be desirable or necessary for a process to be modified to obtain the best and most economical fit between the available alternative techniques and the overall performance requirements of the process.

2-6 ANALYSIS OF ALTERNATIVES

Woods and Davies (1972) developed a system of analysis of process alternatives based on stripping the process down into its basic attributes or functions. The alternative methods for realizing these functions are then considered in isolation, and new and likely combinations of the alternatives are sought. This technique can also result in process synthesis.

As a homey example to demonstrate the principle they use the following objective function. Given a pencil as a sample of a writing instrument, produce an improved writing instrument. After examining the pencil, they decide that the functions of a pencil are marking, metering, storage, and holding. In the pencil, marking is performed by the graphite tip, metering by the grade of graphite, storage by the solid rod of graphite, and holding by the wooden tube.

Each of these functions is now examined, and alternative ways of performing them are generated. The functions and options are then presented in an attribute table, as in Table 2-4. For example, the function of marking can be performed by an inked ball, an inked felt, a split-metal plate, or a hardened tip, not only by a graphite tip. The number of options for each function need not be equal, although they are in Table 2-4, and some of the options in one function may be incompatible with certain options of a different function. A graphite marking option would be incompatible, for example, with viscosity as the metering option.

Now that the attribute table is available, we look for combinations of options that could result in an improved writing instrument. For example, graphite stored in felt could provide an improved writing instrument with unbreakable lead.

Woods and Davies maintain that the problems in using attribute listing arise, in practice, in the sorting of the combination of options and not in the

Table 2-4 Attribute table for a writing instrument

Function			
Marking	Metering	Storage	Holding
Graphite	Hardness	Rod	Tube
Ball	Tolerance	Cylinder	Bottle
Felt	Viscosity	Cartridge	Cylinder
Nib	Pore size	Reservoir	Rod
Hard tip	Capillary	Tube	Case

Source: Woods and Davies (1972), by permission.

generation of options. It would seem likely that an unimaginative engineer faced with the problem of generating options for performing defined functions would in all probability produce a series of conventional options that might result in a new combination, i.e., a process synthesis. It is in the generation of options, however, that the imaginative engineer and scientist should be given free rein. It is doubtful whether the ball-point pen, for example, would have been developed as a result of the use of attribute listing. Not unless, that is, an imaginative engineer had noticed a wet ball leaving a trail as it rolled down a dry floor, remembered this, and thought of it as an option for marking that the engineer would be willing to propose without fear of ridicule.

In a further, more detailed example, Woods and Davies (1972) study a process in which a frozen-food product is to be produced from thermosetting fluid feed. The processing path involves pumping the fluid into a container, where it is allowed to set. The set product is removed from the container, cut into pieces, and then either packed, stored, and frozen or kept apart, boxed, stored, and finally frozen. Each of the functions had several options: the liquid hold could be in trough belt, partial mold, complete mold, bucket chain, or disposable mold; freezing could be in plate, blast, or bed freezer, etc. The functions and options are shown in Table 2-5, along with ratings against cost and feasibility with a rating of 1 to 10, where 1 is the best score and 10 is the worst score. A rating of 0 is for a "not necessary" option. Additional or different criteria can be used, such as reliability or complexity.

It should be pointed out that some of the options are incompatible. A trough belt, option I-11, is self-emptying and is therefore incompatible with options II-21 and II-22. If a trough belt is used, the thermoset product must be cut into pieces; therefore I-11 is also incompatible with III-31.

The process pathways were then examined, sorted, and ranked in accordance with the feasibility criteria and the cost criteria. The details of the two "best" solutions are shown in Table 2-6. These two top solution pathways

Table 2-5 Functions, options, and criteria for frozen-food process

No.	Function	Option	Criteria	
			Cost	Feasibility
I	Liquid hold	11 Trough belt	3	2
		12 Partial mold	5	1
		13 Complete mold	9	6
		14 Disposable mold	6	5
		15 Bucket chain	4	2
II	Remove	21 Air	1	6
		22 Mechanical	1	6
		23 Not necessary	0	0
III	Cut	31 Not necessary	0	0
		32 Gang knife	3	1
		33 Reciprocate	5	1
		34 "Scrapless"	8	2
		35 Wire	1	4
IV	Pack	41 Box	1	1
		42 Hopper	2	1
V	Keep apart	51 Not necessary	0	0
		52 Dust	5	2
VI	Store	61 Buckets	3	1
		62 Conveyors	6	2
		63 Tins	2	1
VII	Box	71 Yes	10	10
		72 No	10	10
VIII	Freeze	81 Plate	9	1
		82 Blast	6	1
		83 Bed	6	2

Source: Woods and Davies (1972), by permission.

would now have to be merged so that a suitable trade-off between feasibility and cost could be obtained.

There is an additional option which can also be pursued. Several of the pathways that are not shown had good cost criteria but poor feasibility criteria. The engineer should explore the possibilities of improving the feasibility of these low-cost pathways by, for example, equipment or process modifications.

This technique, like all the methods for screening and selection between alternatives, forces the engineer to think in detail about the process, its elements, its functions, and its goals. Not only does this help the engineer order the decision-making process and make better decisions, but it can also result in process synthesis.

Table 2-6 Details of best frozen-food-product process pathways

No.	Function	Option pathway	
		A (Cost 23, feasibility 19)	B (Feasibility 17, cost 25)
I	Liquid hold	11 Trough belt	11 Trough belt
II	Remove	23 Not necessary	23 Not necessary
III	Cut	35 Wire cutter	32 Gang knife
IV	Pack	41 Box	41 Box
V	Keep apart	51 Not necessary	51 Not necessary
VI	Store	63 Tins	63 Tins
VII	Box	72 No	72 No
VIII	Freeze	82 Blast	83 Bed

Source: Woods and Davies (1972), by permission.

2-7 SYNTHESIS OF A PROCESS

It is not uncommon for the engineer to be presented with a rather ill-defined problem. In this section we shall present such a problem and sketch some of the considerations that could lead to its solution.

The problem we shall discuss is that of synthesizing a better, cheaper process for drying foods. This problem may appear to young chemical engineers to be a bit remote from the type of problem they think they are prepared and expected to cope with. This choice was intentional because the training and background of chemical engineers prepare them to deal with problems that go beyond the design of purely chemical processes and typical chemical-equipment items such as chemical reactors, heat exchangers, and distillation columns.

We shall now sketch the synthesis of a possible process to solve this problem. This will be presented in the form of several imaginary conferences between engineer I, a group leader, and engineer II, a recent recruit to the group.

ENGINEER I: Now that you've had several days to consider your problem, you'll have learned that dried food products have a number of attractive features. They are easy to ship, provided they are properly packed. They can be stored almost anywhere and have a long shelf life, even without refrigeration. Also, because of their low moisture content, most agents that are active in causing flavor deterioration or spoilage are either inactive or greatly inhibited.

ENGINEER II: Yes, all that is true, but if the drying is carried out at elevated temperatures, flavor, aroma, and nutritional value degrade quite rapidly.

ENGINEER I: The degradation problem can be overcome to a great extent if the food product is first frozen and then dried under vacuum at a low temperature. Of course, drying under vacuum and at low temperature is a rather expensive operation.

ENGINEER II: From several articles I have read on freeze drying I learned that some cellular destruction takes place during freezing. This means that the texture of the reconstituted food is not the same as that of the original. Also, not all foods are amenable to this drying procedure. That is why freeze-drying has been applied commercially to only a limited number of relatively expensive foods and for restricted uses where cost is not a prohibitive factor.

ENGINEER I: I see that you have managed to learn quite a bit about dried foods and food drying. Even though you thought originally when I presented this problem to you that it was in a subject remote from your training, I believe you will now agree with me that you are prepared to deal with it. In fact, your training as a chemical engineer has equipped you with a unique kit of tools, and there is a good probability you'll be able to come up with a possible solution to our problem.

ENGINEER II: I must admit that I am much more confident now than I was several days ago that I shall be able to cope with the problem. From my library research during the past few days I have learned that food drying is not new. It has been practised in one form or another for thousands of years. I think we can reasonably assume that all "conventional" methods for drying solids have been considered in the past.

ENGINEER I: I would be inclined to agree with you. You probably shouldn't look at this problem as being that of drying a food product but to remove water or, alternatively, to separate one material (water) from another material (water-free food). What techniques can you think of that are not conventional for removing water or for separating two materials?

ENGINEER II: My thinking also was along those same general lines, and I suggest we consider the possibility of extracting water from the food product with a dry, volatile solvent. Our problem then would be to remove solvent from the food. This might be a relatively easy problem to solve. Of course, the solvent itself would also have to be dried, probably by distillation, to permit its recycle and reuse.

ENGINEER I: Your proposal sounds good. I suggest that you spend some time on it and you can also do some preliminary laboratory work. Just one more thing. You realized immediately that our original problem, when I presented it to you, was framed in a broad and rather primitive manner—a better, cheaper process. The "better process" could mean a process that will not have the drawbacks of existing processes and that will produce a product that would be acceptable to the consumer and preferable to the products available from present drying techniques. Or it could mean a process that would permit us to dry foods that are not amenable to drying at present. Your proposal may lead to a better process. At this stage of our work I suggest that we don't concern ourselves too much with the "cheaper" part. But if your proposal looks promising after some preliminary research on it, you must not forget that the "cheaper" (when we do get around to it, and this shouldn't be delayed too long) must include all the cost elements involved in the processing operation, including factors such as operating cost, investment, and labor.

Several days later the new recruit presents a verbal progress report to Engineer I.

ENGINEER II: The laboratory work looks promising. I considered a number of possible solvents and selected ethyl acetate for several reasons. It is partially miscible with water, and water can be removed from it by distillation. But there is another important consideration. It will be almost impossible to remove all the solvent from the dried product, and traces are certain to remain. Therefore, whatever solvent is used must be safe as a food additive, and ethyl acetate has been cleared as a food additive if it is present at low enough concentration levels.

ENGINEER I: What processing scheme have you been looking at?

ENGINEER II: A conventional countercurrent solid-liquid contacting operation. The ethyl acetate extracts the water from the food product, which is then dried in a lab vacuum drying oven. The dried product we obtained was easily reconstituted when water was added to it, and the ethyl acetate was easily dehydrated by distillation. Incidentally ethyl acetate and water form a minimum-boiling azeotrope.

ENGINEER I: Doesn't the azeotrope overhead separate into two liquid phases upon condensation? Ethyl acetate and water are partially miscible, and your overhead condensate should produce a water-rich layer and an acetate-rich layer. That means you might be able to combine the extraction step with the solvent-drying step by carrying out the extraction in a vessel that could simultaneously serve as the still boiler. The vapor leaving the still would be at the azeotropic composition, and after condensation and separation the solvent-rich layer, which really acts as a water carrier, could be returned to the boiler. The water-rich layer would have to be distilled to recover the solvent present in it.

ENGINEER II: Your scheme would probably simplify the operation. I will look into it and report back.

Some time later.

ENGINEER II: Your proposal has been incorporated into the processing scheme and we now have a complete flowsheet for the operation. We haven't studied all the elements of the process in the laboratory, only the elements unique to our problem. We have dried raw hamburger, whole strawberries, sliced bananas and potatoes, and even a chunk of 2-in steak. We ran the dehydrator at ambient temperature and a medium vacuum of 0.132 atm. The vacuum dryer, which eliminates solvent from the dried product, was operated at 38°C and 0.000123 atm.

ENGINEER I: What about the dried product; how does it behave?

ENGINEER II: It is easily reconstituted and we didn't get any cellular destruction. The product looks as if it were freeze-dried. Incidentally, after the vacuum-drying operation the solvent is present at levels permitted by the Food and Drug Administration for ethyl acetate as a food additive. We also did a preliminary cost study, and I think our extraction scheme is more attractive than freeze-drying.

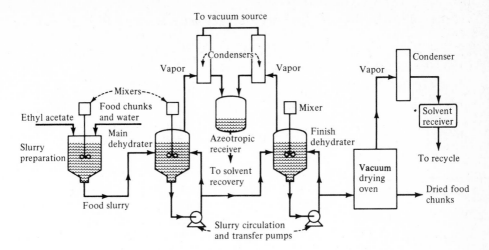

Figure 2-3 Azeotropic dehydration process. [*From Chem. Eng., 75(3):60 (1968).*]

ENGINEER I: You've come a long way since you tackled this problem, but there still is a long way to go. Of course, I am now going to want a more comprehensive cost and economic analysis and a comparison with other food-drying processes. Then there is a long list of questions that will have to be answered. I shall just mention a few. What effect does the solvent extraction have on product flavor? Has the solvent eliminated aroma from the reconstituted product? What products can be dried by the extraction scheme? You will be able to add to this list yourself, and more questions and problems are going to arise as you continue your development work on this process and later on if the company decides to commercialize this project.

This series of meetings was, of course, purely imaginary, but some sequence of discussion, study, and evaluation of alternatives did lead to a solvent-extraction food-drying process as shown in Fig. 2-3.† The process developers claim that the room-temperature dehydration process provides faster drying than freeze-drying and extends the range of foodstuffs amenable to drying. The interested reader is referred to King (1974), who presents a thorough analysis and discussion of alternative processes for food dehydration.

2-8 CONCLUDING REMARKS

The generation of alternative solutions to problems and their evaluation represent some of the most important functions of the chemical engineer. In this chapter we have looked at several facets of this work and gone through examples

† Described in *Chem. Eng.*, **75**(3): 60 (1968).

of the screening of alternatives and on to the synthesis of a plausible and unique solution to a processing problem. We shall return to the creation and examination of alternatives and to process synthesis in Chap. 10 after we have examined and become familiar with a number of tools and techniques of process analysis. Our approach to process synthesis will then be more formal and systematic compared with the preliminary approach taken in this chapter.

The objective of the chemical engineer is, of course, to arrive at the most satisfactory and economic solution to whatever problem is presented. By the very nature of their work, engineers are forced to get to a solution in the face of uncertainty and in the absence of complete information. Perhaps this is one of the reasons why chemical engineering can be such an exciting profession and why the process engineer is rarely bored.

PROBLEMS

2-1 A special-purpose low-molecular-weight polymer is produced in the presence of a SiO_2-supported catalyst. The physical properties of the catalyst resemble those of sand. The raw polymer product contains 1% catalyst, which must be removed before further processing. The polymer melts in the range of 70 to 90°C and is soluble in paraffinic and aromatic solvents. Suggest several schemes for removing the catalyst.

2-2 Sulfur is to be transported to a chemical plant. Propose and discuss several plausible methods for performing this task if (a) several hundred tons per year is to be transported (1) 250 m and (2) 50 mi; (b) 1000 tons/day is to be transported 50 mi.†

2-3 Propose several plausible methods for performing the following tasks:
  (a) A pilot-plant installation is to be purged with inert gas.
  (b) Several hundred cubic feet per hour of inert gas are to be provided for blanketing storage tanks containing flammable liquids.

2-4 List and discuss several methods for effecting separations in (a) gas-solid, (b) solid-solid, and (c) liquid-solid systems. Place special emphasis on the physical or chemical basis for the separation.

2-5 Describe and briefly discuss at least three different routes for the production of acetic anhydride. Write the equations for the chemical reactions involved, calculate the raw material costs for each route, and compare with published prices for acetic anhydride.

2-6 Describe and briefly discuss at least three different routes for the production of aniline. Write the equations for the chemical reactions involved, calculate the raw-material costs for each route, and compare with the published prices for aniline.

2-7 Describe and briefly discuss at least four different routes for the production of phenol. Write the equations for the chemical reactions involved, calculate the raw-material costs for each route, and compare with the published prices for phenol.

2-8 Describe and briefly discuss at least three different routes for the production of benzoic acid. Write the equations for the chemical reactions involved, calculate the raw-material costs for each route, and compare with published prices for benzoic acid.

2-9 The feed to one of the coal-gasification processes is dry coal, which can be no larger than 8-mesh and no smaller than 100-mesh. The raw coal as received is mainly $+\frac{1}{4}$ to $1\frac{1}{2}$ in with small amounts of fines down to 200-mesh. Devise a flow scheme to perform the task. Any undersize material can be burned as pulverized coal in a steam boiler.

---

† This problem is discussed in S. W. Bodman, "The Industrial Practice of Chemical Engineering," MIT Press, Cambridge, Mass., 1968.

2-10 Prepare a rough process flow diagram and propose at least two different types of equipment for each step of the process described below. No additional information concerning the process is available at this stage. A solution of two liquids, A and B, reacts in the presence of a dilute suspension of catalyst C to produce D, a moderately volatile liquid. A is present in large excess. C is a dense, nonfriable solid in the size range of 80 to 100-mesh and can be recycled. It must be removed before the next operation, in which D in solution in A is reacted with gas E. The reaction is rapid and complete, and F is obtained in solution with A. A and F are separated, A is recycled, and F is the marketable, liquid product. Both reactions take place at moderate temperatures and at atmospheric pressure.

2-11 A laboratory synthesis has been worked out for the synthesis of E, in which 1 L of methyl ethyl ketone, a solvent which does not participate in the reactions, is heated to boiling in a round-bottomed flask equipped with a reflux condenser. Then 100 mL of organic liquid A is added to the flask. The organic liquid B is added with agitation in two portions of 75 mL each. The reaction between A and B to produce C is rapid and slightly endothermic. The flask is then cooled to room temperature, and gas D is bubbled through the liquid under agitation. E is obtained as a precipitate. When the reaction, which is exothermic, ceases, the flask contents are filtered and then washed on the filter paper with methyl ethyl ketone. The wet solid is scraped off the filter paper into an evaporating dish and dried in a lab oven. Work out a preliminary, rough, process flow diagram for producing 1 t/day (1 t = 1000 kg) of E as a dry, crystalline solid. List some of the additional information you would need to be able to proceed to a detailed process design.

2-12 The marketing department of your company completed a market survey for E (its laboratory synthesis was described in Prob. 2-11). They believe that a market for 15,000 t/yr can be developed quickly. Prepare a preliminary, rough process design based on the information available to you in Prob. 2-11 to produce 15,000 t/yr of E. List some of the additional information you would need to be able to proceed to a detailed process design.

2-13 Aspirin is produced by the reaction between salicylic acid and acetic anhydride, usually in a solvent medium. The following procedure has been described. A mixture of 250 kg of acetic anhydride, 300 kg of acetic acid, and 250 kg of salicylic acid is heated with agitation to 60°C. Sulfuric acid catalyzes the reaction, and 1 kg of it is added to the reactor as a solution in 2 kg of acetic acid. The temperature is slowly increased to 90°C and maintained for 2 to 3 h. Upon cooling to 20°C the aspirin product crystallizes, and the crystals are filtered, washed with distilled water, and dried. The yield based on salicylic acid is at least 90 percent of theoretical. Work out a process flow diagram to produce aspirin at a rate of 10^6 kg/yr. As much of the process as possible is to be continuous, although a batch reactor will probably be required. Recovery and recycle of unreacted material should be included in the scheme. List some of the additional information you would need to be able to proceed to a final design.

2-14 A typical analysis of cracked butane gas is as follows:

Component	wt %
Hydrogen	0.2
Methane	12.0
Ethylene	30.0
Ethane	10.0
Propylene	20.0
Propane	5.8
Butanes	22.0

Devise a scheme to separate the hydrogen and methane for use as a fuel in the butane cracking reactor, ethylene for use in the manufacture of polyethylene, and the heavier components for recycle to the cracking reactor. Submit the scheme as a neat process flowsheet. Explain your reasoning in the development of your processing scheme.

THREE

ESTIMATION OF THERMOPHYSICAL PROPERTIES

> The sages of Neharde'a used to say: "An estimate can be revised within twelve months."
>
> Said Ameimar: "I am from Neharde'a and it is my opinion that an estimate can always be revised."
>
> *"Babylonian Talmud," Baba Metzia 35ᵃ*

Data and information on the thermodynamic and physical properties of the chemicals with which the chemical engineer is concerned are required in process design, evaluation, and analysis. Although data for many compounds are tabulated in various handbooks and reference sources, the practicing engineer will frequently be confronted with the fact that data for the specific property needed are available neither from the literature nor from company sources. Even when some data are available, they are not always complete and may not cover the range of conditions in which the engineer is interested.

Recourse can be had to one of the numerous methods that exist for estimating the various properties. As we shall see, a complete system of thermodynamic and physical properties can be estimated from only two or three pieces of information about the compound. The value of a method depends on its accuracy, simplicity, and the type and amount of information necessary for its use. Although a property may be predicted quite accurately if sufficient information about the compound is available, in many cases the chemical engineer will know little more about the substance than its chemical formula and normal boiling point. Less reliable procedures would then probably have to be employed and a lesser degree of accuracy of the predicted properties would be expected. The accuracy of the data generated, however, is usually adequate for process evaluation and may also be sufficiently accurate for process design.

Computer packages are available for the calculation of thermodynamic and physical properties. It is wise to know the estimating methods upon which the computer packages are based and the likely accuracies of the predicted thermophysical properties.

3-1 PURE-COMPOUND PROPERTIES

Reid and Sherwood (1958, 1966) presented a critical review of procedures for estimating the properties of gases and liquids. Their pioneering work has been broadened and expanded, and in the third edition by Reid et al. (1977) the literature is reviewed up to 1977. To demonstrate the potential of currently available estimating methods we shall estimate a number of physical and thermodynamic properties of one compound, chlorobenzene. Practically all the properties we shall estimate will be based on only two pieces of information, the structural formula of the compound and its atmospheric pressure boiling point. The properties will be estimated from well-developed and easy procedures, and all the methods used are detailed by Reid et al. (1977).

Chlorobenzene, molecular weight 112.6, can be represented as

Its boiling point at atmospheric pressure is 404.9K (Reid et al., 1977). We shall see that knowledge of only the molecular structure is sufficient to permit the estimation of a number of important properties.

3-2 CRITICAL PROPERTIES

Forman and Thodos (1958) developed a technique to estimate suitable values for the van der Waals constants. The critical points can then be estimated from

$$T_c = \frac{8a}{27bR} \tag{3-1}$$

$$P_c = \frac{a}{27b^2} \tag{3-2}$$

Values for a and b are calculated from group and bond contributions that are available in tabular form as $a^{2/3}$ and $b^{3/4}$. The procedure is to sum up all the individual contributions and then to calculate a and b. The critical properties are then easily estimated.

The compressibility factor at the critical condition Z_c can be estimated by the Garcia-Barcena method (Reid and Sherwood, 1966). This method also utilizes an additive group technique

$$Z_c = 0.293 - \sum_i \Delta Z_i \qquad (3\text{-}3)$$

The method is accurate but not applicable to polar compounds, aliphatic halides, or inorganic materials. Values for ΔZ are presented in tabular form in Reid and Sherwood (1966) for the various groups.

Example 3-1 (*a*) Estimate the critical temperature and critical pressure of chlorobenzene

Group	$\Delta a^{2/3}$	$\Delta b^{3/4}$
Type 3*a* ($\overset{\mid}{C}{-}H$)	5(11,646)	5(5.991)
Type 4*a* ($\overset{\sslash}{C}{-}$)	11,144	1.043
Chloride (—Cl)	17,200	10.88
	86,574	41.878

SOLUTION
$a = (\sum \Delta a^{2/3})^{3/2} = 86{,}574^{3/2} = 25.5 \times 10^6$ (cm$^6 \cdot$ atm)/g mol^2
$b = (\sum \Delta b^{3/4})^{4/3} = 41.878^{4/3} = 146$ cm^3/g mol

$$T_{c,\text{est}} = \frac{8a}{27bR} = \frac{(8)(25.5 \times 10^6)}{(27)(146)(82.06)}$$

$$= 631 \text{ K} \qquad \text{experimental value} = 632.4 \text{ K}$$

$$P_{c,\text{est}} = \frac{a}{27b^2} = \frac{25.5 \times 10^6}{(27)(146)^2}$$

$$= 44.2 \text{ atm} \qquad \text{experimental value} = 44.6 \text{ atm}$$

(*b*) Estimate the critical compressibility factor for chlorobenzene.

SOLUTION Using the Garcia-Barcena method, we find the group contributions to be

Group	ΔZ
Benzene ring	0.0178
—Cl (aromatic)	0.0112
	$\sum \Delta Z = 0.0290$

$Z_{c,\text{est}} = 0.293 - 0.029 = 0.264 \qquad \text{experimental value} = 0.265$

3-3 PRESSURE-VOLUME-TEMPERATURE RELATIONSHIPS

One of the simplest equations of state that the chemical engineer can use is

$$PV = ZRT \tag{3-4}$$

The compressibility factor Z is unity for ideal gases and approaches unity for all gases at low pressures. For temperatures above 2.5 to 5 times the critical temperature Z is quite close to unity up to pressures of about 10 times the critical pressure. Thus, the ideal-gas law is a surprisingly accurate approximation for many cases.

For practical use, values for the compressibility factor are usually calculated on the basis of the law of corresponding states. For the simple nonpolar fluids Z can be represented as a function of two parameters, reduced temperature and reduced pressure

$$T_r = \frac{T}{T_c} \tag{3-5}$$

$$P_r = \frac{P}{P_c} \tag{3-6}$$

A set of compressibility-factor charts due to Nelson and Obert (Obert, 1960) is reproduced here in Figs. 3-1 to 3-3. If these charts are to be used for hydrogen, helium, and neon, the reduced temperatures and pressures should be calculated from

$$T_r = \frac{T}{T_c + 8} \qquad P_r = \frac{P}{P_c + 8} \tag{3-7}$$

where temperature is in kelvins and pressure is in atmospheres.

More accurate results can be attained if a third correlating parameter is also used. Lyderson et al. (1955) developed an extensive correlation with Z_c as the third parameter so that

$$Z = f(T_r, P_r, Z_c) \tag{3-8}$$

They presented tabular values for Z as a function of T_r and P_r for values of Z_c in the range of 0.23 to 0.29 which covers most compounds. Their tabular results cover the liquid as well as the gas range.

The procedure for determining the compressibility factor is obvious, and in the previous section we have already estimated the critical properties based only on knowledge of the structural formula for our compound of interest. We are, therefore, in a position to estimate the pressure-volume-temperature relationships for chlorobenzene.

Example 3-2 Estimate the density of liquid chlorobenzene at its atmospheric boiling point.

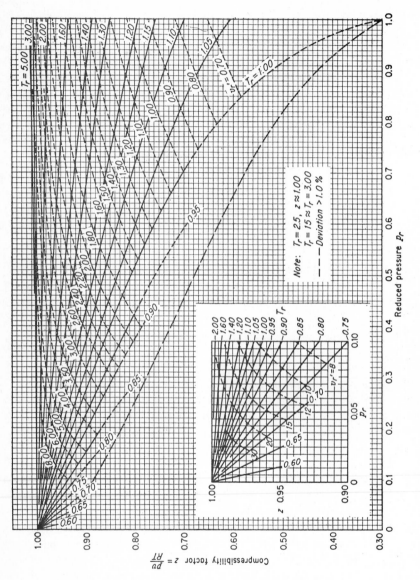

Figure 3-1 Nelson-Obert generalized compressibility chart (low-pressure region). [*From Obert (1960), by permission.*]

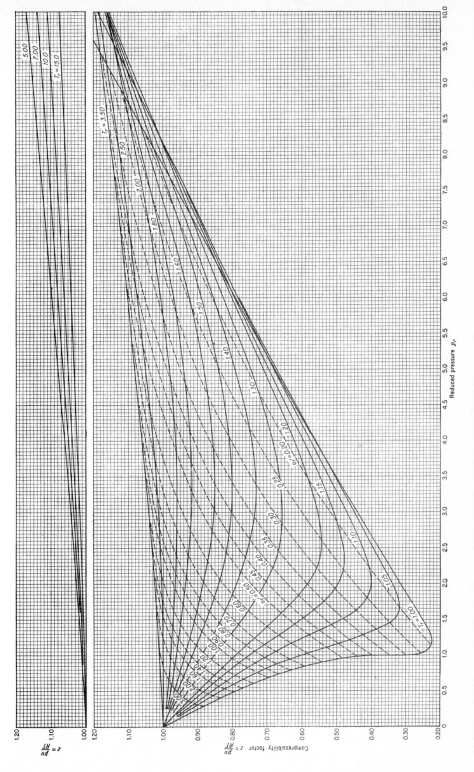

Figure 3-2 Nelson-Obert generalized compressibility chart (medium-pressure region). [*From Obert (1960), by permission.*]

37

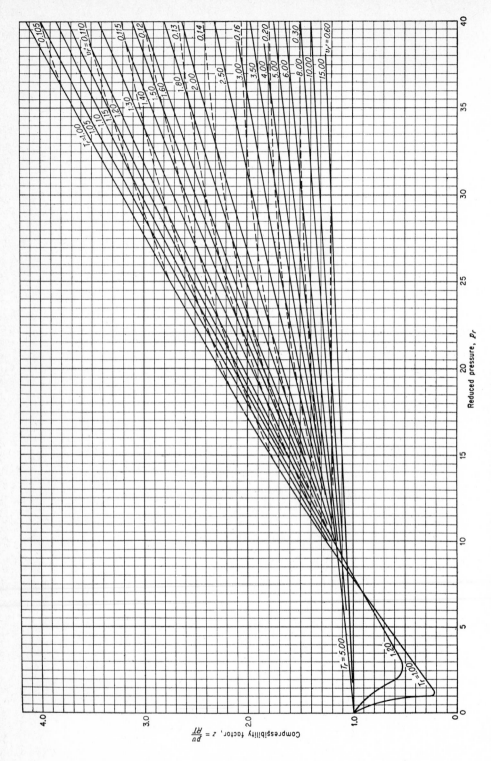

Figure 3-3 Nelson-Obert generalized compressibility chart (high-pressure region). [*From Obert (1960), by permission.*]

38

SOLUTION We shall use values from the tables prepared by Lyderson et al. (1955) for the reduced density of liquids. The density of chlorobenzene at the critical conditions can be calculated from our previously estimated values for T_c, P_c, and Z_c.

$$\rho_c = \frac{P_c}{Z_c R T_c} = \frac{44.2}{(0.264)(82.06)(631)} = 0.00323 \text{ g mol/cm}^3$$

At the atmospheric boiling point

$$T_r = \frac{404.9}{631} = 0.642$$

The reduced density for a saturated liquid whose $Z_c = 0.264$ at a reduced temperature of 0.642 can be obtained from the tabulated values by interpolation. At these conditions, ρ_r is equal to 2.709, and

$$\rho = (2.709)(0.00323) = 0.00875 \text{ g mol/cm}^3$$

The specific volume will be equal to $1/\rho$, or 114.1 cm³/g mol. The experimental value is 115 cm³/g mol (Reid and Sherwood, 1966).

3-4 VAPOR PRESSURE

One of the easiest methods available to estimate the vapor pressure is that proposed by Riedel (1954). This method requires the value of T_c and P_c, which we have already estimated, and the atmospheric boiling point which is part of the input data we permitted ourselves. The Riedel equation permits the estimation of the reduced vapor pressure from

$$\log P_{vr} = -\Phi(T_r) - (\alpha_c - 7)\psi(T_r)$$

where

$$\Phi(T_r) = 0.118\phi(T_r) - 7 \log T_r$$

$$\psi(T_r) = 0.0364\phi(T_r) - \log T_r$$

$$\phi(T_r) = \frac{36}{T_r} + 42 \ln T_r - 35 - T_r^6$$

The functions $\Phi(T_r)$ and $\psi(T_r)$ are available in tabular form in Reid and Sherwood (1966). It is necessary to have one vapor-pressure data point in order to obtain the value of α_c (hence the necessity for the atmospheric boiling point in our data), but any other known value for the vapor pressure is just as suitable.

Example 3-3 Estimate the vapor pressure of chlorobenzene at 245.3°C.

SOLUTION The value for α_c is first obtained from the atmospheric boiling point

$$P_{br} = \frac{1}{44.2}$$

$$T_{br} = \frac{404.9}{631} = 0.642$$

From the tabular values in Reid and Sherwood (1966) at $T_{br} = 0.642$

$$\Phi(T_{br}) = 1.630$$

$$\psi(T_{br}) = 0.280$$

$$\alpha_c = \frac{\log (1/P_{br}) - \Phi(T_{br})}{\psi(T_{br})} + 7$$

$$= \frac{\log 44.2 - 1.630}{0.280} + 7$$

$$= 7.0551$$

Now that we have a value for α_c, we can estimate the reduced vapor pressures at 245.3°C

$$T_r = \frac{245.3 + 273.2}{631} = 0.821$$

From the tables

$$\Phi(T_r) = 0.630$$

$$\psi(T_r) = 0.095$$

$$\log P_{vr} = -0.630 - (7.0551 - 7)(0.095)$$

$$= -0.630 - 0.0052 = -0.6352$$

$$P_{vr} = 0.232$$

and, finally,

$$P_v = (0.232)(44.2) = 10.3 \text{ atm} \qquad \text{experimental value} = 10.0 \text{ atm}$$

3-5 LATENT HEAT OF VAPORIZATION

The simplest method for estimating the latent heat of vaporization at the atmospheric boiling point is that of Giacalone (Reid et al., 1977)

$$\Delta H_{vb} = \frac{R T_b T_c \ln P_c}{T_c - T_b} \qquad (3\text{-}9)$$

The latent heat at any pressure can then be estimated from the vapor-pressure curve, calculable by the Riedel method, by means of the Clausius-Clapeyron equation

$$\frac{dp_v}{dT} = \frac{\Delta H_v}{T \, \Delta V}$$
(3-10)

Alternatively, the Watson correlation can be used

$$\Delta H_{v_2} = \Delta H_{v_1} \left(\frac{1 - T_{r_2}}{1 - T_{r_1}}\right)^{0.38}$$
(3-11)

A nomograph of Eq. (3-11) is available in Reid and Sherwood (1966).

Example 3-4 Estimate the latent heat of vaporization of chlorobenzene at the atmospheric boiling point.

SOLUTION Substituting the appropriate values into Eq. (3-9) gives

$$\Delta H_{vb} = \frac{(1.987)(404.9)(631) \ln 44.2}{631 - 404.9}$$

$$= 8507 \text{ cal/g mol}$$

Experimental value = 8423 cal/g mol

Had we used experimental values for T_c and P_c rather than the estimated values, the estimated value for ΔH_{vb} would have been 8493 cal/g mol.

3-6 GAS HEAT CAPACITY

Ideal-gas heat capacities can be estimated by group-contribution techniques to obtain the values for a, b, c, and d in the heat-capacity equation in the form

$$C_p^0 = a + bT + cT^2 + dT^3$$
(3-12)

Example 3-5 Estimate the heat capacity of chlorobenzene vapor at 298 K by the Rihani-Doraiswamy (1965) group-contribution method.

SOLUTION

Group	a	$b \times 10^2$	$c \times 10^4$	$d \times 10^6$
5HC	5(−1.4572	1.9147	−0.1233	0.002985)
−C	−1.3883	1.5159	−0.1069	0.002659
−Cl	3.0660	0.2122	−0.0128	0.000276

Summing up the contributions yields

$$C_p^0 = -5.6083 + 11.3016 \times 10^{-2}T - 0.7362 \times 10^{-4}T^2 + 0.01786 \times 10^{-6}T^3$$

Assuming that the ideal-gas heat capacity at zero pressure will be equal to the heat capacity of the chlorobenzene at 298 K and substituting the value for T into the equation gives

$$C_p = 22.0 \text{ cal/(g mol} \cdot \text{K)} \qquad \text{experimental value} = 23.2 \text{ cal/(g mol} \cdot \text{K)}$$

It should be pointed out that the accuracy of the Rihani-Doraiswamy method is better at higher temperatures and that the estimated values usually lie within 3 percent of the measured values.

3-7 ENTHALPY CHARTS

We now have developed sufficient information to permit the estimation of enthalpy as a function of pressure and temperature. Generalized charts such as Figs. 3-4 and 3-5 will permit the estimation of the effect of pressure on enthalpy from values for reduced pressure and reduced temperatures. In Example 3-5 we have just developed an expression for the heat capacity as a function of temperature, and Fig. 3-6 will give the isothermal pressure correction for the vapor heat

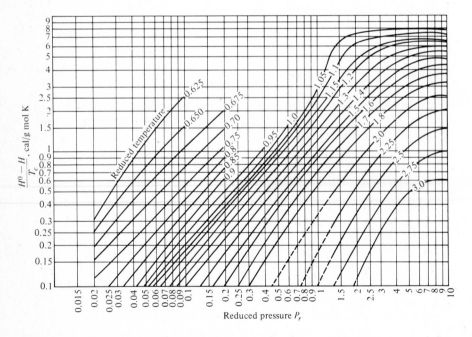

Figure 3-4 Kordbachen-Tien enthalpy correction chart (low- and medium-pressure regions). [*From Kordbachen and Tien (1959), by permission.*]

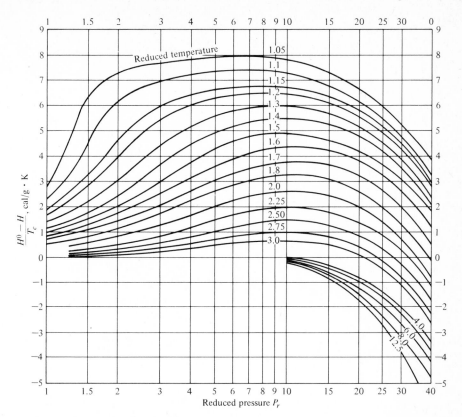

Figure 3-5 Kordbachen-Tien enthalpy correction chart (high-pressure region). [*From Kordbachen and Tien (1959), by permission.*]

capacity. The complete chart of enthalpy as a function of temperature and pressure can therefore be built up once the datum conditions are defined. A pressure-enthalpy diagram for methane is shown in Fig. 3-7.

3-8 STANDARD HEAT OF FORMATION

The standard heat of formation of a compound is defined as the heat absorbed in the reaction in which 1 mol of the compound is formed from its elements at a given temperature and pressure. The reference pressure is usually 1 atm, and the reference temperature is usually 298 K. The standard heat of formation can easily be estimated by atomic-group contributions. The method we shall use for an example is that proposed by Verma and Doraiswamy (1965), who expressed group contributions such that

$$\Delta H_f^\circ = \sum A + \left(\sum B\right)T \tag{3-13}$$

Values for A and B are available in tabular form for each atomic group.

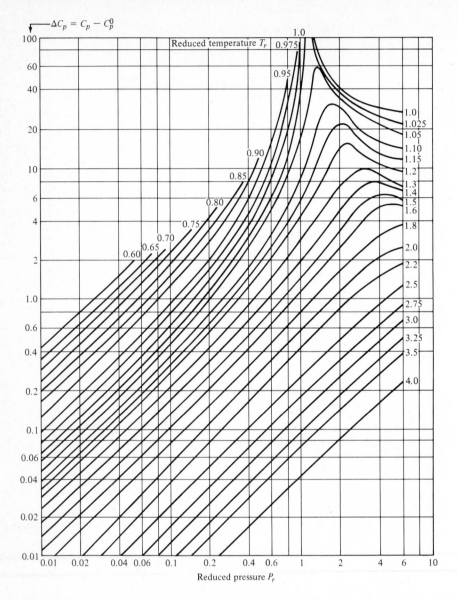

Figure 3-6 Isothermal pressure correction for heat capacity of gases. [*From W. C. Edminster, Petrol. Eng., December 1950, p. C-16, by permission.*]

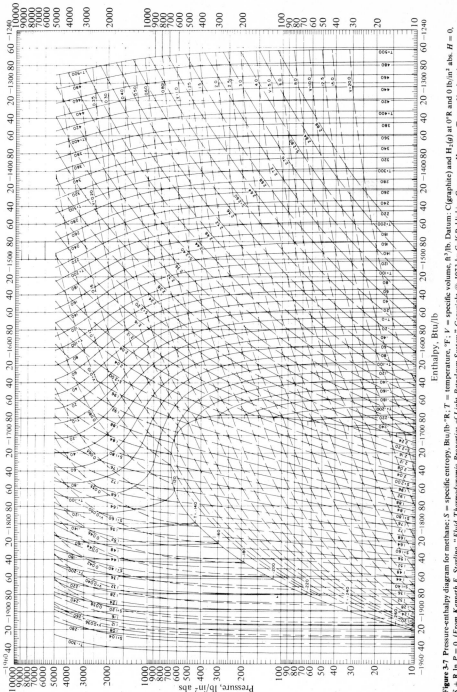

Enthalpy, Btu/lb

Figure 3-7 Pressure-enthalpy diagram for methane; S = specific entropy, Btu/lb·°R; T = temperature, °F; V = specific volume, ft³/lb. Datum: C(graphite) and $H_2(g)$ at 0°R and 0 lb/in² abs. H = 0, S + R ln P = 0. *(From Kenneth E. Starling, "Fluid Thermodynamic Properties of Light Petroleum System." Copyright © 1973 by Gulf Publishing Company, Houston, Texas. Used with permission. All rights reserved.)*

45

Example 3-6 Estimate the standard heat of formation of chlorobenzene at 298 K by the Verma-Doraiswamy procedure.

SOLUTION

Group	A	$B \times 10^2$
5HC⟨	5(3.768)	5(−0.167)
1—C⟨	5.437	0.037
1—Cl	−9.25	0

Summing up the contributions, we get

$$\Delta H_f^\circ = 15.03 - 0.798 \times 10^{-2} T$$

and at 298 K

$$\Delta H_f^\circ = 12.65 \text{ kcal/g mol} \qquad \text{experimental value} = 12.39 \text{ kcal/g mol}$$

It should be pointed out that the Verma-Doraiswamy tables do not give a group contribution for chlorine bonded to an aromatic ring. The value used above was extracted from group contributions for the calculation of ΔG by the Van Krevelin–Chermin (1951) method by use of the thermodynamic relationship $\Delta H_f^\circ = \Delta G_f + T \, \Delta S_f^\circ$.

3-9 FREE ENERGY OF FORMATION

A study of chemical reaction equilibrium requires values for the free energy of formation of the compound participating in the reaction. Again, as in the case of the standard heat of formation, the method of group contributions can be used.

Example 3-7 Estimate the free energy of formation of chlorobenzene at 298 K.

SOLUTION The method of Van Krevelin and Chermin (1951) will be used. They present tabular values for A and B for atomic groups permitting ΔG_f° to be estimated by

$$\Delta G_f^\circ = A + BT \tag{3-14}$$

Group	A	$B \times 10^2$
5HC $\diagup \diagdown$	5(3.047)	5(0.615)
$\diagup$ —C $\diagdown$	4.675	1.150
—Cl	-8.25	0

Summing up the contributions, we have

$$\Delta G_f^\circ = 11.66 + 4.225 \times 10^{-2}T$$

and at 298 K

$$\Delta G_f^\circ = 24.3 \text{ kcal/g mol} \qquad \text{experimental value} = 23.7 \text{ kcal/g mol}$$

Although the error is small, it will be shown in a later chapter that small errors in ΔG_f° can lead to a large error in the predicted equilibrium constant for a reaction and hence to the composition to be expected if thermodynamic equilibrum is achieved.

3-10 ENTROPY OF FORMATION

At this stage of our estimating procedure the entropy of formation of a compound can be predicted by either of two methods. It can be calculated from the estimated values for ΔH_f° and ΔG_f° from the relationship

$$\Delta S_f^\circ = \frac{\Delta H_f^\circ - \Delta G_f^\circ}{T} \tag{3-15}$$

or, alternatively, it can be estimated directly, again by group-contribution methods.

Example 3-8 Estimate ΔS_f° for chlorobenzene at 298 K using (a) the Andersen-Beyer-Watson method (1944) of group contributions and (b) the previously estimated values for ΔH_f° and ΔG_f°.

SOLUTION
(a) The Andersen-Beyer-Watson method estimates entropy of the compound by summing up the group contributions. The entropy of formation is then obtained by subtracting the absolute entropies of the component elements from the absolute entropy of the compound.

Estimate of absolute entropy of chlorobenzene at 298 K

Group	Contribution, cal/(g mol · K)
Benzene base group	64.4
First substitution (methyl)	12.0
Substitution of —Cl for methyl group	0
	76.4

Values for absolute entropy of elements at 298 K

	cal/(g mol · K)
C(graphite)	1.36
$H_2(g)$	31.21
$Cl_2(g)$	53.29

The entropy of formation of chlorobenzene is therefore

$$\Delta S_f^\circ = 76.4 - [(6)(1.36) + (2.5)(31.21) + (0.5)53.29]$$

$$= -36.4 \text{ cal/(g mol · K)}$$

experimental value $= -37.97$ cal/(g mol · K)

(b) Using the previously estimated values obtained for ΔH_f° and ΔG_f°, we get

$$\Delta S_f^\circ = \frac{12,650 - 24,300}{298} = -39.1 \text{ cal/(g mol · K)}$$

It should be borne in mind that the value of any one member of the trio of ΔH_f°, ΔG_f°, and ΔS_f° can be calculated from the other two by Eq. (3-15).

3-11 OTHER PHYSICAL PROPERTIES

A number of additional properties are usually necessary for design and evaluation purposes, the most important being viscosity, surface tension, thermal conductivity, and liquid specific heat. These, too, can be estimated with reasonable accuracy within the constraint of the restricted information we have set for ourselves. We shall now estimate these properties.

Example 3-9 Estimate for chlorobenzene the values of (a) liquid specific heat at 20°C, (b) liquid viscosity at 80°C, (c) liquid thermal conductivity at 80°C, and (d) surface tension of liquid at boiling point.

SOLUTION (a) Johnson and Huang (1955) proposed an additive group method for estimating liquid heat capacities at 20°C. From their tabular values

Group	Contribution, cal/(g mol · K)
C_6H_5-	30.5
$Cl-$	8.6
	$\sum = 39.1$

The liquid heat capacity, as estimated by this method, will therefore be

$$C_{pl} = \frac{39.1}{112.5} = 0.347 \text{ cal/(g · K)} \qquad \text{experimental value} = 0.318 \text{ cal/(g · K)}$$

(b) The Stiel and Thodos (1964) method based on corresponding state relationships can be used. The viscosity can be expressed as

$$\mu_L \xi = f(Z_c, T_r) \qquad \text{where} \qquad \xi = \frac{T_c^{1/6}}{M_c^{1/2} P_c^{2/3}}$$

Values for $\mu_L \xi$ are available in graphical form (Reid and Sherwood, 1966). At 80°C

$$T_r = \frac{353}{631} = 0.559$$

At this value of T_r and for Z_c equal to our estimated value of 0.264, the value for $\mu_L \xi$ read from the graph is 0.010. Therefore

$$\mu_L = \frac{(0.010)(M_c)^{1/2}(P_c)^{2/3}}{(T_c)^{1/6}}$$

$$= \frac{(0.010)(112.5^{1/2})(44.2^{2/3})}{631^{1/6}}$$

$$\mu_L = 0.453 \text{ cP} \qquad \text{experimental value} = 0.441 \text{ cP}$$

(c) Robbins and Kingrea (1962) suggested the following relation for estimating liquid thermal conductivity

$$k_L = \frac{(88.0 - 4.94H) \times 10^{-3}}{(\Delta H_{vb}/T_b) + R \ln (273/T_b)} \frac{0.55}{T_r} C_p \rho^{4/3}$$

In this expression C_p is the molal heat capacity and ρ is expressed as gram moles per cubic centimeter. The units for k_L will be calories per centimeter-second-kelvin. Values for H depend upon molecular structure and are available in Reid et al., 1977. For chlorobenzene H has a value of unity. The other terms in the expression have been estimated previously.

Substitution gives

$$k_L = \frac{(88.0 - 4.94) \times 10^{-3}}{(8540/404.9) + 1.98 \ln (273/404.9)}$$

$$\times \frac{(0.55)(631)}{(353)}(0.347)(112.6)\left(\frac{1}{114.1}\right)^{4/3}$$

$$= 2.84 \times 10^{-4} \text{ cal/(cm} \cdot \text{s} \cdot \text{K)}$$

Experimental value $= 2.66 \times 10^{-4}$

(*d*) The surface tension can be estimated from the Macleod-Sugden expression relating surface tension to the temperature-independent parachor [*P*] and liquid and vapor densities

$$\sigma^{1/4} = [P](\rho_l - \rho_v)$$

The parachor can be estimated by additive contributions as proposed by Quale (1953) and listed in Reid et al. (1977). For chlorobenzene the contributions are 189.6 for C_6H_5 — and 55.2 for Cl—, yielding 244.8. The vapor density will be negligible compared with the liquid density, and using our previously estimated value for ρ_l gives

$$\sigma = \left(\frac{244.8}{114.1}\right)^4 = 21.2 \text{ dyn/cm} \qquad \text{experimental value} = 20.6$$

Had we used the experimental value for the liquid density, the estimated value for σ would have been 20.5 dyn/cm.

We have demonstrated that a complete set of physical and thermodynamic properties can be estimated for pure compounds from a minimum of data by the use of simple procedures that are not too onerous for hand calculation. Most of the procedures permitted the value for a property to be estimated with knowledge only of the molecular structure of the compound in question. Some methods required, in addition, data on the atmospheric boiling point, which is an easily determined property. As Table 3-1 shows, the accuracies were reasonable and in most cases were what would be expected from the various estimating procedures. This does not mean that other compounds might not give poorer accuracies. The accuracies were such, however, that they could be used for preliminary calculations, and in many cases estimated values could also be used for design purposes.

Although the procedures used in this chapter are suitable for hand calculation, they would be time-consuming if it were necessary to obtain a complete map of the properties. Frith (1972) presented a number of methods for estimating physical properties in a form that is easily programmed for computer use.

One should not conclude, however, that estimated properties are preferable to experimental values. Reliable experimental data are always to be preferred

Table 3-1 Comparison of estimated and experimental values for a number of properties of chlorobenzene

Property	Estimated value	Experimental value
T_c, K	631	632.4
P_c, atm	44.2	44.6
Z_c	0.264	0.265
v_{b_1}, cm^3/g mol	114.1	115
Vapor pressure at 245.3°C, atm	10.3	10
ΔH_{vb}, cal/g mol	8507	8423
C_{pv} at 298 K, cal/(g mol · K)	22.0	23.2
ΔH_f° at 298 K, kcal/g mol	12.65	12.39
ΔG_f° at 298 K, cal/g mol	24.3	23.7
ΔS_f° at 298 K, cal/(g mol · K)	−36.4	−37.97
C_{pL} at 20°C, cal/(g · K)	0.347	0.318
μ_L at 80°C, cP	0.453	0.441
k_L at 80°C, cal/(cm · s · K)	2.84×10^{-4}	2.66×10^{-4}
σ at 132.2°C, dyn/cm	21.2	20.7

over values estimated by any of the estimating procedures. When used with good experimental data, the estimating procedures can be a reliable guide for the chemical engineer to permit interpolation, extrapolation, and extension of the data in a logical and accurate manner.

3-12 MIXTURE PROPERTIES

Numerous methods are available for predicting properties of liquid and gas mixtures. Some of the estimating equations yield results of reasonable accuracy over wide ranges of conditions. The equations are usually more complicated than those for pure components, and in many cases their solutions require trial-and-error calculations. Most of the procedures are closely related to the pure-component methods with the invocation of appropriate mixing rules.

The density of mixtures is one of the most accurately predictable properties. The Benedict-Webb-Rubin (1940, 1942) equation has proved to be accurate to within 1 percent when predicting gas or liquid densities at reduced temperatures above 0.6 and reduced densities below 2.0. The Yen-Woods (1966) procedure has been shown to be accurate to within 3 percent for a number of liquid mixtures. Methods available for predicting density include procedures based on tabular or graphical presentations of generalized correlations as well as analytic expressions suitable for computer use.

The enthalpy and heat capacity of mixtures are functions of both temperature and pressure. They can be expressed by values based on ideal-gas behavior plus a term that expresses the departure from ideal behavior. The fact that the PVT behavior of mixtures can be predicted with reasonable accuracy permits the departure term to be calculated.

Transport properties of mixtures can also be estimated by well-tested procedures. Viscosity is the most accurately predictable of the transport properties, and accuracies to within 5 percent are usual. One of the problems facing the development and testing of predictive equations is the scarcity of experimental data covering large temperature and pressure ranges and the large experimental error associated with the data that are available.

For details of the various estimating and predicting equations the reader is referred to Reid et al. (1977), who recommend methods for estimating and correlating properties based on the literature up to 1977.

3-13 VAPOR-LIQUID EQUILIBRIUM

One of the most common operations in the chemical industry is the separation of fluid mixtures into their components by distillation. The design of distillation equipment requires knowledge of vapor-liquid equilibria in multicomponent mixtures. The variety of liquid mixtures in chemical technology is so large that it is extremely unlikely that experimental data will be available for every mixture of interest. It will therefore be necessary to estimate the equilibrium compositions. Such estimates should be based, whenever possible, on reliable data for the particular mixture, fragmentary as the data may be. Techniques are available to reduce and correlate the limited data that might be available and to make well-based interpolations and extrapolations.

The basic thermodynamics of vapor-liquid equilibrium will first be reviewed. We assume that a liquid mixture at temperature T and pressure P is in equilibrium with a vapor mixture at the same temperature and pressure. We are interested in the relationship existing between temperature, pressure, and composition of the two phases. At equilibrium the thermodynamic criterion will be

$$f_i^V = f_i^L \tag{3-16}$$

and our problem is to relate the vapor and liquid fugacities to temperature, pressure, and composition.

For the vapor phase we can introduce the fugacity coefficient

$$\phi_i = \frac{f_i^V}{y_i P} \tag{3-17}$$

where, by definition, $\phi_i \rightarrow 1$ as $P \rightarrow 0$. The fugacity coefficient can be calculated if an equation of state for the mixture is available. A virial equation of state is an especially convenient one for the purpose (see, for example, Van Ness, 1964). At low pressure ϕ_i can usually be assumed to have a value of unity.

For the liquid phase an activity coefficient γ_i is introduced

$$\gamma_i = \frac{f_i^L}{x_i f_i^\circ}$$

where the standard-state fugacity, f_i°, is usually taken as the fugacity of pure liquid i at the system temperature and pressure.

A number of models and equations are available (see Reid et al., 1977) for calculating activity coefficients. Among the models are the Wilson equation, the NTRL equation, and the UNIQUAC model, all of which have a number of features in common, namely, that multicomponent vapor-liquid equilibria can be predicted from experimental information on binary systems; their use requires equilibrium data for all possible binary combinations of the multicomponent mixture and assumes that extrapolation with respect to temperature is possible.

It is not possible within the limitations of this section to present a complete review of the subject of activity-coefficient prediction, which is one of the keys to the prediction of vapor-liquid equilibrium. It is possible, however, to outline one approach to the problem, an approach which can utilize experimental data when available but which can also predict when no experimental information is at hand. One such approach can be based on the group-contribution concept, which we have already seen and used in our earlier predictions of properties of pure compounds.

The basic idea of group contributions is that the thousands of chemical compounds of interest to the chemical industry can all be represented by a small number of appropriately constituted functional groups. By examining and summing the contributions made by the various groups we can develop a possible technique for correlating the properties of a large number of compounds in terms of a small number of functional groups. In addition, we open the possibility for predicting properties in systems for which no experimental data are available.

The group-contribution concept is used in the UNIFAC (*UNI*QUAC *f*unctional-groups *a*ctivity *c*oefficients) model for predicting activity coefficients, as developed and presented by Fredenslund et al. (1977). In the UNIQUAC model the expression for the activity coefficient contains two parts, a combinatorial part, due essentially to differences in size and shape of the molecules comprising the mixture, and a residual part, due essentially to energetic interactions. In the UNIFAC model the combinatorial part depends on the group sizes and shapes, and the residual contribution depends on the group areas and group interactions.

UNIFAC can be applied to nonelectrolytic binary and multicomponent mixtures at conditions removed from the critical region and where all components are condensable. The method correlates more than 70 percent of the published vapor-liquid equilibrium data. From the user's point of view UNIFAC provides a number of advantages in that it is flexible, simple, reliable and has a large range of applicability. It is also capable of predicting phase equilibrium in systems where no experimental equilibrium data exist.

In their work, which is intended for the process design engineer, Fredenslund et al. (1977) describe the UNIFAC group-contribution method for estimating activity coefficients and present computer programs for determining the parameters needed for the model and for predicting binary and multicomponent vapor-liquid equilibria. For the latter purpose they also provide programs for estimating vapor-phase fugacity coefficients with virial coefficients predicted by the Hayden and O'Connell (1975) method. They also present a complete program in which the UNIFAC and fugacity-coefficient programs are incorporated into a distillation-column design program for the separation of multicomponent mixtures.

3-14 DATA SOURCES

It appears appropriate to conclude this chapter with mention of several collections and compilations of thermodynamic and physical properties and vapor-liquid equilibrium data and of physical properties programs and data banks. In addition to the sources mentioned here, every practicing engineer will build up a personal library of data, sources, estimating procedures and computer programs. A number of sources of thermodynamic and physical properties data are presented in Chap. 4 in the problems section. Several sources of vapor-liquid equilibrium data are shown below.

Air Liquide in "Gas Encyclopedia," Elsevier, Amsterdam, 1976.
Chu, Ju-Chin, et al.: "Distillation Equilibrium Data," Reinhold, New York, 1950.
————: "Vapor-Liquid Equilibrium Data," Edwards, Ann Arbor, Mich., 1956.
Gmehling, J., and U. Onken: "Vapor-Liquid Equilibrium Data Collection," vol. I, DECHEMA, Frankfurt, 1977.
Ohe, S.: "Computer-Aided Data Book of Vapor Pressure," Data Book Publishing Company, Tokyo, 1976.
Wichterle, I., J. Linek, and E. Hala: "Vapor-Liquid Equilibrium Data Bibliography," Elsevier, Amsterdam, 1973, and Supplement I, 1976.

As mentioned earlier, a number of computer packages are available for the calculation and estimation of thermodynamic and physical properties. Such programs are usually incorporated into the chemical-plant simulators discussed in Chap. 10. Most of these packages also include data banks. A survey of computer-stored data banks available in Europe was presented by Rose (1978). The American Institute of Chemical Engineers recently announced the establishment of the Design Institute for Physical Property Data (DIPPR). The DIPPR

Table 3-2 Data banks of thermophysical properties

Bank name	Source
APPES	American Institute of Chemical Engineers, New York
DSD DECHEMA	DECHEMA, Frankfurt, Germany
FLOWTRAN	Monsanto Company, St. Louis, Missouri
NEL-APPES	National Engineering Laboratory, Glasgow, Scotland
PPDS	Institution of Chemical Engineers, Rugby, England
PROPDAD	University of Connecticut, Storrs, Connecticut
TRC/API	Texas A & M University, College Station, Texas
TRL	Washington University, St. Louis, Missouri

program will include the evaluation and correlation of available property data and the development of additional data and of correlative and predictive equations and procedures. The results are to be computer-accessible. Some of the more comprehensive data banks presently available are listed in Table 3-2.

PROBLEMS

3-1 Estimate the critical properties of ethylene and propylene and compare with experimental values from the literature.

3-2 Estimate the critical properties of acetone and isopropanol and compare with experimental values.

3-3 Suppose that 10,000 lb of ethylene is to be stored in an uninsulated pressure vessel. The design temperatures are $-40°F$ minimum and $120°F$ maximum. What volume is required, and what design pressure should be specified based on the appropriate estimated and observed physical properties of ethylene?

3-4 If 2000 lb/h of saturated chlorobenzene vapor at atmospheric pressure is to be condensed and cooled to $120°F$, estimate the heat-transfer duties for the condenser and the cooler.

3-5 Monoethanolamine at $95°C$ is to be cooled to $40°C$ before storage. Estimate the heat-transfer duty for an exchanger to cool 1500 lb/h.

3-6 What is $\Delta G_{298}°$ for the production of chlorobenzene by the chlorination of benzene? Compare the value obtained by a group-contribution estimation procedure with that obtained from literature values for the free energies of formation.

3-7 Methylamine is produced by reaction between methanol and ammonia

$$CH_3OH + NH_3 \rightleftharpoons CH_3NH_2 + H_2O$$

Calculate $\Delta G_{298}°$ and $\Delta H_{298}°$ for this reaction using estimated values for the properties of methanol and methylamine. Compare with the values obtained using literature data.

3-8 Acetone can be produced by the oxidation of isopropanol produced separately by the hydrolysis of propylene. The overall reaction is

$$CH_3CH = CH_2(g) + \tfrac{1}{2}O_2(g) \longrightarrow CH_3COCH_3(g)$$

The question whether this reaction can be carried out in one step rather than in two steps of hydrolysis followed by oxidation is to be considered.

    (*a*) Estimate $\Delta G_{298}°$ for the one-step reaction and compare with value obtained from published data.

    (*b*) Estimate $\Delta G°$ as a function of temperature.

    (*c*) Estimate ΔH as a function of temperature.

3-9 The Prandtl number is necessary for the calculation of heat-transfer coefficients. Estimate the value of the Prandtl number for isopropyl alcohol at $20°C$. Use one of the procedures for estimating the Prandtl number directly. Compare with the Prandtl numbers obtained with estimated and with experimental values of C_p, k, and μ.

3-10 Estimate the value of the Prandtl number for acetone at $20°C$ using (*a*) one of the estimating procedures for the Prandtl number, (*b*) estimated values of C_p, k, and μ, and (*c*) experimental values for C_p, k, and μ.

3-11 The analysis of a process proposed for the one-step production of acetone by the oxidation of propylene (Prob. 3-8) requires information concerning the transport properties of acetone in the gaseous phase at 400 K. Estimate the value of C_p, k, and μ at 400 K and 1 atm. Compare with experimental values.

3-12 If you have access to a thermodynamic and physical properties package, use it to calculate the properties requested in some of the preceding problems. Compare the package results with values you obtained by your calculations and with experimental values.

FOUR

REACTION EQUILIBRIUM ANALYSIS

Sulfur etc.: Etched silver vessels are smoked in sulfur fumes until the designs darken and become visible.

Rabbi Solomon Itzhaki (Rashi, 1040–1105)
Commentary on " Babylonian Talmud,"
Sabbath, 18ª

Knowledge of the effects of temperature, pressure, and reactor feed composition on equilibrium reaction-mixture composition is essential for the determination of optimum reactor operating conditions. The possibility of simultaneous reactions must also be considered. These effects can be calculated by appropriate thermodynamic considerations. Although no attempt at an extensive treatment of the thermodynamics of chemical reaction equilibrium will be made in this chapter, they will be briefly reviewed. This review material will then serve as the framework for the study and analysis of the effects of changes in the various operating parameters on reaction equilibrium. Thermodynamic considerations sometimes indicate that only a small fractional conversion of reactants into products can be expected at equilibrium conditions. Methods and approaches for overcoming or bypassing such unfavorable thermodynamics will also be presented and discussed.

4-1 CHEMICAL EQUILIBRIUM

A reacting system maintained at constant temperature and pressure will spontaneously change in the direction of increasing total entropy and will reach equilibrium conditions when the entropy can no longer increase. This equilibrium condition is related to the Gibbs free-energy change by

$$\Delta G° = -RT \ln K \tag{4-1}$$

The equilibrium composition is calculated from the equilibrium constant K, obtainable from the standard-free-energy change for the reaction, which is a function only of temperature. Since information for $\Delta G°$ is usually not available at the temperature at which the reaction is to be studied, the first problem to be solved is how to obtain the value of $\Delta G°$ or, alternatively, the value of the equilibrium constant at the temperature of interest to us.

By definition,

$$G = H - TS \tag{4-2}$$

which by differentiation will yield

$$dG = V \, dP - S \, dT \tag{4-3}$$

The pressure, however, is defined by the standard conditions and does not vary; hence

$$dG° = -S° \, dT \tag{4-4}$$

for a single component. For the reacting mixture, therefore,

$$\frac{d(\Delta G°)}{dT} = -\Delta S° \tag{4-5}$$

Since

$$\Delta G° = \Delta H° - T \, \Delta S° \tag{4-6}$$

we can obtain an expression for the change in the standard-free-energy change for the reaction as a function of temperature

$$\frac{d(\Delta G°)}{dT} = \frac{\Delta G° - \Delta H°}{T} \tag{4-7}$$

By differentiation of Eq. (4-1)

$$\frac{d(\Delta G°)}{dT} = \frac{d(-RT \ln K)}{dT} = -R \ln K - RT\frac{d(\ln K)}{dT} \tag{4-8}$$

and, upon combining Eqs. (4-1), (4-7), and (4-8), we get

$$\frac{d(\ln K)}{dT} = \frac{\Delta H°}{RT^2} \tag{4-9}$$

We have reached our goal of an expression relating the equilibrium constant to temperature through the standard heat of reaction.

To permit Eq. (4-9) to be integrated, however, we must relate $\Delta H°$ to temperature. Heat capacity can generally be expressed as a function of temperature by an equation of the form

$$C_p = a + bT + cT^2 \tag{4-10}$$

and, because $dH = C_p\, dT$, the standard heat of reaction becomes

$$\Delta H_T° = I + \Delta a\, T + \tfrac{1}{2}\Delta b\, T^2 + \tfrac{1}{3}\Delta c\, T^3 \tag{4-11}$$

where Δa, Δb, and Δc are the sums of the respective coefficients in Eq. (4-10) for each compound in the reaction multiplied by its stoichiometric coefficient. The constant of integration I can be obtained from one known value of $\Delta H°$.

The value for K can now be determined from Eq. (4-9) at any temperature from a value of $\Delta G°$ at one temperature, usually 298 K, a value for $\Delta H°$, also usually at 298 K, and equations for the heat capacity of the reactants and products.

The composition to be expected at equilibrium can be calculated once K is known. Let us consider the general reaction:

$$a\text{A} + b\text{B} + \cdots \quad \rightleftharpoons \quad r\text{R} + s\text{S} + \cdots$$

and let us also assume a gaseous system in which the standard state of the component gases is the ideal-gas state at 1 atm pressure. Let us further assume that the components form an ideal solution; i.e., the activity coefficients are unity. For this case

$$K = \frac{x_R^r\, x_S^s \cdots}{x_A^a\, x_B^b \cdots} \frac{\gamma_R^r\, \gamma_S^s \cdots}{\gamma_A^a\, \gamma_B^b \cdots} P^{\Sigma\, \alpha_j} \tag{4-12}$$

where x_j = mole fraction of component j
$\quad \gamma_j = f_j/P$
$\quad f_j$ = fugacity of component j
$\quad \sum \alpha_j$ = sum of stoichiometric coefficients
$\quad P$ = total pressure

Since we normally would be interested in calculating the number of moles of product material that could be obtained at equilibrium under given conditions of temperature, pressure, and feed composition, Eq. (4-12) is modified to the more convenient form

$$K = \frac{n_R^r\, n_S^s \cdots}{n_A^a\, n_B^b}\, K_\gamma \left(\frac{P}{\sum n_j}\right)^{\Sigma\, \alpha_j} \tag{4-13}$$

where n_j is the number of moles of component j in the reacting system which also contains n_I moles of inert gas. Thus

$$x_j = \frac{n_j}{n_A + n_B + \cdots + n_R + n_S + \cdots + n_I} \tag{4-14}$$

K_y is defined, at a given temperature and pressure by

$$K_\gamma = \frac{\gamma_R^r \, \gamma_S^s \cdots}{\gamma_A^a \, \gamma_B^b \cdots} \qquad (4\text{-}15)$$

We can now look at the effect of the various properties and parameters on the equilibrium yield or conversion. It is apparent from Eq. (4-13) that the larger the value for K the larger the yield of the product species at equilibrium.

One major factor over which we have no control is the value of $\Delta G°$ at a given temperature. If $\Delta G°$ is negative, the values of K will be positive and a large equilibrium conversion of reactants into products can be expected. Dodge (1944) laid down the following rough guide to whether a reaction was thermodynamically promising at a given temperature:

$\Delta G° < 0$	reaction promising
$0 < \Delta G° < +10{,}000$	reaction of doubtful promise but warrants further study
$\Delta G° > +10{,}000$	very unfavorable; would be possible only under unusual circumstances

These are approximate criteria that are useful in preliminary exploratory work, and it will shortly be shown that in many situations it is possible to overcome or get around this thermodynamic obstacle.

Equation (4-13) can be rearranged to read

$$\frac{n_R^r \, n_S^s \cdots}{n_A^a \, n_B^b \cdots} = \frac{K}{K_\gamma} \left(\frac{\sum n_j}{P} \right)^{\sum \alpha_j} \qquad (4\text{-}16)$$

It is apparent that any operating condition that will increase the value of the right-hand side of Eq. (4-16) will increase the ratio of products to reactants at the equilibrium condition and hence will increase the reactant conversion.

4-2 EFFECT OF DATA INACCURACIES

It was shown in Chap. 3 that it is possible to estimate a large number of physical and thermodynamic properties from a minimum of data. The estimated values generally are of sufficient accuracy to permit preliminary evaluations and in many cases are sufficiently accurate to permit engineering design. The estimation of chemical equilibrium composition, however, can require data of a high order of accuracy.

The fractional error in the value for an estimated equilibrium constant can be written as $\Delta K/K$. The allowable error in the standard free energy that will result in this fractional error in K at any given temperature can be calculated from a slight manipulation of Eq. (4-1) yielding

$$\Delta(\Delta G°) = -RT \ln \left(\frac{\Delta K}{K} + 1 \right) \qquad (4\text{-}17)$$

At a temperature of 400 K, for example, an error of only 0.076 kcal/g mol would be sufficient to produce a 10 percent error in the calculated value for the equilibrium constant. Because $\Delta G° = \Delta H° - T \Delta S°$, this error could arise from an error in the estimation of either $\Delta H°$ or $\Delta S°$.

The heat of reaction is usually calculated from heats of combustion. The heats of combustion of organic compounds generally exceed 200 kcal/g mol, so that an error of less than 0.04 percent would result in a 10 percent error in the calculated equilibrium constant. This same error could be produced by an error of 0.19 entropy unit for the standard entropy change of the reaction.

Data of a high degree of accuracy are necessary if accurate equilibrium compositions are to be calculated when K values are in the vicinity of unity. A procedure that can yield reasonably accurate results can be based on a single accurate experimental equilibrium measurement made within the range of interest. Equilibrium constants at other conditions can then be calculated with the aid of Eqs. (4-1), (4-6), and (4-9). The estimated value for $\Delta S°$, which generally is reasonably accurate, would be used in Eq. (4-6) to yield a value for $\Delta H°$ consistent with the estimated value for $\Delta S°$. Estimated values for heat capacity could then be used in Eq. (4-11) to extend the data with reasonable accuracy over a limited range of conditions.

Example 4-1 A highly active catalyst has been developed for the gas-phase reaction

$$A + B \ \rightleftharpoons \ R$$

and it is expected that equilibrium conditions can be closely approached with short residence times in the reactor. The reactor will operate at atmospheric pressure, and you are asked to predict the equilibrium composition at 500 K. Reliable values for the standard free energy of formation of A and B at 500 K can be calculated from available literature data as $-20,820$ and -6075 cal/mol, respectively. It was necessary to estimate the value of $\Delta G_f°$ at 500 K for R, and the estimated value is $-27,500 \pm 500$ cal/g mol.

SOLUTION The uncertainty in the value of $\Delta G_f°$ will have to be taken into consideration. For the reaction

$$\Delta G_{500}° = (-27,500 \pm 500) - (-20,820 - 6075) = -605 \pm 500$$

The equilibrium constant can be calculated from Eq. (4-1). For $\Delta G_{500}° = -605$

$$\ln K = -\frac{-605}{(1.987)(500)} = 0.609$$

$$K = 1.839$$

Taking into account the ± 500 cal/g mol uncertainty in the estimated value for the free energy of formation of R, we have

$$\ln K_{\max} = -\frac{-605 - 500}{(1.987)(500)} = 1.112$$

$$K_{\max} = 3.041$$

$$\ln K_{\min} = -\frac{-605 + 500}{(1.987)(500)} = 0.106$$

$$K_{\min} = 1.111$$

The equilibrium composition can now be calculated. We shall assume that the reactor feed is an equimolar mixture of A and B and that n mol of R is produced per 2 mol of feed. The product stream will contain:

Component	Moles	Mole fraction
A	$1 - n$	$\dfrac{1-n}{2-n}$
B	$1 - n$	$\dfrac{1-n}{2-n}$
R	n	$\dfrac{n}{2-n}$
Total	$2 - n$	

For gas-phase reaction at atmospheric pressure $K_\alpha = 1$, and, from Eq. (4-12),

$$K = \frac{x_R}{x_A x_B} = \frac{n(2-n)}{(1-n)^2}$$

Solving for n yields

$$n = 1 - \frac{1}{\sqrt{K+1}}$$

Substituting the numerical values obtained for K, we find that

$$n_{\max} = 0.503 \qquad n = 0.407 \qquad n_{\min} = 0.312$$

The expected equilibrium compositions can now be calculated:

Component	\multicolumn{3}{c}{Mole fraction at equilibrium}
	$K_{\max}$	K	$K_{\min}$
R	0.336	0.255	0.185
A	0.332	0.372	0.408
B	0.332	0.372	0.408

The uncertainty of only 500 cal/g mol in the free energy resulted in significant effects on the predicted equilibrium composition. These differences could have important effects on the design of downstream separation and recycle equipment.

If the estimated value of the free energy of formation of R had been, say, $-37,500 \pm 500$, the effects of the 500 cal/mol uncertainty would have been negligible, as we shall see. For this latter case

$$\Delta G^{\circ}_{500} = -10,605 \pm 500$$

and $\qquad K_{max} = 71,514 \qquad$ yielding $\qquad n_{max} = 0.996$

$\qquad\qquad K = 43,234 \qquad$ yielding $\qquad n = 0.995$

and $\qquad K_{min} = 26,137 \qquad$ yielding $\qquad n_{min} = 0.994$

The equilibrium composition will be 0.99 mole fraction R for all three values of K.

4-3 EFFECT OF OPERATING PARAMETERS ON EQUILIBRIUM CONVERSION

Temperature

The effect of temperature on equilibrium composition can be calculated with the help of Eq. (4-9). Since the standard heat of reaction is negative for an exothermic reaction, an increase in temperature will result in a decrease in K and also in a decrease in conversion. An exothermic reaction should therefore be carried out at as low a temperature as possible consistent with reaction rates, which, unfortunately, decrease with temperature. For an endothermic reaction ΔH° is positive, and K increases with an increase in temperature, as does the equilibrium conversion. The endothermic reaction should therefore be carried out at an elevated temperature.

Pressure

The equilibrium constant is independent of pressure with the standard states that were chosen. The effect of pressure can be felt in Eq. (4-16) in two places, however, in K_y and, of course, in the term that explicitly includes the pressure. K_y is usually relatively insensitive and may either increase or decrease slightly with pressure. It can be calculated from charts of f/p as a function of reduced temperature and pressure. Such charts are shown in Figs. 4-1 and 4-2. For ammonia synthesis from nitrogen and hydrogen, K_y decreases with pressure, and as a result of this factor the equilibrium conversion would increase with pressure.

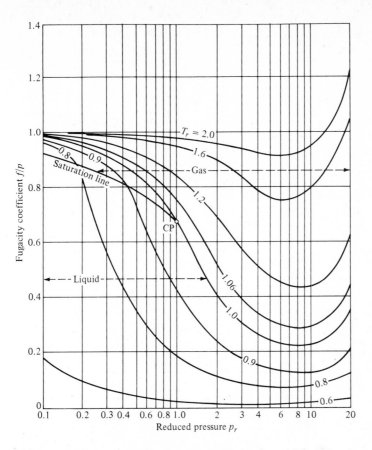

Figure 4-1 Fugacity coefficient of gases and liquids ($Z_c = 0.27$) as function of reduced pressure. [*From Hougen et al. (1959), p. 600, by permission.*]

It is apparent that the effect of pressure on the term in which it appears explicitly is dependent on the value of $\sum \alpha_j$. If $\sum \alpha_j$ is positive, i.e., there is an increase in the number of moles as a result of the reaction, an increase in pressure will result in a decrease in the equilibrium conversion. If the converse is true, i.e., the number of moles decreases as a result of reaction and $\sum \alpha_j$ is negative, an increase in pressure will increase the equilibrium yield. In ammonia synthesis the reaction results in a decrease in the number of moles; thus an increase in pressure causes an increase in equilibrium conversion because of this factor. This, of course, is in addition to the favorable change in K_y (and hence conversion) resulting from an increase in pressure.

Example 4-2 Your group leader was not too pleased at the results obtained (Example 4-1) for the expected equilibrium composition for the reaction

$$A + B \rightleftharpoons R$$

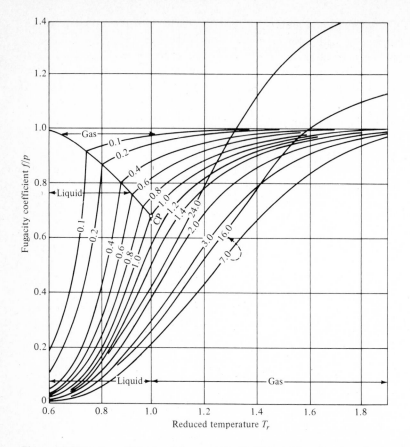

Figure 4-2 Fugacity coefficients of gases and liquids ($Z_c = 0.27$) as function of reduced temperature. [*From Hougen et al. (1959), p. 600, by permission.*]

and asks you to do a preliminary exploration of the effects of relatively mild changes in operating temperature and pressure with the aim of finding conditions that will result in high equilibrium conversion. The temperature range that can be considered is from about 400 K, the lowest temperature at which reaction kinetics are still sufficiently rapid, to 650 K, at which temperature catalyst activity begins to decrease rapidly. Operation at subatmospheric pressure is ruled out because of the danger of air leakage into the reactor. The reaction is mildly exothermic, -7 kcal/mol, and the heat capacities of the products and reactants are approximately equal. Thus, Δa, Δb, and Δc equal zero.

SOLUTION This reaction proceeds with a decrease in the number of moles, $\sum \alpha_j = -1$. Thus, operating at elevated pressure should result in an increase in equilibrium conversion. Since your group leader was interested in relatively mild changes, you assume that operation at 5 atm is a relatively mild change compared with operation at 1 atm.

Because the reaction is slightly exothermic, an increase in equilibrium conversion can be expected if we decrease the reactor temperature. The minimum permitted temperature is 400 K, and we therefore choose to predict the equilibrium conversion at 400 K and 5 atm. By integration of Eq. (4-9) and remembering that Δa, Δb, and $\Delta c = 0$, we get

$$\ln \frac{K_{400}}{K_{500}} = -\frac{\Delta H^\circ}{R}\left(\frac{1}{400} - \frac{1}{500}\right)$$

$$\ln \frac{K_{400}}{1.839} = \frac{7000}{1.987}\left(\frac{1}{400} - \frac{1}{500}\right)$$

$$K_{400} = 10.704$$

From Eq. (4-12)

$$K = \frac{x_R}{x_A x_B} K_\gamma P^{\Sigma \alpha_j}$$

and, assuming that $K_\gamma = 1$, a reasonable assumption for operation at 5 atm and especially in view of the preliminary nature of this study, we have

$$\frac{x_R}{x_A x_B} = (10.704)(5) = 53.52$$

or

$$\frac{n(2 - n)}{(1 - n)^2} = 53.52$$

Solving for n yields $n = 0.8646$.

The expected equilibrium composition will therefore be

$$x_R = 0.761 \qquad x_A = 0.119 \qquad x_B = 0.119$$

This represents a dramatic improvement over the equilibrium conversion obtained in Example 4-1. You would, of course, recommend to your group leader that operation at a pressure higher than 5 atm should be studied. Possible nonideal behavior of the reacting and product species would have to be considered.

Inert Gas

Inspection of Eq. (4-16) shows that the presence of inert gas corresponds to an increase in the $\sum n_j$ term and that the effect would be opposite that produced by pressure. Hence, the effect of inert gas depends on $\sum \alpha_j$, and if the total number of moles is increased as a result of reaction, dilution by inert gas would result in increased equilibrium conversion of the reactants. Of course, equipment size would have to be increased, and product separation and purification might be complicated by the presence of inert components.

Effect of Feed Composition

An increase in the number of moles of reactants other than reactant i in the feed mixture will increase the degree of conversion of reactant i, which would then become the limiting reactant. In the event that one component is expensive or difficult to separate from the product stream, it may be advisable to make it the limiting reactant since an excess of one reactant tends to increase the equilibrium conversion of the other reactant.

If product compounds are present in the feedstream, the equilibrium conversion of the reactant species is reduced. This factor should be borne in mind in recycle operations.

4-4 OPTIMUM FEED COMPOSITION

The first step in determining the optimum feed composition to a reactor is to decide what the objective function or goal of the optimizing operation is. If the goal is to obtain the maximum yield of product species for an isothermal chemical reaction in an ideal solution system, Dodge (1944) and others have shown that the initial molar concentrations of the reactant species should be proportional to their respective stoichiometric ratios in the chemical reaction taking place; however, the initial distribution of mole fractions of reactants required to obtain maximum yield for the case of an adiabatic reaction can differ substantially from the classic distribution according to the stoichiometric coefficients, as obtained for the isothermal case. For some purposes the attainment of the highest possible equilibrium temperatures is the practical goal. Again, the optimum reactants distribution would not be according to the stoichiometric coefficients.

Pings (1965) derived the appropriate expressions for the optimum initial distribution of mole fractions of reactants for several nonisothermal cases. As an exercise he examined the oxidation of SO_2 with air and assumed that the reaction proceeds adiabatically to equilibrium. For the initial temperature conditions assumed, he calculated that the optimum initial ratio of SO_2 to O_2 should be $6.34:1$ in order to maximize the yield of SO_3 rather than the $2:1$ ratio that would be indicated for an isothermal reaction. The concentration of SO_3 at equilibrium was greater than 12 percent for the optimum initial composition as calculated for the adiabatic reaction case, compared with 10.4 percent if the reactants were fed to the reaction according to the stoichiometric coefficients.

In a further example, he calculated the optimum initial reactant composition required to yield the maximum temperature. For this case the optimum initial ratio of SO_2 to O_2 should be $4.22:1$. The equilibrium temperature reached was 14°C higher than that obtained for the stoichiometric mixture.

In commercial operations the goal is to maximize the economic return. As hinted in the previous section, the distribution of initial mole fractions that will maximize the product yield may not be identical with the distribution that will yield the maximum profit. The net result of a chemical reaction is, in general, an equilibrium system containing product species coexisting with reactants. If the equilibrium conditions strongly favor the product species the reactant species might be present in such low concentration that they can remain with the product as contaminants or impurities. In other cases the unreacted species would have to be separated from the product in some kind of purification operation, either to be discarded as waste or recycled. Costs can easily be assigned to these possibilities so that the economic factors can be considered in addition to the thermodynamic aspects. When the economic considerations are included, we cannot expect that the distribution of initial mole fractions that will yield the maximum profit will necessarily be identical to that required for maximum yields.

Example 4-3 In the process being developed for producing R (examples 4-1 and 4-2) the separation of unreacted A from R is an easy task, and recovered A can be recycled to the reactor. Not so for B, where separation from R turns out to be extremely difficult. Our marketing people say that they can sell the product provided that it contains at least 97.5 mol % R. Some B can be present as an impurity.

   You are requested to find suitable reaction conditions that could meet the marketing requirements without requiring separation of unreacted B as part of product purification. The reaction is to be carried out at 400 K and 5 atm.

SOLUTION Our problem is to calculate the ratio of A to B in the reactor feed that could result in practically all the B, as the limiting reactant, being consumed at equilibrium conditions. If we assume that n mol of R is produced from 1 mol of A and m mol of B, the reactor product stream will contain:

Component	Moles	Mole fraction
R	n	$\dfrac{n}{1+m-n}$
A	$1-n$	$\dfrac{1-n}{1+m-n}$
B	$m-n$	$\dfrac{m-n}{1+m-n}$
Total	$1+m-n$	

At equilibrium at 400 K and 5 atm from Example 4-2

$$KP^{-\Sigma \alpha_j} = 53.52$$

and

$$\frac{n(1+m-n)}{(1-n)(m-n)} = 53.52$$

To solve this equation we need an expression relating m to n, and this can be obtained from the product purity requirement. To meet the purity requirements

$$\frac{n}{n + (m - n)} = 0.975$$

Therefore, $\qquad\qquad\qquad m = 1.02564n$

We can now solve for n and m, yielding

$$n = 0.2664 \qquad m = 0.2732$$

Thus, the feed to the reactor should be in the ratio of 0.2732 mol of B per mole of A. This will result in a composition at equilibrium of

$$x_R = 0.2646 \qquad x_A = 0.7287 \qquad x_B = 0.00678$$

After complete separation and removal of A the product will contain 0.025 mol % of B as an impurity. Thus, a separation step for the removal of B is unnecessary. This benefit is bought at the price of a larger reactor than would be necessary if the reactor were fed at the stoichiometric ratio.

4-5 OVERCOMING POSITIVE $\Delta G°$

In the preceding paragraphs we have examined several ways of adjusting various operating parameters to increase the equilibrium conversion in the face of possible unfavorable thermodynamics as represented by a large positive or a small negative value of $\Delta G°$. With sufficient originality and inventiveness, however, it is possible to overcome a positive $\Delta G°$ and to produce the desired product by reaction sequences that are characterized by negative (and hence favorable) values of $\Delta G°$. It is the purpose of this section to examine several such procedures.

Reaction Coupling

This procedure will be demonstrated by reference to the production of acetylene, and the general procedure will then be described. In principle, methane can be cracked to yield acetylene and hydrogen according to

$$2CH_4 \rightleftharpoons C_2H_2 + 3H_2$$

This endothermic reaction has a large, positive $\Delta G°$ that becomes negative and hence thermodynamically favorable only at very high temperatures, but the cracking reaction can be coupled with the reaction

$$H_2 + \tfrac{1}{2}O_2 \rightleftharpoons H_2O$$

which has a highly negative $\Delta G°$. The desired product, acetylene, can then be produced by the summed partial-oxidation reaction

$$2CH_4 + \tfrac{3}{2}O_2 \;\rightleftharpoons\; C_2H_2 + 3H_2O$$

which has a negative $\Delta G°$ and therefore favorable equilibrium conversion.

The principle of reaction coupling involves the addition of a new reactant species that will react with one of the product species produced by the thermodynamically unfavorable reaction. The additional coupled reaction should be one for which $\Delta G°$ has a large negative value. Where R is the desired product species, the general sequence would be

$$aA + bB \;\rightleftharpoons\; rR + sS \qquad \Delta G_1° > 0$$

The product species S reacts in an additional coupled reaction with a new reactant C according to

$$sS + cC \;\rightleftharpoons\; tT \qquad \Delta G_2° < 0$$

The overall reaction would be the sum of these coupled reactions

$$aA + bB + cC \;\rightleftharpoons\; rR + tT$$

The overall $\Delta G°$ for this reaction should be negative or a small positive number.

An additional industrial example of the use of a coupled reaction is the reduction of iron oxide to elemental iron

$$FeO \;\rightleftharpoons\; Fe + \tfrac{1}{2}O_2 \qquad \Delta G_{1000\,K}° = 47.55 \text{ kcal}$$

$$C + \tfrac{1}{2}O_2 \;\rightleftharpoons\; CO \qquad \Delta G_{1000\,K}° = -47.94 \text{ kcal}$$

giving the overall reaction

$$FeO + C \;\rightleftharpoons\; Fe + CO \qquad \Delta G_{1000\,K}° = -0.39 \text{ kcal}$$

Staging Reactions

The overall reaction for the production of methanol can be written

$$3CH_4 + CO_2 + 2H_2O \;\rightleftharpoons\; 4CH_3OH$$

This is an endothermic reaction with an extremely large positive $\Delta G°$, and the thermodynamic equilibrium is unfavorable. The reaction can be carried out in two stages, however, the first stage being the production of synthesis gas, a mixture of CO and H_2, at a high temperature where the thermodynamic conditions are favorable, followed by the exothermic production of methanol at a lower temperature, again with a favorable thermodynamic situation prevailing.

The staged reactions can be written in idealized form as

$$3CH_4 + CO_2 + 2H_2O \;\rightleftharpoons\; 4CO + 8H_2$$

followed by

$$4CO + 8H_2 \;\rightleftharpoons\; 4CH_3OH$$

The first reaction is carried out at approximately 800°C. The second reaction is carried out, depending on the particular process, in the range of 250 to 390°C.

Solvay Clusters

In 1860 Ernest Solvay discovered a cluster of six chemical reactions which proceed rapidly to the right under reasonable industrial operating conditions. The net result of these reactions is the conversion of salt and limestone into calcium chloride and soda ash, a reaction that cannot be made to proceed under industrial conditions

$$2NaCl + CaCO_3 = CaCl_2 + Na_2CO_3 \qquad \Delta G^\circ_{298} = 9600$$

The cluster of reactions that yields this net result and the approximate temperatures at which they proceed are

$$CaCO_3 = CaO + CO_2 \qquad\qquad 1000°C$$
$$CaO + H_2O = Ca(OH)_2 \qquad\qquad 100°C$$
$$Ca(OH)_2 + 2NH_4Cl = CaCl_2 + 2NH_3 + 2H_2O \qquad 120°C$$
$$2NH_3 + 2H_2O + 2CO_2 = 2NH_4HCO_3 \qquad\qquad 60°C$$
$$2NH_4HCO_3 + 2NaCl = 2NaHCO_3 + 2NH_4Cl \qquad 60°C$$
$$2NaHCO_3 = Na_2CO_3 + H_2O + CO_2 \qquad\qquad 200°C$$

The sum of these reactions is the Solvay soda-ash process, the classic example of the closed-cycle use of a cluster of reactions to arrive at an important but otherwise infeasible reaction.

The Solvay cluster must meet several conditions: the cluster must have stoichiometric integrity in that each intermediate material must be produced and consumed in identical amounts. The reactions must each move to the right at reasonable operating conditions and at a sufficiently high rate. The equilibrium conditions must guarantee formation of reaction products with sufficient yield. Some flexibility can exist in the thermodynamic conditions, as we have seen earlier, with respect to the temperature and pressure at which reactions can be carried out, but this flexibility is limited.

A number of commercial reactions in addition to the Solvay soda-ash process are driven by Solvay clusters, e.g., the production of carbon tetrachloride and of acetaldehyde. For carbon tetrachloride the reaction clusters are

$$3CH_4 + 6S_2 = 3CS_2 + 6H_2S$$
$$2CS_2 + 6Cl_2 = 2CCl_4 + 2S_2Cl_2$$
$$CS_2 + 2S_2Cl_2 = 3S_2 + CCl_4$$
$$6H_2S + 3O_2 = 3S_2 + 6H_2O$$

$$\overline{3CH_4 + 6Cl_2 + 3O_2 = 3CCl_4 + 6H_2O}$$

and, for acetaldehyde

$$C_2H_4 + H_2O + PdCl_2 = CH_3CHO + 2HCl + Pd$$

$$Pd + 2CuCl_2 = 2CuCl + PdCl_2$$

$$2HCl + 2CuCl + \tfrac{1}{2}O_2 = H_2O + 2CuCl_2$$

$$C_2H_4 + \tfrac{1}{2}O_2 = CH_3CHO$$

These processes, like other commercial Solvay-cluster-driven reactions, are characterized by the fact that the intermediate chemicals and reactions seem to have little relation to the net reaction. The development of these reaction sequences is poorly understood, and until recently there have been no principles to guide the development of closed-cycle sequences of reactions. May and Rudd (1976) addressed themselves to this problem and showed that the development of thermodynamically feasible Solvay clusters is a problem in pattern recognition. The following discussion is based on their treatment.

The three factors that must be considered in the development of a Solvay cluster are the stoichiometry of the reactions, the thermodynamic equilibrium, and the reaction kinetics. We shall deal only with the thermodynamic aspect. The necessary condition for a reaction to be thermodynamically feasible is that

$$\Delta G^\circ \leq \varepsilon$$

where, as we have seen earlier, ε is of the order of 10,000 cal/g mol.

Let us now examine a three-step cluster. We shall see that a common difference can be found for each of the reactions in the cluster. The net reaction is

$$A + B + C = Z$$

where $\Delta G^\circ_{net} \geq \varepsilon$ and the reaction is therefore thermodynamically infeasible. This reaction can also be written as

$$Z - B - C = A$$

where A will be the common difference. The reactions in the cluster are:

No.	Reaction	Common difference
(1)	$A + L = N$	$N - L = A$
(2)	$N + K = L + P + Q$	$P + Q - K = N - L = A$
(3)	$B + C + P + Q = K + Z$	$P + Q - K = Z - B - C = A$

For the Solvay cluster to be feasible we must have $\Delta G^\circ \leq \varepsilon$ at the reaction conditions for each of the three reactions; that is,

$$\Delta G^\circ_1 = G_N - (G_A + G_L) \leq \varepsilon$$

$$\Delta G^\circ_2 = (G_L + G_P + G_Q) - (G_N + G_K) \leq \varepsilon$$

$$\Delta G^\circ_3 = (G_K + G_Z) - (G_B + G_C + G_Q + G_P) \leq \varepsilon$$

and for the net reaction

$$\Delta G^\circ_{net} = G_Z - (G_A + G_B + G_C) \geq \varepsilon$$

These expressions can be rewritten as

$$G_Z - G_B - G_C \geq G_A + \varepsilon$$

at all reaction conditions for the net reaction and

$$G_N - G_L < G_A + \varepsilon \qquad \text{at conditions of reaction (1)}$$

$$G_N - G_L + \varepsilon \geq G_P + G_Q - G_K \qquad \text{at conditions of reaction (2)}$$

$$G_Z - G_B - G_C \leq G_P + G_Q - G_K + \varepsilon \qquad \text{at conditions of reaction (3)}$$

These criteria are shown in Fig. 4-3, and we see that the thermodynamic conditions are satisfied for reaction (1) at $T \geq T_1$, for reaction (2) at $T \leq T_2$, and for reaction (3) at $T \geq T_3$.

It should be noted that a ladderlike plot is formed when the thermodynamic criteria are satisfied. It does not involve the free energy of any reaction but instead involves the difference of the free energy of formation of the species that enter into the reactions. Climbing the ladder, i.e., the jumps from line to line, corresponds to executing reactions, and the jumps can occur only if the lines are within the ε distance. The general trend is that of executing the net reaction by

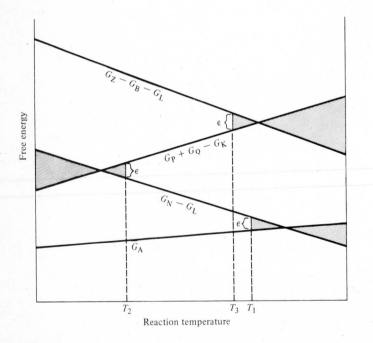

Figure 4-3 Solvay-cluster ladder pattern.

climbing from reactants to products along the ladder defined by the common-difference equation. If a proposed Solvay cluster does not exhibit the proper geometric pattern on the free-energy-reaction condition diagram, the cluster is not thermodynamically feasible.

As an example, May and Rudd (1976) demonstrate the synthesis of thermodynamically feasible Solvay clusters for the infeasible reaction

$$2HCl = H_2 + Cl_2$$

The free-energy diagram is shown in Fig. 4-4. The heavy lines identify one feasible ladder pattern corresponding to the following cluster that yields the net reaction

$$2HCl + 2CrCl_2 = H_2 + 2CrCl_3$$

$$2MnCl_3 + 2CrCl_3 = 2CrCl_2 + 2MnCl_4$$

$$2MnCl_4 = 2MnCl_3 + Cl_2$$

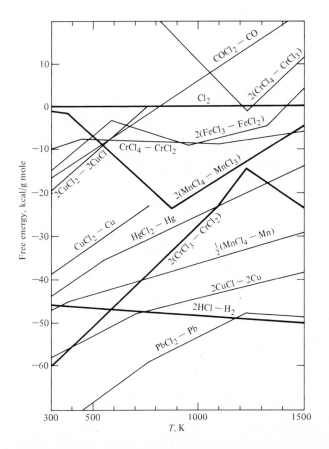

Figure 4-4 Ladder patterns yielding $H_2 + Cl_2$. [*From May and Rudd (1976), by permission.*]

4-6 ADIABATIC REACTION PATH

In laboratory or small-scale pilot-plant equipment it is relatively easy to maintain or at least approach isothermal conditions in a chemical reactor. As the
equipment size increases and the surface-to-volume ratio decreases, it becomes
more and more difficult to provide sufficient heat-transfer surface to maintain
even moderately endothermic or exothermic reactions at conditions approximat-
ing isothermal conditions. In many practical situations the reaction will be
carried out under conditions which will be much closer to the adiabatic than to
isothermal conditions.

For any given initial temperature of the reactants, the temperature attained
in an adiabatic reaction will be a function of the extent of conversion, and the
equilibrium extent of conversion reached, in turn, is dependent on the temperatures. The equilibrium temperature can easily be determined graphically by plotting the equilibrium extent of conversion for an isothermal reaction as a function
of temperature and on the same coordinates plotting the adiabatic-reaction
temperature. The intersection of these two curves will give the adiabatic equilibrium temperature and conversion. The characteristic behavior of these curves is
shown in Fig. 4-5 for endothermic and exothermic reactions. In the exothermic
case, the adiabatic reaction path is upward and to the right, and it will continue
until it intersects the equilibrium curve, giving the equilibrium conversion and
temperature. In the endothermic case, the adiabatic reaction path is upward and
to the left, until it intersects the equilibrium curve. In both cases isothermal
operation would have resulted in a higher equilibrium conversion than is possible with adiabatic operation.

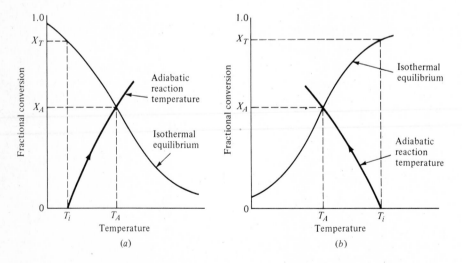

Figure 4-5 Adiabatic equilibrium conversion: (*a*) exothermic and (*b*) endothermic reaction.

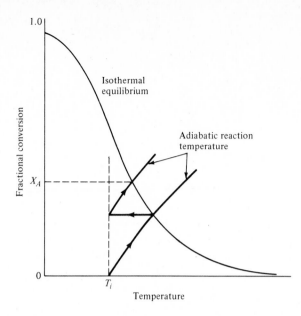

Figure 4-6 Adiabatic reaction with intercooling.

Even though it may not be possible to maintain other than adiabatic conditions in the reactor, it is possible to increase equilibrium yields by operating with several reactors in series with intermediate cooling between reactors. This is shown graphically in Fig. 4-6 for one stage of cooling for an exothermic reaction.

4-7 REPRESENTATION OF SIMULTANEOUS REACTIONS

We are generally accustomed to writing a chemical reaction according to the convention that the reactants appear on the left-hand side of the expression and the products appear on the right-hand side. The production of methanol from synthesis gas would be written

$$CO + 2H_2 = CH_3OH$$

But we can adopt another convention whereby the same reaction could be written

$$CH_3OH - CO - 2H_2 = 0$$

The numbers 1, -1, and -2 are, of course, the stoichiometric coefficients.

This method of writing a chemical reaction can be easily extended to simultaneous reactions. Let us assume that there are R chemical reactions involving S

chemical species. This set of simultaneous reactions can be expressed compactly by

$$\sum_{j=1}^{S} \alpha_{ij} A_j = 0 \qquad i = 1, 2, ..., R \tag{4-18}$$

Consider the two reactions

$$CO + 2H_2 \rightleftharpoons CH_3OH$$

$$CH_4 + H_2O \rightleftharpoons CO + 3H_2$$

Let us identify the chemical species as

$$A_1 = CH_3OH \qquad A_2 = CO \qquad A_3 = H_2 \qquad A_4 = CH_4 \qquad A_5 = H_2O$$

The reactions can now be written as

$$A_1 - A_2 - 2A_3 \qquad\qquad = 0$$

$$A_2 + 3A_3 - A_4 - A_5 = 0$$

For this set of reactions $R = 2$ and $S = 5$ and the stoichiometric coefficients are

$$\alpha_{11} = \alpha_{22} = 1 \qquad \alpha_{12} = \alpha_{24} = \alpha_{25} = -1 \qquad \alpha_{13} = -2 \qquad \alpha_{23} = 3$$

and

$$\alpha_{14} = \alpha_{15} = \alpha_{21} = 0$$

Now consider the reaction

$$CH_4 + H_2O \rightleftharpoons CH_3OH + H_2$$

Using the same species identification as before, we can write this reaction as

$$A_1 + A_3 - A_4 - A_5 = 0$$

and this clearly is the sum of the preceding two reactions.

In any reaction-equilibrium analysis in which simultaneous reactions can be involved it is important to know how many independent reactions there are in the system. We shall now present a procedure by which the number of independent equations and therefore the number of independent reactions can be determined. There are several possibilities by which a set of R reactions would not all be independent. One reaction could, for example, be a multiple of another reaction, or one reaction could be linear combination of two or more other reactions. Cases of these kinds could then be eliminated from the thermodynamic considerations.

From linear algebra we know that the linear equations

$$\sum_{j=1}^{S} \alpha_{ij} A_j = 0 \qquad i = 1, 2, ..., R \tag{4-19}$$

are independent if no set of multipliers λ_i, where $i = 1, 2, ..., R$, can be found other than the trivial set $\lambda_i = 0$ such that

$$\sum_{i=1}^{R} \lambda_i \alpha_{ij} = 0 \qquad j = 1, 2, ..., S \tag{4-20}$$

This is equivalent to saying that the rank of the matrix whose element in the ith row and jth column is α_{ij} is R. Thus the rank of the matrix of the stoichiometric coefficients will give us the number of independent reactions in the set.

A simple procedure for finding the number of independent reactions will be presented by means of an example.

Example 4-4 A number of reactions can be written involving C, O_2, H_2, CO_2, CO, and H_2O. One possible group of simultaneous reactions is

$$C + O_2 \rightleftharpoons CO_2$$

$$C + CO_2 \rightleftharpoons 2CO$$

$$H_2O + C \rightleftharpoons CO + H_2$$

$$2H_2 + O_2 \rightleftharpoons 2H_2O$$

$$CO + H_2O \rightleftharpoons CO_2 + H_2$$

How many of these reactions are independent?

SOLUTION Six chemical species are present. We shall identify them as follows:

$$A_1 = C \quad A_2 = O_2 \quad A_3 = H_2 \quad A_4 = CO \quad A_5 = CO_2 \quad A_6 = H_2O$$

The reactions are

$$-A_1 - A_2 \qquad\qquad + A_5 \qquad\quad = 0$$

$$-A_1 \qquad\qquad + 2A_4 - A_5 \qquad\quad = 0$$

$$-A_1 \qquad + A_3 + A_4 \qquad - A_6 = 0$$

$$-A_2 - 2A_3 \qquad\qquad + 2A_6 = 0$$

$$+ A_3 - A_4 + A_5 - A_6 = 0$$

The stoichiometric coefficients are now written in matrix form

$$\begin{bmatrix} -1 & -1 & 0 & 0 & 1 & 0 \\ -1 & 0 & 0 & 2 & -1 & 0 \\ -1 & 0 & 1 & 1 & 0 & -1 \\ 0 & -1 & -2 & 0 & 0 & 2 \\ 0 & 0 & 1 & -1 & 1 & -1 \end{bmatrix}$$

The first step is to treat the first row so that the leading number will be $+1$. This row is then used to make the leading element of the remaining rows equal to zero. In our case we can make the leading element of the first row equal to $+1$ by multiplying the row by -1. The leading element of the next two rows can be made zero by adding the new first row to the second and third rows. The final two rows already have zeros in their leading elements. The resulting matrix will be

$$\begin{bmatrix} 1 & 1 & 0 & 0 & -1 & 0 \\ 0 & 1 & 0 & 2 & -2 & 0 \\ 0 & 1 & 1 & 1 & -1 & -1 \\ 0 & -1 & -2 & 0 & 0 & 2 \\ 0 & 0 & 1 & -1 & 1 & -1 \end{bmatrix}$$

Treatment of the first row and column is now complete, and the same procedure is continued on the second row and column, but this time, to produce a value of 1 in location (2, 2) with zeros in the second column of the subsequent rows. This procedure is continued to produce 1s along the diagonal as far as they will go. The matrices that will appear for our example are, for the next two stages

$$\begin{bmatrix} 1 & 1 & 0 & 0 & -1 & 0 \\ 0 & 1 & 0 & 2 & -2 & 0 \\ 0 & 0 & 1 & -1 & 1 & -1 \\ 0 & 0 & -2 & 2 & -2 & 2 \\ 0 & 0 & 1 & -1 & 1 & -1 \end{bmatrix} \quad \text{and} \quad \begin{bmatrix} 1 & 1 & 0 & 0 & -1 & 0 \\ 0 & 1 & 0 & 2 & -2 & 0 \\ 0 & 0 & 1 & -1 & 1 & -1 \\ 0 & 0 & 0 & 0 & 0 & 0 \\ 0 & 0 & 0 & 0 & 0 & 0 \end{bmatrix}$$

and we can go no farther.

The number of independent reactions will be equal to the number of 1s appearing on the diagonal, or, alternatively, it will be equal to the number of reactions R minus the number of rows that are entirely zero in the final array.

In our thermodynamic analysis we can work with an independent subset of the original reaction set, and it usually is convenient to do so. In the example the independent reactions are

$$A_1 + A_2 \quad\quad - A_5 \quad\quad = 0$$
$$A_2 + 2A_4 - 2A_5 \quad\quad = 0$$
$$A_3 - A_4 + A_5 - A_6 = 0$$

or, in terms of chemical species,

$$CO_2 \rightleftharpoons C + O_2$$
$$2CO_2 \rightleftharpoons O_2 + 2CO$$
$$CO + H_2O \rightleftharpoons H_2 + CO_2$$

All the reactions in the original scheme can be generated from this set by appropriate linear combinations.

4-8 EQUILIBRIUM IN SIMULTANEOUS REACTIONS

In practice, and particularly in organic reactions, one must be concerned with a number of reactions occurring simultaneously. Many reactions are possible for a given set of reactants. Consider, for example, a possible sequence of reactions that could develop with CO and H_2, as the reactants:

$$CO + H_2 \rightleftharpoons HCOH$$

$$CO + 2H_2 \rightleftharpoons CH_3OH$$

$$CO + 3H_2 \rightleftharpoons CH_4 + H_2O$$

$$2CO + 5H_2 \rightleftharpoons C_2H_6 + 2H_2O$$

$$CH_4 + 2H_2O \rightleftharpoons CO_2 + 4H_2$$

and so forth.

The analysis of such a system would require knowledge of the product distribution to be expected at thermodynamic equilibrium. The same principle is involved in solving problems of equilibrium composition, regardless of whether a single reaction or simultaneous reactions are involved. The principle is to find the composition that simultaneously satisfies the thermodynamic-equilibrium requirements and the mass-balance requirements. The classical approach to this problem is to produce an independent set of reactions and then to examine the various equilibria involved and eliminate those reactions which are extremely unfavorable from the thermodynamic viewpoint. Then the set of nonlinear simultaneous equations involving the equilibrium constants, composition, and pressure is set up along with the elemental mass balances. The set of rather formidable equations that may remain is then solved.

Dodge (1944) discusses a process for producing hydrogen by passing methane and steam over a catalyst. The question posed is how much methane will be decomposed and what the composition of the off-gas will be if 5 mol of steam per mole of CH_4 is passed over a catalyst at 600°C and 1 atm and equilibrium is attained. Many reactions as possible. The set proposed by Dodge is

$$CH_4 + H_2O \rightleftharpoons CO + 3H_2 \tag{1}$$

$$CH_4 + 2H_2O \rightleftharpoons CO_2 + 4H_2 \tag{2}$$

$$CH_4 \rightleftharpoons C(s) + 2H_2 \tag{3}$$

$$CO + H_2O \rightleftharpoons CO_2 + H_2 \tag{4}$$

$$CO_2 + C(s) \rightleftharpoons 2CO \tag{5}$$

$$CO_2 \rightleftharpoons C(s) + O_2 \tag{6}$$

$$2CO \rightleftharpoons 2C(s) + O_2 \tag{7}$$

$$H_2O + C(s) \rightleftharpoons H_2 + CO \tag{8}$$

$$H_2O \rightleftharpoons H_2 + \tfrac{1}{2}O_2 \tag{9}$$

$$2CH_4 \rightleftharpoons C_2H_6 + H_2 \tag{10}$$

$$C_2H_2 \rightleftharpoons C_2H_4 + H_2 \tag{11}$$

. . . .

This situation looks extremely difficult, but it can be simplified. Not all the reactions shown are independent, and by applying the procedure described earlier reactions (1), (2), (7), and (8) can be eliminated as being linear combinations of other reactions. Next, an examination of the values for the standard-free-energy change shows that reactions (6), (9), and (10) can be neglected because the standard-free-energy change is highly positive. If reaction (10) can be neglected, ethane will not be produced and hence reaction (11) and the subsequent reactions can also be neglected.

Only three reactions survive this preliminary screening

$$CH_4 + H_2O \rightleftharpoons CO + 3H_2 \tag{1}$$

$$CO + H_2O \rightleftharpoons CO_2 + H_2 \tag{4}$$

$$CO_2 + C(s) \rightleftharpoons 2CO \tag{5}$$

For the time being reaction (5) will be neglected, and now only two simultaneous reactions remain. These reactions will give two simultaneous equations to solve that arise from $\Delta G°$ and the resulting value for K, the equilibrium constant. Expressions for K are then set up, careful attention being paid to the required mass and element balances.

When these equations are solved, the off-gas equilibrium composition shown in Table 4-1 is obtained.

The problem is not yet completely solved, as we must now go back to determine whether we were justified in neglecting reaction (5). For this reaction at 600°C

$$K = K_p = 0.086$$

$$K_p = \frac{\bar{P}_{CO}^2}{\bar{P}_{CO_2}}$$

For the composition shown in Table 4-1, however,

$$\frac{\bar{P}_{CO}^2}{\bar{P}_{CO_2}} = \frac{0.0315^2}{0.0848} = 0.0117$$

Table 4-1 Equilibrium off-gas composition

Component	volume %
CH_4	1.16
H_2	43.36
H_2O	43.85
CO	3.15
CO_2	8.48

This is smaller than 0.086, the value for the equilibrium constant; therefore no solid carbon can be present. In fact, if carbon were present, it would be consumed by reaction (5) in an attempt to increase the value of P_{CO}^2/P_{CO_2} to satisfy the equilibrium requirements.

4-9 FREE-ENERGY-MINIMIZATION METHOD

The procedure described above is tedious and requires laborious hand computation. The problem of obtaining an equilibrium composition is really a problem of minimizing the free energy of the system. When the system is at equilibrium, it is at its minimum free energy. White et al. (1958) suggested the free-energy-minimization method in 1958, and the technique has since been programmed for use on computers. In this method the individual reactions are not considered. Instead, all possible chemical species that could appear are noted, and the distribution of these species, consistent with material and elemental balance considerations, that will minimize the free energy of the system is computed.

Oliver et al. (1962) described one such program that includes the elemental balances, the constraint of nonnegative concentrations, and also expressions for the free energy of the various chemical species normally encountered in combustion reactions. Their procedure calculates the free energy of the system for an assumed concentration and then uses an efficient procedure to search for the composition that will minimize the free energy. The operator must guess at an initial composition to begin the iterative procedure.

They applied their program to solve the problem discussed above using two approaches. In the first approach they assumed, just as Dodge had done in his solution, that only six species would be present at equilibrium, CH_4, CO, H_2O, H_2, CO_2, and $C(s)$. The same off-gas concentration as obtained by Dodge was reached as the solution to the problem after six iterations.

In the second approach, they postulated the possible presence of 20 chemical species at equilibrium. Again, the same off-gas composition as in the first solution was found, but this time 22 iterations were required to reach the solution.

If the analysis of a given reaction situation will require the exploration of a broad temperature and pressure field, the free-energy-minimization method programmed for a computer will prove to be a time-saving procedure.

Dluzniewski and Adler (1972) considered nonidealities in the vapor phase and also included the liquid phase in their analysis. This permitted an artifice by which the equilibrium of complex nonreacting systems could be calculated in addition to the equilibrium of reacting systems. They present several examples in which equilibrium compositions were calculated by free-energy minimization and compared with experimental results.

In one example, they estimate the equilibrium composition for a steam-reforming operation at 1730°F and 73 lb/in^2 abs at a steam-to-carbon ratio of 0.95. The experimental results obtained from pilot-plant operation are presented in Table 4-2 along with the calculated results. The comparison is very favorable.

Table 4-2 Comparison of calculated and experimental values for steam-reforming-reaction products

	Composition, mole fraction	
Compound	Calculated	Experimental
Methane	0.035	0.035
Carbon dioxide	0.004	0.006
Carbon monoxide	0.232	0.231
Water	0.018	0.021
Hydrogen	0.711	0.707

Source: From Dluzniewski and Adler, (1972), by permission.

In addition, the free-energy-minimization model predicted carbon deposition. Dluzniewski and Adler were pleased to point out that the pilot-plant reactor actually plugged with carbon before the scheduled termination of the run.

Free-energy minimization for predicting equilibrium is thermodynamically sound, and complete faith can be placed in the calculated results provided the thermodynamic data are accurate. The equilibrium predicted is the true thermodynamic equilibrium, but its reliability is limited by the accuracy of the input thermodynamic data. If predicted and experimental equilibrium results are not in agreement, either the input data or the experimental results are in error.

PROBLEMS

The thermodynamic properties necessary for solution of most of these problems are not included in the problem statements. The following sources of property data are suggested:

American Petroleum Institute: "Selected Values of Physical and Thermodynamic Properties of Hydrocarbons and Related Compounds," Project 44, Carnegie Press, Pittsburgh, 1953.

Bichowsky, F. R., and F. D. Rossini: "Thermochemistry of Chemical Substances," Reinhold, New York, 1936.

Canjar, L. N., and F. S. Manning: "Thermodynamic Properties and Reduced Correlation for Gases," Gulf Publishing, Houston, Tex., 1967.

"International Critical Tables," McGraw-Hill, New York, 1926.

Lange, N. A., "Handbook of Chemistry," various editions, Handbook Publishers, Sandusky, Ohio.

Parks, G. S., and H. M. Huffman: The Free Energies of Some Organic Compounds, *A.C.S. Monog.* 60, Reinhold, New York, 1932.

Perry, J. H.: "Chemical Engineer's Handbook," various editions, McGraw-Hill, New York.

Reid, R. C., J. M. Prausnitz, and T. K. Sherwood: "The Properties of Gases and Liquids," 3d ed., McGraw-Hill, New York, 1977.

Rossini, F. D., et al.: Selected Values of Chemical Thermodynamic Properties, *NBS Circ.* 500, 1952.

Stull, D. R., and H. Prophet: "JANAF Thermochemical Tables," 2d ed., NSRDS-NBS 37, Washington, 1971.

————, E. F. Westrum, and G. C. Sinke: "The Chemical Thermodynamics of Organic Compounds," Wiley, New York, 1969.

Weast, R. C. (ed.): "Handbook of Chemistry and Physics," various editions, CRC Press, West Palm Beach, Fla.

4-1 Recent researches have shown that certain zeolites are able to convert methanol into products in the gasoline range. The process is carried out in two reactors operating in series. In the first reactor, operated at 400°C and 20 atm, the dehydration of methanol occurs according to

$$2CH_3OH \; \rightleftharpoons \; CH_3OCH_3 + H_2O$$

This is followed by the conversion reactor, operated at about 450°C and 18 atm, in which the dimethyl ether and the unreacted methanol are further dehydrated and converted into paraffins and aromatics

$$nCH_3OH \; \rightleftharpoons \; n(-CH_2-) + nH_2O$$

$$mCH_3OCH_3 \; \rightleftharpoons \; 2m(-CH_2-) + mH_2O$$

The dehydration of methanol to dimethyl ether is believed to reach equilibrium very quickly. On the assumption that this is the only reaction occurring in the first reactor, estimate the composition of its exit gas.

4-2 Aniline is produced by the continuous catalytic vapor-phase reduction of nitrobenzene

$$C_6H_5NO_2(g) + 3H_2(g) \; \rightleftharpoons \; C_6H_5NH_2(g) + 2H_2O(g)$$

The reaction is carried out at about 270°C and 2 atm in a fluidized catalyst bed. The ratio of hydrogen to nitrobenzene in the feed to the catalyst bed is about 3 times the stoichiometric ratio.

(a) Estimate the equilibrium conversion of nitrobenzene to aniline.

(b) Why is an excess of hydrogen used?

4-3 Your company has received an order to supply Ag_2CO_3 to a catalyst manufacturer. The purity requirements are extremely strict, and the purchase contract includes a penalty clause in the event that the Ag_2O content exceeds 0.001 percent. The final stages of the production process include washing Ag_2CO_3 crystals followed by drying at 110°C. Is there any danger that Ag_2O could be formed during the drying step by $Ag_2CO_3 \rightleftharpoons Ag_2O + CO_2$? (Air contains 0.03% CO_2.) If so, what steps would you recommend to prevent the decomposition reaction?

4-4 An inert gas is required at 400°C. It is proposed that the gas be produced by burning CH_4 with a slight deficiency of air so that all the oxygen will be consumed. It is expected that 95 percent of the C in the CH_4 would burn to CO_2 and 5 percent to CO. The gas will be dried to remove the water vapor and then reheated to 400°C. Would you expect to have any problems with carbon deposition in the heater due to $2CO(g) \rightleftharpoons C + CO_2(g)$?

4-5 Butadiene can be produced by the dehydrogenation of butylene according to

$$C_4H_8(g) \; \rightleftharpoons \; C_4H_6 + H_2$$

For this reaction assume that $\Delta G^\circ_{298} = 21.6$ kcal/g mol; $\Delta H^\circ_{298} = 28.6$ kcal/g mol, and $\Delta C_p = 0$. Ideal-gas behavior may also be assumed.

(a) Estimate the equilibrium conversion to be expected at 600°C and 1 atm assuming that only the dehydrogenation reaction takes place.

(b) Estimate the equilibrium conversion if 10 mol of steam is fed as inert diluent per mole of butylene.

(c) If steam were not used, what operating pressure would be required to obtain the same equilibrium conversion as in (b)? Is this a desirable operating condition?

4-6 For each mole of chlorine consumed in the chlorination of hydrocarbons 1 mol of HCl is produced. The chlorine value in the HCl can be recovered by the Deacon process

$$4HCl(g) + O_2(g) \rightleftharpoons 2Cl_2(g) + 2H_2O$$

The reaction can be carried out at temperatures in the range of 450 to 650°C over a copper(II) chloride catalyst.

(a) What conversion of HCl to Cl_2 would you expect if air is used?

(b) Would it be advantageous to use oxygen instead of air?

(c) Sketch a flowsheet for a process to produce dry Cl_2 from HCl produced in a hydrocarbon chlorination reaction.

4-7 By oxychlorination HCl can be used directly for the chlorination of organic compounds instead of proceeding via the Deacon process to Cl_2. In the oxychlorination of benzene, for example, the overall reaction is

$$C_6H_6 + HCl + \tfrac{1}{2}O_2 \rightleftharpoons C_6H_5Cl + H_2O$$

This can be considered as being the sum of two reactions

$$2HCl + \tfrac{1}{2}O_2 \rightleftharpoons Cl_2 + H_2O$$

and

$$C_6H_6 + Cl_2 \rightleftharpoons C_6H_5Cl + HCl$$

The Raschig process produces chlorobenzene by the above reaction in the vapor phase at about 200°C and with a ratio of benzene to HCl of about 8 : 1. The chlorobenzene is then hydrolyzed in the vapor phase at about 475°C and with an excess of steam to produce phenol

$$C_6H_5Cl + H_2O \rightleftharpoons C_6H_5OH + HCl$$

Thus, the overall reaction is the production of phenol by oxidation of benzene

$$C_6H_6 + \tfrac{1}{2}O_2 \rightleftharpoons C_6H_5OH$$

(a) What equilibrium conversions would you expect for the oxychlorination reaction and the hydrolysis reaction?

(b) Why would a high ratio of benzene to HCl be used in the oxychlorination reaction? Why use an excess of steam in the hydrolysis reaction?

4-8 In the contact process for the oxidation of SO_2 to SO_3 the reactor consists of several beds of catalyst in series with gas intercooling between the adiabatic catalyst beds. The design is such that equilibrium is very nearly reached at the exit of each bed. The feed gas to a certain contact converter is from a sulfur burner, and the analysis is 11% SO_2, 10% O_2, balance N_2. The gas enters the first contact stage and essentially reaches equilibrium. It is cooled to 410°C and enters the second stage. The gas leaving the second stage enters a primary absorber, in which the SO_3 is absorbed in sulfuric acid. The exit gas from the absorber is reheated to 400°C and enters the third catalyst bed.

$$SO_2(g) + \tfrac{1}{2}O_2 \rightleftharpoons SO_3(g)$$

$$\log K_p = \frac{4760}{T} - 4.474$$

where the temperature is in kelvins and pressure is in atmospheres.

(a) What will be the temperature of the gas leaving the first catalyst contact stage and the composition of the gas leaving the second stage?

(b) What will be the composition on an SO_3-free basis of the gas leaving the third contact stage? What is the overall conversion of SO_2 to SO_3?

4-9 Carbon tetrachloride is one of the important halogenated paraffins. One method of production is by the chlorination of methane, which produces methyl chloride, methylene chloride, and chloroform in addition to carbon tetrachloride. Carbon tetrachloride can also be produced by the following cluster of reactions (approximate reaction conditions are shown alongside each reaction):

$$2CH_4(g) + 4S_2(g) \rightleftharpoons 2CS_2(g) + 4H_2S(g) \qquad \text{gas phase, 700°C, 2 atm}$$

$$4H_2S(g) + 2O_2(g) \rightleftharpoons 4H_2O(g) + 2S_2(g) \qquad \text{gas phase, 500°C, 1 atm}$$

$$CS_2(l) + 3Cl_2(g) \rightleftharpoons CCl_4(l) + S_2Cl_2(l) \qquad \text{liquid phase, 30°C, 1 atm}$$

$$CS_2(l) + 2S_2Cl_2(l) \rightleftharpoons 6S(c) + CCl_4(l) \qquad \text{liquid phase, 60°C, 1 atm}$$

(a) Show that this is a Solvay cluster and that the thermodynamic conditions are favorable for each reaction at the conditions at which it is executed. (Although sulfur in the vapor phase is present as an equilibrium mixture of S_2, S_6, and S_8 molecules, assume that all vapor-phase S is present as S_2.)

(b) Equilibrium considerations show that the product composition from methane chlorination depends on the initial ratio of chlorine to methane. Discuss the effects to be expected on product distribution as the initial ratio of chlorine to methane is increased. What would be the effect of recycle of the various product species? Analyze the effect of the competing reaction

$$CH_4(g) + 2Cl_2(g) \rightleftharpoons C(s) + HCl(g)$$

4-10 Synthesis gas is a mixture of CO and H_2 which is used in the synthesis of ammonia, methanol, and oxo products. When natural gas (assume 100% CH_4) is the raw material, the first stage in synthesis-gas preparation is primary reforming, in which the following reactions take place:

Reforming: $\qquad CH_4 + H_2O \rightleftharpoons CO + 3H_2$

Shift: $\qquad CO + H_2O \rightleftharpoons CO_2 + H_2$

Carbon can deposit on the reforming catalyst according to

$$2CO \rightleftharpoons C + CO_2$$

The ratio of steam to natural gas in the feed to a typical primary reformer would be 5 : 1, and the reformer would operate at 790°C and 20 atm. At 790°C, K_p for the reforming reaction is 128; it is 1 for the shift reaction and 0.17 for carbon deposition.

(a) Assuming ideal-gas behavior and neglecting the carbon deposition reaction, what would the equilibrium composition of the gas leaving the primary reformer be?

(b) Would you expect carbon deposition to occur?

(c) Although the shift reaction is independent of pressure (if ideal-gas behavior is assumed), the conversion of methane in the reforming reaction is favored by a low pressure. In practice, however, elevated pressures are used, and the trend is to even higher pressures. Explain.

4-11 Methylamine is produced by the vapor-phase reaction between methanol and ammonia over an active catalyst. All three methylamines are produced, according to

$$CH_3OH + NH_3 \rightleftharpoons CH_3NH_2 + H_2O$$

$$2CH_3OH + NH_3 \rightleftharpoons (CH_3)_2NH + 2H_2O$$

$$3CH_3OH + NH_3 \rightleftharpoons (CH_3)_3N + 3H_2O$$

The reactor operates at 350°C and 3 atm. The sales forecast for the three methylamines is that they will be sold in the ratio (molar basis) of 2 tri- : 3 mono- : 5 di-. Assuming that equilibrium is reached, that only these three reactions take place, and that ideal-gas behavior obtains, what should the ratio of CH_3OH to NH_3 in the reactor feed be? Discuss briefly the effect of changes in the methanol-to-ammonia ratio on the equilibrium product distribution.

4-12 Consider the first eight reactions shown in the set of reactions (1) to (11) in Sec. 4-8. How many of them are independent?

CHAPTER FIVE

NONREACTING PROCESS ANALYSIS

> Glass makers heat their furnace for seven consecutive days and nights. The intensity of the fire produces a creature similar to a spider which is called a salamander.
>
> *"Midrash Tanhuma," on Vayeshev, 3*

In Chap. 4 we dealt with the thermodynamic analysis of a number of the factors that affect the equilibrium conversion in a reaction process. Although the chemical reactor and its operation represent the core and heart of a chemical process, the reactor represents only one part of the entire complex of equipment and operations in a chemical plant. The object of this chapter is to show how thermodynamics can be applied to evaluate and analyze other parts of the chemical plant.

Much of our use of thermodynamics is based on the assumption of ideal conditions or reversible processes. By comparing the actual situation to the ideal one it is possible to analyze operations, to calculate their efficiencies, and to help make a rational choice between alternatives.

5-1 LOST-WORK CONCEPT

Let us consider a case where an amount of heat Q at the temperature T_1 is being transferred in a heat exchanger to cooling water at T_0. The total change in entropy that takes place as a result of this operation is the sum of the changes that occur in the system, the heat exchanger, and in the surroundings, the cooling water,

$$\Delta S_{surr} + \Delta S_{sys} = \frac{Q}{T_0} - \frac{Q}{T_1} = \frac{Q(T_1 - T_0)}{T_i T_0} = \Delta S_{tot} \tag{5-1}$$

This heat-transfer operation was carried out in an irreversible manner. It could have been carried out reversibly, however, by the following procedure. The heat Q could have been transferred at temperature T_1 to a reversible heat engine. This engine could have converted part of the heat into work and rejected the remainder at the temperature T_0. The maximum work that could have been obtained is

$$W = Q \frac{T_1 - T_0}{T_1} \tag{5-2}$$

The energy was not produced but was wasted. This *lost work* is the result of the irreversibility inherent in the original heat-transfer scheme.

An expression for lost work can be obtained for this case by eliminating Q from Eqs. (5-1) and (5-2)

$$W_l = T_0 \, \Delta S_{tot} \tag{5-3}$$

We now show that this equation is general and not specific to the particular case selected. We shall consider the nonflow situation first. According to the first law of thermodynamics, for a nonflow process

$$dW = dQ - dE \tag{5-4}$$

If all heat transfer between the system and the surroundings took place at T_0, and if the process were reversible, we would have

$$dW_{rev} = T_0 \, dS - dE \tag{5-5}$$

where dS and dE represent the changes in entropy and internal energy of the system. The difference between Eqs. (5-5) and (5-4) will be the lost work

$$dW_l = dW_{rev} - dW = T_0 \, dS - dQ \tag{5-6}$$

The heat transferred to the system dQ is equal to (minus) the heat absorbed by the surroundings $-dQ_0$, but the entropy change of the surroundings must be dQ_0/T_0; therefore,

$$dW_l = T_0 \, dS + dQ_0 = T_0 \, dS + T_0 \, dS_{rev} \tag{5-7}$$

and $\qquad dW_l = T_0 \, dS_{tot} \tag{5-8}$

The development for a flow system is analogous. For this case, neglecting kinetic- and potential-energy terms, we have

$$dW = dQ - dH \tag{5-9}$$

and $\qquad dW_{rev} = T_0 \, dS - dH \tag{5-10}$

leading to

$$dW_l = T_0 \, dS - dQ = T_0 \, dS + T_0 \, dS_{\text{surr}} \tag{5-11}$$

and to the same final result as for the nonflow case

$$dW_l = T_0 \, dS_{\text{tot}} \tag{5-12}$$

For a finite process if T_0 is constant, integration of Eq. (5-8) or (5-12) leads to

$$W_l = T_0 \, \Delta S_{\text{tot}}$$

as we found in Eq. (5-3) for our example. For practical problems T_0 is usually taken as the temperature at which heat can be conveniently discarded to the atmosphere or to cooling water.

Because T_0 is a constant temperature, and because the process which is the most efficient is the one with the smallest lost work, the reader will immediately realize that the process producing the smallest total increase in the entropy is the most efficient. This means that in many cases the thermodynamic comparison of alternative processing schemes requires merely the computation and comparison of the entropy increase of the competing alternatives.

Example 5-1 A plant is located where two-thirds of its cooling-water requirements are to be supplied by recirculation from a cooling tower that will provide cooling water at 90°F. The remaining one-third is to be supplied on a once-through basis from wells that produce water at 60°F. Should these streams be mixed, or should the water systems be kept separate?

SOLUTION The answer to this problem from the thermodynamic viewpoint should be obvious; since mixing is an irreversible process, the two streams should be kept separate. We shall go through the calculations, however, in order to demonstrate that the total entropy will increase if the two cooling-water streams are mixed. The basis for the calculations will be 3 lb of water, total.

If unmixed, the total entropy will be that of 2 lb of recirculating water at 90°F and 1 lb of well water at 60°F

$$S_{\text{tot}} = (2)(0.1115) + (1)(0.0555) = 0.2785 \text{ Btu/°R}$$

If mixed, the combined stream will be at 80°F, and the total entropy will be

$$S_{\text{tot}} = (3)(0.0932) = 0.2796 \text{ Btu/°R}$$

The entropy of the mixed streams has increased; hence, the thermal efficiency of the system has decreased. From the thermodynamic viewpoint the alternative of mixing the two streams is undesirable, but some practical means must also be available to make use of the energy that would be potentially available if the cooling-water streams were kept separate.

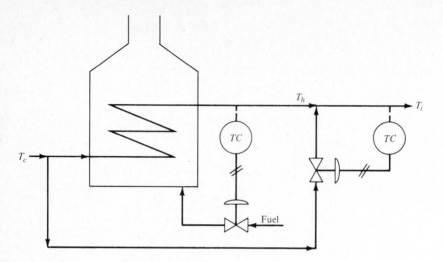

Figure 5-1 Mixing of hot and cold streams for temperature control.

The thermodynamic viewpoint is one of a number that should be considered. If the cooling-water streams are kept separate, there is a possibility, for example, that the heat-transfer surface could be reduced. A thorough study of the operations might indicate that 90°F cooling water would be sufficient for some of the streams. Only an overall study which would involve not only the thermodynamic aspects but also other aspects, such as economic, process, and plant operation, could resolve the question whether it was desirable to have one or two cooling-water systems.

In the above example cold well water was blended with recirculating cooling water to provide a sufficient quantity of cooling water, but hot and cold fluids are quite frequently mixed to provide responsive temperature control. Such a system is illustrated in Fig. 5-1 and is in common use when hot oil is provided by a furnace. Although the same net amount of heat is supplied to the fluid with or without the blending, the efficiency of the overall operation is reduced compared with that expected if T_i were produced directly by the furnace rather than the higher temperature T_h. The furnace will have to operate at a higher temperature, so that heat losses will be increased, as will the exit-flue-gas temperature, compared with the nonblending case.

In both these cases, no practical means for using the lost work exist. The thermodynamic evaluation, however, raised some warning flags, and the resulting more complete examination can disclose previously unrecognized energy losses or the possibility of reduction of capital expenditure.

5-2 AVAILABILITY CONCEPT

The *availability function* was first introduced by Gibbs and later employed by Keenan. It is defined as the maximum amount of useful work that could be performed when a substance in a given state passes to the "dead" state, which is usually taken as the pressure and temperature of the surroundings. The change in the availability can then be used to compare the amount of work used or produced in a given process with the minimum amount that could be used or the maximum that could be produced. It also can lead to a convenient means of calculating irreversibilities, thereby permitting thermodynamically inefficient process elements to be pinpointed.

For a nonflow process the maximum work obtainable when the system passes from an initial state to the dead state will be

$$B = (E_1 + P_0 v_1 - T_0 S_1) - (E_0 + P_0 v_0 - T_0 S_0) \tag{5-13}$$

The change in availability when the system passes from state 1 to a final state 2 will be

$$\Delta B = (E_2 + P_0 v_2 - T_0 S_2) - (E_1 + P_0 v_1 - T_0 S_1) \tag{5-14}$$

and Eq. (5-14) represents the net work performed upon moving from state 1 to state 2. The quantity $E + P_0 v - T_0 S$ is a property, and its value is fixed by the state of the system for any given value of P_0 and T_0.

For a flow system if changes in kinetic or potential energy are neglected, the expression for availability becomes

$$B = (H_1 - T_0 S_1) - (H_0 - T_0 S_0) \tag{5-15}$$

The change in availability between state 1 and state 2, including changes in kinetic and potential energy that can be completely converted into work, is

$$\Delta B = (H_2 - T_0 S_2) - (H_1 - T_0 S_1) + \frac{u_2^2}{2g_c} - \frac{u_1^2}{2g_c} + \frac{g z_2}{g_c} - \frac{g z_1}{g_c} \tag{5-16}$$

or $$\Delta B = \Delta H - T_0 \, \Delta S + \frac{\Delta u^2}{2g_c} + \frac{g \, \Delta z}{g_c} \tag{5-17}$$

A decrease in availability represents energy that is potentially available for performing useful work. The difference between the change in availability and the useful work performed represents work that was lost because of inversibilities in the system. The term $T_0 \, \Delta S$ in Eq. (5-17) represents energy that is unavailable for performing useful work. It is the heat that would be discarded to the surroundings at T_0 by a Carnot engine.

Example 5-2 Nitrogen at 230°C, 10 atm abs pressure, is expanded through a turbine, leaving it at 35°C and 1.5 atm. It is then discharged to the atmosphere. The turbine produces 1.8 hph of shaft work per kilogram mole of nitrogen. Assuming that the lowest convenient temperature at which heat can be discharged to the surroundings is 20°C, (a) calculate the maximum useful work that could be obtained from the nitrogen and (b) perform a lost-work analysis.

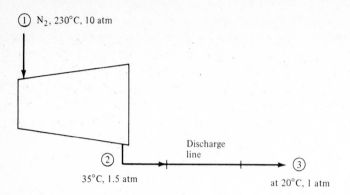

① N_2, 230°C, 10 atm

Discharge
line

②

35°C, 1.5 atm

③

at 20°C, 1 atm

Figure 5-2 Nitrogen expander.

SOLUTION The system is shown schematically in Fig. 5-2. The availability at 1 will be calculated on the assumption that N_2 behaves as an ideal gas. Its constant-pressure heat capacity will therefore be $\frac{7}{2}R$.

Availability at 1

$$B_1 = (H_1 - H_0) - T_0(S_0 - S_1)$$

$$= \int_{T_0}^{T_1} C_p \, dT - T_0 \left(\int_{T_0}^{T_1} \frac{C_p \, dT}{T} - \int_{P_0}^{P_1} R \frac{dP}{P} \right)$$

Substituting $T_0 = 293$ K, $T_1 = 503$ K, $P_0 = 1$ atm, $P_1 = 10$ atm and integrating gives

$$B_1 = 1699.8 \text{ kcal/kg mol}$$

Similarly for B_2 with $T_2 = 308$ K and $P_2 = 1.5$ atm, we get

$$B_2 = 238.7 \text{ kcal/kg mol}$$

(a) The maximum useful work that can be obtained from the nitrogen is equal to the availability at 1, or 1699.8 kcal/kg mol.

(b) The lost work will first be calculated, and then its allocation will be determined. The useful work produced is 1.8 hph/kg mol, or

$$W_u = (1.8)(641.7) = 1155.1 \text{ kcal/kg mol}$$

The lost work is therefore

$$W_1 = 1699.8 - 1155.1 = 544.7 \text{ kcal/kg mol}$$

The allocation of this lost work is easily determined. The nitrogen exiting from the expander had an availability of 238.7 kcal/kg mol, but none of this potential for performing useful work was realized. This unrealized availability represents lost work in the turbine discharge line. The remainder of the lost work is in the expander. The work allocations are shown in Table 5-1.

The analysis could have been carried out in a slightly different manner. It could have been argued, for example, that the difference in availability between the nitrogen entering and leaving the turbine was 1461.1 kcal/kg mol. Out of this ΔB the turbine managed to produce 1155.1 kcal; in other words, 79.1 percent of the ideal-work potential.

If we assume that the turbine discharge pressure of 1.5 atm is a process requirement, then what the turbine is asked to do is to expand nitrogen at 230°C and 10 atm to 1.5 atm and to produce work from this operation. The question that should be explored, therefore, is somewhat different. What is the maximum work one could obtain from an adiabatic, reversible expansion, and how does the actual performance compare with this ideal?

For nitrogen entering at 230°C and 10 atm and leaving at 1.5 atm

$$W_{\text{ideal}} = -\frac{kRT_1}{k-1}\left[\left(\frac{P_2}{P_1}\right)^{(k-1)/k} - 1\right]$$

$$= -\frac{(1.4)(1.987)(503)}{1.4-1}\left[\left(\frac{1.5}{10}\right)^{0.4/1.4} - 1\right]$$

$$= 1463.7 \text{ kcal/kg mol}$$

The turbine produces 1155.1 kcal/kg mol, and its efficiency is therefore $(1155.1/1463.7)(100) = 78.9$ percent.

One must exercise caution in this last analysis. In the above case, the nitrogen that would leave the reversible, adiabatic turbine would be at 19.5°C, a temperature lower than T_0. The thermodynamic analysis would require that the Carnot work necessary to bring this nitrogen up to the dead temperature be deducted from the ideal work. The thermodynamics bookkeeping would then be satisfied. In practice, however, any engineer would be pleased if he could discharge a stream at a temperature lower than T_0 to the surroundings when by doing so that stream is producing more work than would otherwise be obtained.

Table 5-1 Work allocation

	kcal/kg mol	Percentage of total
W_u	1155.1	68.0
W_l, in turbine	306.0	18.0
In discharge line	238.7	14.0
W_{ideal}	1699.8	100

Availability in Energy Accounting

Chemical plants and petroleum refineries need electrical as well as thermal energy. Electricity and process heat can be supplied from a power plant as coproducts by generating steam at a pressure higher than is necessary for process heat. The steam is let down to the pressure required for process heat through a backpressure or extraction turbine, which supplies the motive power for electricity generation. A cost-accounting problem can now arise: What portion of the power-plant fuel cost should be charged to electricity generation, and what portion should be charged to process steam?

The power-plant engineer can claim that the steam is being generated primarily for the provision of process heat; therefore, the work delivered by the turbine should be considered a by-product. Thus, the charge against electricity production should be only the excess of cost over that when only process steam is produced. After all, each pound of steam provided to the process engineer will provide about 900 Btu of latent heat, whereas the change in enthalpy through the turbine will be only several hundred Btu per pound. This, the power-plant engineer can claim, is the appropriate ratio to use in the energy accounting.

The process engineer can claim, on the other hand, that the steam is being generated primarily for the production of electricity and the steam withdrawn from the turbine for process heat is by-product. The charge against process steam should therefore be only the excess of cost over that when power is produced without process steam. The fuel charge to the processing units should be based on the loss in power per pound of steam due to steam being extracted at process steam pressure rather than at a lower pressure consistent with the cooling-water temperature.

Thermodynamics and the application of the availability concept can help us resolve this issue, as demonstrated in the following example.

Example 5-3 Electricity and process steam are produced as coproducts from our power plant. Steam is generated at 650 lb/in² abs and 700°F and enters a backpressure turbine to supply motive power to a generator. The steam leaves at 100 lb/in² abs, saturated, to supply process heat. Condensate may be assumed to leave as a saturated liquid at 100 lb/in² abs. Cooling water is available at 80°F. From availability considerations, how would you recommend that power-plant fuel costs be allocated between electricity and process steam?

SOLUTION We can recommend the appropriate accounting ratio by considering the loss in steam availability during its passage through the turbine and during condensation as process steam.

Using 80°F and atmospheric pressure as our basis, we can calculate the availabilities from Eq. (5-15):

At turbine entrance:
$$B_1 = (1348.0 - 48.02) - 540(1.5767 - 0.0932)$$
$$= 498.89 \text{ Btu/lb}$$

At turbine exit:

$$B_2 = (1187.02 - 48.02) - 540(1.6026 - 0.0932)$$
$$= 324.10 \text{ Btu/lb}$$

Condensate:

$$B_3 = (298.4 - 48.02) - 540(0.4740 - 0.0932)$$
$$= 44.75 \text{ Btu/lb}$$

$$\Delta B_{\text{turb}} = 324.10 - 498.89 = -174.79 \text{ Btu/lb}$$

$$\Delta B_{\text{cond}} = 44.75 - 324.10 = -279.35 \text{ Btu/lb}$$

$$\Delta B_{\text{tot}} = 44.75 - 498.89 = -454.14 \text{ Btu/lb}$$

The fuel charges should therefore be allocated as follows:

To power:
$$\frac{174.79}{454.14} \times 100 = 38.49\%$$

To process steam:
$$\frac{279.35}{454.14} \times 100 = 61.51\%$$

It should be noted that the loss in availability due to condensation is less than the process heat supplied because of the $T_0 \Delta S$ term in the expression for availability. If condensation takes place at T_0, there is no loss in availability because energy at the dead temperature is unavailable for work production.

5-3 SEPARATION AND PURIFICATION

In most chemical plants the majority of the operations are devoted to separation and purification of feed and product streams. Generally these operations, such as distillation, evaporation, and drying, are also the largest consumers of energy. In spite of this large expenditure of energy the products of these operations have almost the same energy content as the feed materials. The products differ from the feedstocks only in being purer.

A certain amount of energy, either as heat or work, must be expended to effect separation of a solution into its component parts. In most distillation processes heat energy is expended, but in vapor recompression, for example, work is expended to effect the separation. A measure of the thermodynamic efficiency of the process can be obtained once we know the least possible amount of heat or work required for the separation. A series of isothermal reversible steps can be postulated that, in summary, result in the desired separation and permit us to calculate the minimum work of separation.

If one wants to calculate the minimum work required to separate a binary liquid solution of components A and B and composition x_i into two products of composition x_0 and x_b the process could be carried out as follows:

1. Vaporize the two components from the initial solution separately through semipermeable membranes at their respective equilibrium pressures but in the correct proportions to form the product x_0.
2. Compress the two vapors to the equilibrium pressure of the components corresponding to the composition x_0.
3. Condense the vapors through a semipermeable membrane into the solution of composition x_0.

Repeat steps 1 to 3 for solution of composition x_b.

The work terms for all the steps are then added, giving the minimum work of separation. If ideal gases are assumed and it is further assumed that liquid volumes are negligible compared with gas volume, the minimum work required per mole of solution of composition x_i is

$$W_{min} = -RT\left[\frac{x_0(x_i - x_b)}{x_0 - x_b} \ln \frac{\bar{P}_{A_0}}{\bar{P}_{A_i}} + \frac{(1 - x_0)(x_i - x_b)}{x_0 - x_b} \ln \frac{\bar{P}_{B_0}}{\bar{P}_{B_i}}\right.$$
$$\left. + \frac{x_b(x_0 - x_i)}{x_0 - x_b} \ln \frac{\bar{P}_{A_b}}{\bar{P}_{A_i}} + \frac{(1 - x_b)(x_0 - x_i)}{x_0 - x_b} \ln \frac{\bar{P}_{B_b}}{\bar{P}_{B_1}}\right] \quad (5\text{-}18)$$

For separation into pure components this equation reduces to

$$W_{min} = -RT\left[x_i \ln \frac{P_{A_i}}{\bar{P}_{A_i}} + (1 - x_i) \ln \frac{P_{B_i}}{\bar{P}_{B_i}}\right] \quad (5\text{-}19)$$

and if ideal solutions are assumed so that $\bar{P}_A = xP_A$, then

$$W_{min} = RT[x_i \ln x_i + (1 - x_i) \ln (1 - x_i)] \quad (5\text{-}20)$$

The decrease in entropy of an ideal solution when it is separated into n pure components is easily obtained from Eq. (5-20) as

$$\Delta S = R \sum_{i=1}^{n} x_i \ln x_i \quad (5\text{-}21)$$

Example 5-4 Calculate, analyze, and discuss the minimum work of separation of a binary mixture of 50 mol % benzene, 50 mol % toluene at 20°C. This binary mixture can be assumed to form ideal solutions and to act as an ideal gas in the vapor phase. Saturated steam at 21 lb/in² abs is available, and the surroundings are at 20°C. Compare the minimum work requirements to produce distillate and bottom products containing 90, 95, 99, and 100 percent benzene and toluene, respectively.

SOLUTION The minimum isothermal work of separation can be calculated from Eq. (5-18). For the 90% benzene distillate product, 90% toluene bottom product

$$x_0 = 0.9 \qquad x_b = 0.1 \qquad \text{and} \qquad x_i = 0.5$$

Substituting into Eq. (5-18) and remembering that for the ideal solution $\bar{P}_i = xP_i$, we obtain -214 kcal/kg mol as the minimum work requirement. The minimum work requirements for all of the separations are:

Products, mol %	W_{min}
90	-214
95	-288
99	-371
100	-404

We now continue the analysis for the 90 percent distillate and bottom products. The availability of 21 lb/in² abs steam with respect to surroundings at 20°C is

$$B = \Delta H - T_0 \, \Delta S$$

$$= (1157 - 36) - (528)(1.7286 - 0.0708)$$

$$= 245.7 \text{ Btu/lb} = 136.5 \text{ kcal/kg}$$

The steam requirements to supply the minimum work of separation are therefore

$$\frac{214}{136.5} = 1.57 \text{ kg steam/kg mol feed}$$

Let us look into this a bit further. To produce the overhead product by a distillation process would require that a minimum of $\frac{1}{2}$ mol of distillate be vaporized and condensed at a molar latent heat of approximately 7500 kcal. The thermal requirements for the process are therefore at least 3750 kcal per mole of feed. This will, of course, be supplied by the steam, which at 21 lb/in² abs will have a condensing temperature approximately equal to the boiling point of the bottom product and a latent heat of 532 kcal/kg. The steam requirement for vaporization will be

$$\text{Steam for vaporization} = \frac{3750}{532} = 7.05 \text{ kg/kg mol feed}$$

This quantity of steam could provide, in accordance with its availability, $(7.05)(136.5) = 962.3$ kcal of work, compared with the minimum work requirement for this separation of 214 kcal. The efficiency of the operation is easily calculated:

$$\text{Efficiency} = \frac{214}{962.3} \times 100 = 22.2\%$$

This does not end the analysis, however. The minimum steam requirement for the distillation was based on a zero reflux ratio. The fact that the separation in a fractionation column will require operation at a reflux ratio above some minimum value greater than zero was not considered. The actual efficiency of a distillation column will generally be less than 10 percent. Distillation is one of the least efficient operations, from a thermodynamic viewpoint, that take place in a chemical plant.

5-4 BLENDING

The operation of blending is the opposite of separation and purification, and it always imposes an energy penalty. It is sometimes used intentionally to meet a purity specification by blending overpurified with underpurified product. It can occur unintentionally in storage and intermediate surge tanks when the composition of the stream fed to the tank varies with time. Blending can also occur unintentionally in a distillation operation if the column is not fed on the optimum tray. Feed would then be mixed on the tray with a mixture of different composition. Feed to a nonoptimum tray is reflected on the McCabe-Thiele or Ponchon-Savarit diagram by an increase in the number of trays required to effect the separation at a given reflux ratio or by an increase in the required reflux ratio at a constant number of plates.

The operation of blending is always accompanied by an increase in entropy and lost work. Where n pure components are blended and form an ideal solution, the change in entropy, in accordance with (minus) Eq. (5-21), will be

$$\Delta S = -R \sum_{i=1}^{n} x_i \ln x_i \tag{5-22}$$

Lost work due to blending can be calculated from the equations developed in the preceding section.

At first glance, lost work due to blending may appear to be so small that it can be neglected. It should not be forgotten, however, that distillation is an extremely inefficient operation. Bojnowski et al. (1975) reported on a spectacular reduction in steam consumption upon the redesign of a multitower distillation operation. Much of the reduction in steam consumption from 6 to 2 kg per kilogram of product was due to the elimination of blending operations. In the original operation overhead products from the final four stills in the five-tower train were combined and recycled to the first still, where they were blended with the feedstream. In the redesigned and revised operation practically no recycle was used, and two columns were eliminated. In addition to the reduction in steam consumption, the redesigned system, while maintaining product quality, exhibited improved product-recovery efficiency.

PROBLEMS

5-1 In Example 5-1 we saw that it was thermodynamically desirable to keep separate the two cooling-water streams that were at different temperatures. Compare the heat-transfer-surface requirements for two schemes presented below and estimate the percentage difference in exchanger costs.

The cost of heat exchangers can be assumed to be proportional to $(\text{surface})^{0.6}$. For this problem you may assume that the overall heat-transfer coefficients are equal to 100 Btu/(h · ft^2 · °F). In both schemes two streams are to be cooled, stream A from 145 to 105°F in one exchanger by the transfer of 1×10^6 Btu/h and stream B from 135 to 95°F in another exchanger by the transfer of 1×10^6 Btu/h.

Scheme 1: A mixed cooling-water stream at 80°F is to be used for both heat exchangers. The cooling water will be returned to the cooling tower at 120°F.

Scheme 2: The cold-water stream at 60°F will first cool stream B and will leave the exchanger at 90°F. It will then cool stream A and will leave at 120°F for circulation to the cooling tower.

5-2 In a petroleum refinery operation a kerosene stream is to be cooled from 500 to 150°F, a naphtha stream from 400 to 130°F, and a circulating reflux stream from 350 to 200°F. As much as possible of this heat is to be used to preheat an oil stream from 100°F. Cooling water, which is to be returned to the cooling tower at no more than 130°F, is available at 80°F. Minimum approach temperatures are limited to 50°F for oil-oil exchangers and 40°F for oil-water exchangers. For the purposes of this problem assume that the specific heats of all of the oil streams are constant and equal to 0.6. The amount of the circulating reflux is equal to that of the oil to be preheated, while the kerosene and naphtha streams are equal to 20 and 30 percent, respectively, of the oil stream.

Calculate the exit temperatures and heat duties of the exchangers per pound of oil for the two schemes in Fig. P5-2. Where necessary, add cooling-water–oil exchangers and their duties.

5-3 A heat-exchanger train is to be designed to deal with the streams shown in the table. Propose a scheme in which no cooling water will be required and which will satisfy the duty requirements. Any additional heat requirements are to be provided by steam at 250 lb/in^2 gauge. The streams shown in the table are light and heavy distillates in a petroleum refinery. Assume that the heat capacities are constant and equal to 0.55 Btu/(lb · °F). Minimum approach temperatures are 40°F for oil-oil exchangers and 35°F for steam-oil.

Stream	Flow rate, lb/h	Temperature, °F	
		Inlet	Outlet
A	60,000	460	250
B	30,000	400	150
C	25,000	200	400
D	40,000	100	360
E	40,000	150	400

5-4 A limited quantity of cooling water is available at 70°F for condensing exhaust steam from a twin steam-turbine installation. The question has arisen whether the cooling water should pass through both condensers in series or whether the stream should be split, half of the water going through each condenser. The steam enters the turbines at 160 lb/in^2 abs and 600°F. To make the comparison assume for both alternatives that the heat-transfer surfaces are equal, the condensing loads are equal, and the overall heat-transfer coefficients are equal. For both alternatives the log mean temperature difference between the condensing steam and the cooling water is 25°F. The cooling water is returned to the cooling tower at 110°F.

(a) Which alternative will produce the higher power output?

(b) Justify your conclusions by thermodynamic reasoning.

5-5 Distillate and bottom products containing no less than 97 mol% of benzene and toluene, respectively, are to be produced by distillation of a binary equimolar mixture of benzene and toluene. The feed will enter the tower as a saturated liquid, and a reflux ratio 1.2 times the minimum is to be used.

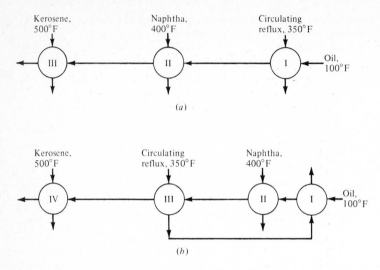

(a)

(b)

Figure P5-2 (a) Straight-through and (b) looping scheme.

The tray efficiency is expected to be 80 percent.

(a) Calculate the number of actual trays to be used and the feed-tray location.

(b) The tower was built with the same number of trays as calculated in part (a), and three feed nozzles were provided, one on the predicted feed tray, one two trays above, and one two trays below. During initial operation with the feed entering the predicted tray the tower performed as predicted. Later, by mistake, the wrong valve was opened, and the feed entered the tower through the lower feed nozzle. What effect would this error have on the operating conditions required to maintain product specifications?

5-6 In Prob. 5-5 the condenser was designed to operate with cooling water entering at 80°F and leaving at 120°F. Excessive fouling caused the overall heat-transfer coefficient to drop by 10 percent compared with the design coefficient. What effect will this have on the cooling-water requirements? Although the tower would, in practice, operate at a pressure slightly above atmospheric, you may assume operation at atmospheric pressure.

5-7 Ethyl alcohol can be produced by fermentation of starches or sugars. Typically, the product from a fermenter would contain 10 wt% alcohol. A product containing 90 wt% alcohol is required. Assuming that the product leaving the fermenter is essentially only alcohol and water at 20°C:

(a) Calculate the minimum work required to separate 1 mol of fermentation product into 90 wt% alcohol and essentially pure water at 20°C.

Data for ethanol-water system

Weight fraction ethanol	Partial pressure mmHg	
	H_2O	C_2H_5OH
0	17.5	0
10	16.8	6.7
90	7.5	35.8

(b) Calculate the thermodynamic efficiency of a distillation operation to perform the separation required in part (a). A reflux ratio of 4 mol of liquid per mole of distillate product will be used. Steam is available at 50 lb/in² gauge and cooling water at 20°C (T_0). Repeat for steam at 25 lb/in² abs.

5-8 In Prob. 5-7 consider the possibility of supplying heat to the reboiler by using the overhead vapor from the tower. This would require the vapor to be compressed to a pressure such that the condensing vapor would be able to supply heat to the reboiling liquid. Assume a minimum ΔT of 30°F. Estimate the compressor work requirements per pound mole of feed to the tower. Compare with the results obtained in Prob. 5-7. For the purposes of this problem assume that pure ethanol enters the compressor, that ideal-gas behavior is obtained, and that $C_p/C_v = 1.13$. Neglect any possible desuperheating problems.

5-9 Energy consumption in distillation can be reduced by reusing the vapor in a manner reminiscent of multiple-effect evaporation. The reuse of vapor requires that a train of stills be used, each one operating at a lower temperature and pressure than the preceding one. In one realization of this principle a split tower is used. The first tower operates at an elevated pressure, and the vapor from the top of this tower is condensed in the reboiler of the second one. Design a split tower to produce 70.0 mol% NH_3 overhead products from a 5 mol% NH_3 aqueous liquid feed. The bottom products are to be essentially pure water. The first tower is to operate at 10 atm abs and the second column at 1 atm.

(a) Compare the steam requirements for the split tower with those for a single tower operating at 1 atm.

(b) What effect does elevated pressure have on tower size? Compare elevated-pressure operation with atmospheric-pressure operation.

Vapor-liquid equilibrium data for NH_3–H_2O system

Pressure, mmHg	NH₃, mol%		Temperature, °C
	Liquid	Vapor	
760	2.1	23.2	91.5
	4.1	42.1	84.8
	6.1	55.5	79.0
	8.2	67.0	73.5
	10.2	73.9	68.4
	16.0	85.1	57.1
7600	2.1	14.0	172.8
	4.1	26.0	166.5
	6.1	36.7	160.5
	8.2	45.0	154.7
	10.2	54.4	149.3
	16.0	70.0	135.8
	21.0	80.9	123.7

5-10 Zdonik (*Chem. Eng.*, July 4, 1977, p. 99) claims that substantial energy savings can be realized by minimizing the pressure drop in the interstage coolers of multistage compressors. Hydrogen is to be compressed from 15 lb/in² abs and 100°F to 3000 lb/in² abs in a four-stage compressor. The compression ratio is the same in all stages. Assuming ideal-gas behavior, adiabatic compression, and perfect intercooling, calculate the minimum work requirement per pound mole of hydrogen if pressure drop in the interstage coolers is (a) zero, (b) 4 lb/in², and (c) 2 lb/in².

5-11 A chemical plant that is now under design will require 25,000 lb/h of saturated steam at 150 lb/in^2 gauge, 12,000 lb/h of saturated steam at 20 lb/in^2 gauge, and 1500 hp of mechanical drives. Using current prices of fuel and electricity, estimate the fuel and purchased electricity costs for the following alternatives:

(a) All electricity is purchased. Steam is generated at 150 lb/in^2 gauge, and the 20 lb/in^2 gauge steam is provided by throttling the required amount of 150 lb/in^2 gauge steam. (Recall that throttling will result in superheated steam that must be desuperheated by the addition of water.)

(b) The throttling valve is replaced by a steam turbine that provides part of the mechanical-drive horsepower.

(c) Steam will be generated at 585 lb/in^2 gauge and 600°F. All the mechanical-drive power requirements will be provided by backpressure turbines operating at 150 and 20 lb/in^2 gauge discharge pressures. Any additional process steam requirements will be provided by throttling the high-pressure steam.

5-12 Ethylene is produced by the thermal cracking of a paraffin feed material. The lighter components of the cracked gas are H_2, CH_4, C_2H_6, C_2H_8, C_3H_6, and C_3H_8; their separation is performed at low temperatures, refrigeration usually being provided by ethylene and propylene refrigeration cycles. In many modern plants the demethanizer tower operates at 475 lb/in^2 gauge, and the overhead methane is used as fuel at a pressure of about 50 lb/in^2 gauge. Assume that the demethanizer produces a pure CH_4 stream and that 20 lb of methane is produced per 100 lb of ethylene.

One scheme that has been proposed to conserve energy involves preheating the demethanizer overhead product by heat exchange with the demethanizer tower feedstream. The preheated methane is passed through a turboexpander and used as a substitute for part of the $-150°F$ ethylene refrigeration. It is finally recompressed to 50 lb/in^2 gauge for delivery to the fuel-gas system.

(a) What should be the methane preheat temperature if the formation of liquid methane in the turboexpander is to be avoided? (A Mollier diagram for CH_4 would be a reasonable aid.)

(b) Suggest a reasonable exhaust pressure for the turboexpander. How much work would be produced per pound of ethylene?

(c) How much refrigeration as Btu per pound of ethylene could be produced?

(d) What would be the net horsepower of the methane turboexpander compressor set for an ethylene plant producing 10^9 lb/yr and operating 8600 h/yr?

5-13 A number of alternatives are being considered for supplying power and process heat to a chemical plant now under design. The scheme you are to study is as follows. A gas-turbine-driven generator will be used to supply part of the power requirements. The gas will leave the turbine at 850°F and will be used in a waste-heat boiler to raise superheated steam at 600 lb/in^2 and 750°F, the gas leaving the boiler at 450°F. The steam will enter an extraction turbine to generate additional power. One-third of the steam will leave at 65 lb/in^2 abs to be used for process heating. The remaining two-thirds will be withdrawn at 150 lb/in^2 abs to be used for operating turbine-drive pumps and will exhaust to a condenser at 120°F. The condensate (assume no subcooling) will be returned to the waste-heat boiler along with makeup 80°F boiler feedwater. The isentropic efficiency of the backpressure turbine is expected to be 85 percent and that of the pump turbines 80 percent. Cooling water is available at 80°F. The heat capacity of the gas-turbine exhaust gas is $6.24 + 1.3 \times 10^{-3}T$ Btu/(lb mol · °R). Heat losses and condensate and boiler-feedwater pumping power requirements are neglible.

(a) Sketch a flowsheet for this scheme.

(b) Calculate the lost work in each part of the scheme and actual work produced.

(c) Perform a thermodynamic analysis of the scheme. Can you recommend any changes in the scheme?

REACTOR DESIGN AND ANALYSIS

And the streams shall be turned into pitch and the soil into brimstone.

Isaiah, 34: 9

The amount of capital investment in the reactor of a chemical plant usually represents only a small fraction of the total investment, especially in the case of continuous processing. The capital invested in feed-preparation and product-purification sections, tankage, piping, control, and ancillaries will generally far exceed the investment in the reactor. The economic and operating success of the plant, however, depends upon the successful design and operation of the reactor. In this chapter we concern ourselves with the analysis of a number of factors that affect the selection of the reactor type and size and its operating conditions.

6-1 REACTOR MATERIAL BALANCE

The general reactor material-balance equation for a reactant species can be written as

Moles entering reactor = moles leaving reactor

+ moles reacting

+ change in number of moles in reactor   (6-1)

or   Input = output + disappearance + accumulation

In this section we shall concern ourselves with the "moles reacting" term.

One component of our system can be selected as the key species and the material-balance equation can be written in terms of this component. The key component can either be a reactant species or a product species because all the components can be related to this key species through the reaction stoichiometry and the stoichiometric coefficients. The rate of change in the number of moles of the key component will be dN_i/dt.

A number of different expressions can be written to relate a reaction rate to the rate of change in the number of moles of the key component. In terms of the volume of the reacting fluid we could write

$$r_i = \frac{1}{v}\frac{dN_i}{dt} \tag{6-2}$$

where r_i will be the number of moles of component i formed per unit volume of reacting fluid per unit of time. Or if the reactor volume differs from the volume of the reacting fluid, the rate can be written as

$$r_i' = \frac{1}{V_r}\frac{dN_i}{dt} \tag{6-3}$$

where r_i' is the number of moles of i formed per unit volume of reactor per unit of time. In the case of a heterogenous reaction in which the interfacial area could be of importance the reaction rate could be expressed as

$$r_i'' = \frac{1}{S}\frac{dN_i}{dt} \tag{6-4}$$

where r_i'' is the number of moles of i formed per unit of interfacial surface per unit of time. This equation could apply to two-phase reacting systems.

We could continue this list of reaction-rate expressions almost ad infinitum. No matter how the reaction rate is expressed, however, its functional form will be the same because there is a functional relationship between the variables of the system. Only the various constants of proportionality and the dimensions will change as one switches from one reaction-rate expression to another. The equation to use for any particular case is a matter of choice, and the choice should be made on the basis of ease of use and convenience for the particular situation involved.

6-2 CHEMICAL KINETICS

What we are interested in at this point is how this reaction rate will be affected by the various parameters over which we have some control. For the purposes of this discussion we shall consider only homogeneous reactions. We can write, in general, that

$$r_i = f(\text{temperature, pressure, composition})$$

But as a result of phase rule and thermodynamic considerations, there will be a relationship between temperature, pressure, and composition, so that the reaction rate can be expressed as

$$r_i = f(\text{temperature, composition})$$

We now look at the reaction

$$A + B \rightarrow R + S$$

A molecule of R and of S can form only when there is a binary collision between molecules of A and B. The reaction rate r_R (rate of appearance of species R) will therefore depend on this collision frequency, which, from the kinetic theory of gases, will be proportional to the product of the concentration of the reactants expressed as moles per unit volume and to the square root of the absolute temperature

$$r_R \propto C_A C_B \sqrt{T} \tag{6-5}$$

But the number of molecules reacting per unit time is smaller than the number of binary collisions between A and B, and, in addition, the temperature is known to have a much greater effect on the reaction rate than one would expect from $T^{1/2}$. One can therefore postulate that for the binary collisions between A and B to result in a reaction, the collision must involve energies of translation, vibration, etc., that are in excess of an energy E called the *activation energy*. The fraction of collisions having energies in excess of E can be represented by $e^{-E/RT}$, and the rate of reaction can now be written as

$$r_R \propto e^{-E/RT} C_A C_B \sqrt{T} \tag{6-6}$$

Since the effect of temperature in the term $\sqrt{T}$ is small compared with its effect in $e^{-E/RT}$, the former term can be lumped with the proportionality constant, yielding

$$r_R = k_0 e^{-E/RT} C_A C_B \tag{6-7}$$

or, in general

$$r_i = f_1(\text{temperature}) f_2(\text{composition})$$

and at a given temperature

$$r_i = k f_2(\text{composition})$$

where

$$k = k_0 e^{-E/RT} \tag{6-8}$$

where k = reaction rate constant or velocity constant
  k_0 = frequency factor
  E = activation energy

This equation is referred to as the *Arrhenius law*.

Many homogeneous reactions can have their kinetics expressed by

$$r_R = k C_A^\alpha C_B^\beta \cdots C_S^\sigma$$

The order of the reaction is equal to the sum of exponents on the concentrations, $\alpha + \beta + \cdots + \sigma$ or, more explicitly, is of order α with respect to species A, of order β with respect to species B, ... and of order σ with respect to species S.

The rate expression discussed so far applies only to the case of "irreversible" reactions, reactions which go substantially to completion at equilibrium. This would be the situation prevailing in reactions for which the equilibrium constant is large. For reactions that do not proceed virtually to completion at equilibrium it is necessary to include the kinetics of the reverse reaction in the rate equation. The net rate equation can then be written as

$$r_R = k_F C_A^\alpha C_B^\beta \cdots - k_R C_R^\rho C_S^\sigma \cdots \tag{6-9}$$

where k_F is the reaction-rate constant for the forward reaction and k_R is the reaction-rate constant for the reverse reaction. At thermodynamic equilibrium the net reaction rate would be zero.

6-3 TYPES OF CHEMICAL REACTIONS

Simple Reactions

Simple reactions are those in which only a single reaction is considered to take place. The reaction considered earlier

$$A + B \longrightarrow R + S$$

is a simple reaction. Other common simple reactions are

$$A \longrightarrow R$$
$$A + B \longrightarrow R$$
$$A \longrightarrow R + S$$
$$2A \longrightarrow R$$

Some industrially important examples of simple reactions are given in Table 6-1.

Multiple Reactions

Parallel, or simultaneous, reactions are those in which more than one product can be obtained from the same reactants by different reactions. Some generalized sets of simultaneous reactions are

$$\begin{cases} A + B \longrightarrow R \\ A + B \longrightarrow S \end{cases} \qquad \begin{cases} A + B \longrightarrow R \\ A \longrightarrow S \end{cases}$$

An industrial example of parallel reactions is the nitration of toluene, where substitutions can appear in the ortho, meta, and para positions

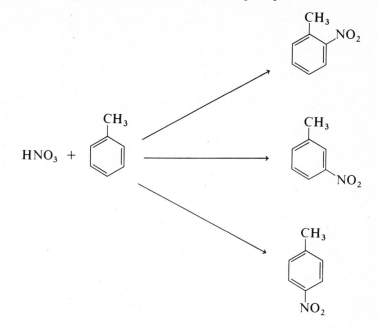

Table 6-1 Simple reactions that are industrially important

Reaction	Example
$A + B \rightarrow R + S$	Ethyl acetate production:
	$CH_3COOH + C_2H_5OH \rightarrow CH_3COOC_2H_5 + H_2O$
$A \rightarrow R$	Isomerization of n-pentane to isopentane:
	$H_3CCH_2CH_2CH_2CH_3 \rightarrow H_3CCHCH_2CH_3$
	$\qquad\qquad\qquad\qquad\quad\;\; CH_3$
$A \rightarrow R + S$	Ethylene production:
	$C_2H_6 \rightarrow C_2H_4 + H_2$
$2A \rightarrow R$	Dimerization of isobutene:
	$\qquad\qquad\qquad\qquad CH_3 \;\; CH_3$
	$2H_3CC{=}CH_2 \rightarrow H_2C{=}CCH_2CCH_3$
	$\qquad\quad CH_3 \qquad\qquad\quad CH_3$

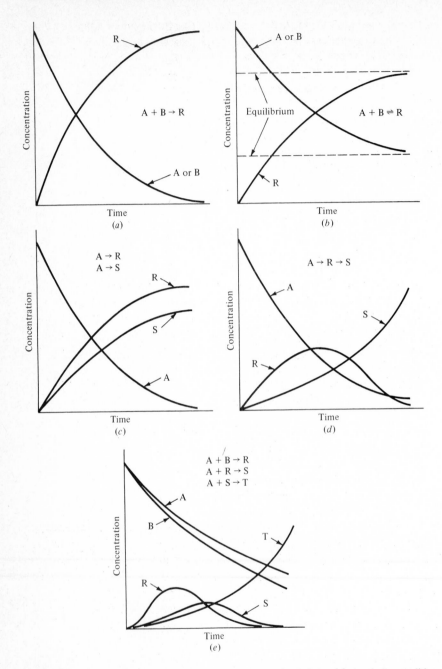

Figure 6-1 Concentration-time curves for several types of reactions: (*a*) simple irreversible reaction, (*b*) simple reversible reaction, (*c*) parallel reaction, (*d*) series reaction, (*e*) complex reaction set.

Series Reactions

Series, or sequential, reactions are those in which the product of the reaction can react further. A generalized series reaction is

$$A \longrightarrow R \longrightarrow S$$

and industrially important instances occur in substitution processes, such as

$$CH_4 + Cl_2 \longrightarrow CH_3Cl + HCl$$

$$CH_3Cl + Cl_2 \longrightarrow CH_2Cl_2 + HCl$$

Series reactions also frequently occur in oxidation processes, where the desired product may oxidize further. Maleic anhydride is produced, for example, by the oxidation of benzene, but the maleic anhydride can be oxidized further to carbon dioxide and water.

Parallel and series reactions and combinations of these two general types, termed *parallel-series* or *complex-series reactions*, occur frequently in reacting systems, especially in organic reactions. In operations involving multiple reaction possibilities, the questions of selectivity, i.e., the yield of the desired product, and of conversion are of overriding importance in the selection of reactor type and operating conditions. Conversion, defined as the moles of reactant disappearing due to reaction, will obviously not be the same as yield, defined as the moles of desired product formed per mole of reactant disappearing, except for a simple reaction.

Concentration-time curves that would be obtained for several types of reactions are shown in Fig. 6-1. These curves indicate how the concentrations of reactants and products change with time in a batch reactor.

6-4 EFFECT OF TEMPERATURE ON REACTION RATES

The effect of temperature on the reaction rate can be easily evaluated from the Arrhenius's law. Equation (6-8) can be written

$$\ln k = (\ln k_0) + \left(\frac{-E}{RT}\right) \tag{6-10}$$

Experimental data for reaction rate as a function of temperature therefore should yield a straight line when plotted as $\ln k$ against $1/T$ if Arrhenius's law is applicable. The slope of the line will be $-E/R$. The temperature dependency of the reaction rate is shown in this form as Fig. 6-2.

The temperature dependency is also presented in a somewhat different manner in Tables 6-2 and 6-3. In Table 6-2 the temperature rise needed to double the rate of reaction is presented for several values of the base temperature

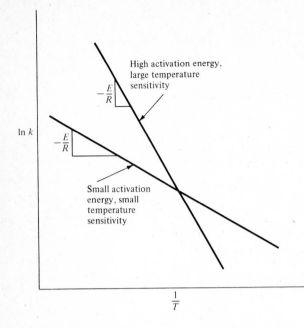

Figure 6-2 Reaction-rate–temperature dependence.

and the activation energy. The relative rates of reaction for several values of temperature and activation energy are presented in Table 6-3.

From these tables and Fig. 6-2 we can see that the reaction rate depends not only on the activation energy but also on the temperature level. Any given reaction will be more temperature-sensitive at a low temperature level than at a high temperature level. Reaction rates for reactions with high activation energies are very temperature-sensitive, whereas the rate of reactions with low activation energies are relatively insensitive to temperature. We shall return later to these general conclusions concerning the temperature dependency of chemical reactions.

Table 6-2 Temperature rise to double rate of reaction

Temperature, K	Activation energy, kcal		
	10	20	40
300	13	6	3
600	54	26	13
1000	160	74	36

Table 6-3 Relative reaction rate

Temperature, K	Activation energy, kcal		
	10	20	30
300	1	52×10^{-9}	2.7×10^{-15}
600	4.4×10^3	1	228×10^{-6}
900	72×10^3	268	1

6-5 BATCHWISE AND CONTINUOUS REACTIONS

In general, chemicals that are produced in rather small quantities are made batchwise. Some chemicals in this category are pharmaceuticals, dyestuffs, and some insecticides. The capital cost for batchwise operation is usually less than for continuous operation, especially for low production rates. Some chemical products, however, are produced in batchwise operation on a large scale, an outstanding example being the polymerization of vinyl chloride.

Continuous operation is generally used in large-scale production for a number of reasons, e.g., lower unit labor costs, greater ease of automatic control, and the ability to maintain constant reaction conditions and product quality. The kinetics principles involved in batchwise compared with continuous operation are the same because the molecular changes are the same, but the macroscopic conditions involved may be vastly different because of the state of flow in continuous operation. This can result in different molecules having histories of different residence time in the reactor, different concentration, and different temperature, compared with batchwise operation. These factors must be considered in an appropriate manner in the design of chemical reactors, which is usually based on laboratory experiments performed under batchwise conditions.

6-6 TYPES OF IDEALIZED CHEMICAL REACTORS

Three idealized chemical reactors are shown schematically in Fig. 6-3. The *batch reactor* is shown in Fig. 6-3a as a stirred vessel. It is charged with reactants, and the reaction is permitted to proceed. When the desired conversion is achieved, the products are discharged from the reactor. It is assumed that the contents are perfectly mixed, so that the concentrations and temperatures at all points in the reactor at any time are uniform. A *semibatch* variant of batchwise operation occurs when one of the reactants is fed continuously to the reactor as it is depleted during reaction or in order to prevent too violent a reaction. This is shown in Fig. 6-4a. Typical cases of this type would be the continuous addition of chlorine during a chlorination reaction or the continuous regulated addition of

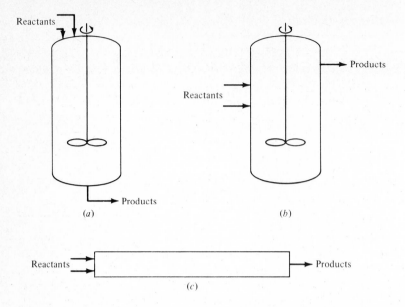

Figure 6-3 Schematic representation of idealized chemical reactors: (*a*) Batch; reactants charged at beginning, products removed at end. (*b*) CSTR; reactants fed continuously, products removed continuously. (*c*) Plug flow or tubular flow; reactants fed continuously, products removed continuously.

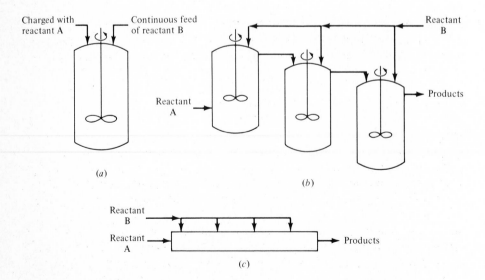

Figure 6-4 Divided-feed schemes: (*a*) semibatch, (*b*) CSTR series operation with divided feed, and (*c*) plug-flow reactor with cross-feed.

nitric acid during a nitration reaction to prevent an uncontrolled temperature rise.

The *continuous stirred-tank reactor* (CSTR) is shown in Fig. 6-3b. The design of a CSTR, or backmix reactor as it is sometimes called, is based on the assumption that the contents are perfectly mixed. As soon as the reactants enter the reactor, they are mixed. A portion of the fresh feed could therefore leave immediately in the product stream. To reduce this bypassing effect a number of stirred-tank reactors in series are frequently used, thereby reducing the probability that a reactant molecule entering the reactor will immediately find its way to the leaving product stream. Stirred-tank reactors in series can be operated with divided feed in a manner analogous to semibatch operation by feeding one of the reactants into each reactor, as shown in Fig. 6-4b.

The *tubular-flow reactor*, or *plug-flow reactor*, is shown in Fig. 6-3c. In the idealized case there would be no movement of the fluid in a direction other than from the reactor inlet to its outlet, and no attempt is made to induce mixing of the elements of the fluid at different points along the direction of flow. The idealized behavior is that of a plug of fluid moving through the reactor, hence the name plug flow. A divided feed or cross-feed can be used in this case also, and it is shown schematically in Fig. 6-4c.

6-7 BASIC DESIGN EQUATIONS FOR A BATCH REACTOR

The time required to attain a given conversion is the first objective in the design of a batch reactor. The volume of reactor required for a given production rate can then be found by simple scale-up.

The reaction time t_r is obtained by applying the material-balance equation (6-1). The calculations can be based on one of the reactant species A which undergoes a fractional conversion X_A. The number of moles of A initially charged will be taken as N_{A_0}. At time t the number of moles of A remaining will be $N_{A_0}(1 - X_A)$, and the rate of change will be

$$\frac{d}{dt}[N_{A_0}(1 - X_A)] = -N_{A_0}\frac{dX_A}{dt}$$

The flow terms in the material balance equation will be zero for the batch case so that the rate of change equals the accumulation and

$$-r_A V = -N_{A_0}\frac{dX_A}{dt} \tag{6-11}$$

Integration of the expression to a final conversion X_{A_f} yields the basic design equation

$$t_r = N_{A_0}\int_0^{X_{A_f}} \frac{dX_A}{r_A V} \tag{6-12}$$

For many liquid-phase reactions the volume of the reaction mixture can be assumed to be constant, so that Eq. (6-12) can be rewritten as

$$t_r = \frac{N_{A_0}}{V} \int_0^{X_{A_f}} \frac{dX_A}{r_A} = C_{A_0} \int_0^{X_{A_f}} \frac{dX_A}{r_A} \tag{6-13}$$

where C_{A_0} would be the initial concentration of reactant A in moles per liter. The final working equation resulting from Eq. (6-12) or (6-13) would depend upon the form of rate equation, upon whether the reaction is reversible or irreversible, and upon the relationship between temperature and conversion. If isothermal operation is assumed, the rate term r_A would be a function only of concentrations.

Example 6-1 The dimerization reaction

$$2A \longrightarrow R$$

is carried out in the liquid phase under isothermal conditions in the presence of an inert solvent at 80°C. The equilibrium constant at this temperature is a large positive number, so that the reaction can be considered irreversible. The reaction rate constant k_A equals 2L/(g mol · h), and the reaction is second order with respect to A, whose density is 6 g mol/L. What reaction time is required for 75 percent of the reactant to be converted into the dimer if the reactor is charged with a feed containing 50 vol % A, the remainder being inert solvent?

SOLUTION The basis for the calculation will be chosen as 1 L, and it will be assumed that no change in volume occurs as a result of the reaction.

The initial volume of A is 0.5 L, which is equivalent to $(0.5)(6) = 3$ mol of A charged to the reactor. C_{A_0} is therefore 3 g mol/L. The number of moles of A present at any instant will be $N_{A_0}(1 - X_A)$, and the concentration will be $3(1 - X_A)/1$ g mol/L. The reaction rate will be

$$r_A = kC_A^2 = 2[3(1 - X_A)]^2$$

Substituting into Eq. (6-13) will yield the reaction time

$$t_r = 3 \int_0^{0.75} \frac{dX_A}{18(1 - X_A)^2} = \frac{1}{6} \frac{1}{1 - X_A}\bigg|_0^{0.75} = \frac{1}{2} \text{ h}$$

The reactor volume required for any specified production rate for R can be easily calculated since we now know that $(0.75)(\frac{3}{2}) = 1.25$ mol of R will be produced from 1 L of feed in 0.5 h.

6-8 BASIC DESIGN EQUATIONS FOR A CSTR

The important basic assumption in the design of a CSTR is that the contents are well mixed. This means that the compositions are uniform everywhere in the reactor and that the product stream is at the same concentration as the mixture in the reactor. When steady-state conditions have been reached, the concentrations of the various chemical species will be constant in the reactor and easily calculable on the basis of the desired conversion. The reactor volume can then be calculated by applying the general material-balance equation. The accumulation term will be equal to zero because there will be no change in the concentration of the various components with time at the steady-state condition.

We shall first consider the general case in which the density of the mixture is not constant. The flow rate of A into the reactor is F_A, and the fractional conversion of A to product species will be X_{A_f} in a reactor of volume V. The material-balance equation becomes

$$F_A = r_A V + F_A(1 - X_{A_f}) \qquad (6\text{-}14)$$

The reaction rate must be evaluated at the concentration prevailing in the reactor, which, in accordance with the assumption of ideal mixing, will be the same as in the product stream.

In most cases where stirred-tank reactors are used the reactions are in the liquid phase, and the volume changes due to reaction are negligible. If we assume that density is constant, that the concentration of A in the feedstream is C_{A_0}, and that the volumetric feed rate is v, the concentration of A in the product stream will be $F_A(1 - X_A)/v$ or $C_{A_0}(1 - X_{A_f})$. Equation (6-14) can be expressed in terms of concentrations for this case as

$$vC_{A_0} = r_A V + C_{A_0}(1 - X_{A_f})v$$

or the reactor volume will be

$$V = \frac{C_{A_0} X_{A_f} v}{r_A} \qquad (6\text{-}15)$$

Example 6-2 In the preceeding batch example we showed that 75 percent of the A in 1 L of a feed material containing 3 g mol of A would be converted into product in $\frac{1}{2}$ h. In other words, 2.25 g mol of A would disappear by reaction per $\frac{1}{2}$ h in a 1-L reactor, or if we could instantaneously dump and recharge the reactor, a 1-L reactor would consume 4.50 g mol/h of reactant. What volume would be required for a CSTR to perform the same duty?

SOLUTION The feed rate for equivalent duty would be 2 L/h of feed containing 6 mol of A. The number of moles of A leaving at 75 percent conversion would be $1.50/2 = 0.75$ g mol/L, which would also be the concentration of A in the reactor. The reaction rate would therefore be $(2)(0.75^2)$. Substituting into Eq. (6-15), we have

$$V = \frac{(3)(0.75)(2)}{(2)(0.75^2)} = 4 \text{ L}$$

In other words, the perfectly mixed reactor would require 4 times the volume of the batch reactor to carry out the same duty. This is the result of the reactions being carried out only at the final concentration of reactant. In the batch reactor the reaction begins at the highest rate corresponding to the high concentration of reactant in the feed and gradually decreases as the reaction proceeds. Only at the end will the reaction rate be the same as the low value that prevails during the entire period in the CSTR.

In this example the reaction rate could be expressed as a function of concentration, and a simple algebraic solution of the problem was possible. In real solutions it may not be possible to fit the data to any simple kinetic expression. The rate data can then be presented graphically and a simple graphical method employed to find the CSTR volume. Such a graphical solution is presented in Fig. 6-5 and is based on the following considerations. The material balance expressed in Eq. (6-14) can be written as

$$v C_{A_0} = r_A V + C_A v \tag{1}$$

where C_A is the reactant concentration in the stirred tank, or

$$C_A = C_{A_0} - \frac{V}{v} r_A \tag{2}$$

If, as in Fig. 6-5, the reaction rate is plotted as a function of reactant concentration, a simple geometric construction can be used to determine

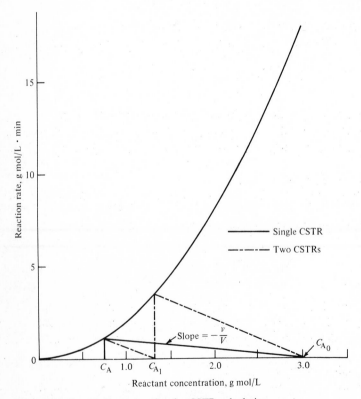

Figure 6-5 Graphical construction for CSTR calculation.

V/v. The reaction rate r_A in Eq. (2) is the rate obtained at C_A, the reactant concentration prevailing in the CSTR. A straight line from (C_A, r_A) to C_{A_0} will therefore have a slope of $-v/V$. Thus, with v known, V is immediately calculable from the measured slope.

The equation developed above can easily be extended to a series of CSTR reactors. The feedstream entering reactor j will have the same composition as the product stream from reactor $j - 1$. The design equation for reactor j will therefore be

$$V_j = \frac{C_{A_{j-1}} X_{A_j} v}{r_{A_j}} \tag{6-16}$$

where X_{A_j} is the fractional conversion in reactor j of reactant A.

Example 6-3 Repeat the previous calculation for the case of two equal-volume CSTRs in series.

SOLUTION The conversions of component A in each reactor are unknown, but they can be related to the overall fractional conversion by an overall material balance. The concentrations leaving each vessel will be

$$C_{A_1} = C_{A_0}(1 - X_{A_1})$$

$$C_{A_2} = C_{A_1}(1 - X_{A_2}) = C_{A_0}(1 - X_{A_f})$$

After simple algebraic manipulation the expression relating the individual conversions to the overall conversion becomes

$$1 - X_{A_f} = (1 - X_{A_1})(1 - X_{A_2})$$

The volume of each reactor, from Eq. (6-16) will be

$$V_1 = \frac{C_{A_0} X_{A_1} v}{r_{A_1}} \qquad V_2 = \frac{C_{A_1} X_{A_2} v}{r_{A_2}}$$

where $r_{A_1} = k(C_{A_1})^2$ and $r_{A_2} = k(C_{A_2})^2$. The specifications were for equal-volume reactors; hence,

$$\frac{C_{A_0} X_{A_1} v}{k(C_{A_1})^2} = \frac{C_{A_1} X_{A_2} v}{k(C_{A_2})^2}$$

Values for X_{A_1} and X_{A_2} can be obtained by trial and error or successive approximations. The values obtained are $X_{A_1} = 0.564$ and $X_{A_2} = 0.426$. The reactor volumes can now be calculated

$$V_1 = \frac{(3)(0.564)(2)}{(2)[(3)(1 - 0.564)]^2} = 0.99 \text{ L}$$

$$V_2 = \frac{(3)(1 - 0.564)(0.426)(2)}{(2)(0.75^2)} = 0.99 \text{ L}$$

By using two reactors in series the total volume has been reduced to 1.98 L compared with the 4 L required for the single CSTR. The reduction in volume is the result, of course, of more than half the conversion taking place at an intermediate concentration of A and hence at a faster rate than in the single reactor. It can easily be shown that as the number of reactors in series is increased, the total reactor volume will decrease and, in the limit of an infinite number of reactors, will reach the value required for a plug-flow reactor.

The graphical construction shown in Example 6-2 for a single CSTR is easily extended to a series of CSTRs. The construction for two CSTRs in series is shown in Fig. 6-5. The problem in the present example is to get to C_{A_2} from C_{A_0}, so C_{A_1} can be chosen at the designer's discretion. Thus, a straight line is drawn from (C_{A_2}, r_{A_2}) to C_{A_1}, and V_2 is determined from the slope $-v/V_2$. The slope of the straight line from (C_{A_1}, r_{A_1}) to C_{A_0} will determine the value of V_1. A trial-and-error solution is required if the constraint of $V_1 = V_2$ is imposed, as it is in this example.

6-9 BASIC DESIGN EQUATIONS FOR TUBULAR-FLOW REACTORS

As mentioned earlier, there is no movement of the fluid in a tubular-flow reactor except in a direction from the reactor inlet to its outlet. The idealized picture is that of a plug of fluid moving through the reactor. The basic design is, as usual, based on the general material-balance equation.

We shall assume a differential element of reactor volume of cross section S and length dL, as shown in Fig. 6-6. The general material balance will be written for the reactant species A for the steady-state condition; i.e., the accumulation term is zero

$$N_A = (N_A + dN_A) + r_A S \, dL \tag{6-17}$$

The number of moles of A present will be $N_{A_0}(1 - X_A)$, and dN_A will therefore equal $-N_{A_0} \, dX_A$. Substituting into Eq. (6-17) and integrating from the inlet to

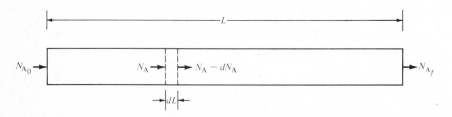

Figure 6-6 Differential element of tubular-flow reactor.

the outlet condition yields the reactor volume

$$V_r = SL = N_{A_0} \int_0^{X_{A_f}} \frac{dX_A}{r_A}$$ (6-18)

Integration of the right-hand side of Eq. (6-18) can be performed either explicitly or numerically once the reaction rate is related to the conversion.

Example 6-4 Repeat Example 6-2 for a tubular-flow reactor.

SOLUTION The feed rate of 2 L/h contains 6 mol/h of A, corresponding to $C_{A_0} = 3$ mol/L. The concentration of A as a function of fractional conversion will be $C_{A_0}(1 - X_A)$. Equation (6-18) becomes

$$V_r = N_{A_0} \int_0^{X_{A_f}} \frac{dX_A}{k(C_{A_0})^2} = 6 \int_0^{0.75} \frac{dX_A}{2[3(1 - X_A)]^2} = 1 \text{ L}$$

The volume of the tubular-flow reactor, for this case, is identical to the volume required for a batch reactor. This result should not be surprising if we imagine that each element of feed was encapsulated in a tiny batch reactor and then moved through the tubular-flow reactor during a period of time equal to t_r as calculated for the batch case. This visualization is valid, however, only if there is no change in the density of the fluid stream during the course of the reaction.

6-10 COMPARISON OF REACTORS

Simple Reactions

Although the treatment presented above is by no means complete or exhaustive, a number of factors have emerged and can be codified concerning the three reactor types discussed. The volumes of a batch reactor and idealized tubular-flow reactor will be identical for the same duty, provided there is no volume change due to reaction. This statement must be modified in practice because time must be provided to charge and discharge the batch reactor. This shutdown time would require that the actual batch reactor volume be larger than that calculated only on the basis of the reaction kinetics.

The volume of a CSTR will always be larger than that required for the tubular-flow reactor performing the same duty except for a zero-order reaction. The difference in volume requirements becomes more marked at high conversions. For the zero-order reaction the volume will be the same. If a series of CSTRs is used, the total volume of reactors required will decrease and approach the volume required by a tubular-flow or plug-flow reactor. The resulting complexity in piping and operation must be considered, however, if a series of reactors is contemplated.

It can be easily shown for a first-order reaction that the minimum total volume for CSTRs in series will be obtained when the tanks are of equal volume. This is not true, however, if the reactions are of order other than 1. The optimum ratio of volumes will be a function not only of the kinetics but also of the conversion to be attained.

A series of CSTRs permits operation of the individual reactors at different temperatures. There is no particular advantage in this possibility with a simple reaction, but in parallel or series reactions there may be significant advantages if reactors can be operated at different temperatures.

Parallel Reactions

A generalized set of parallel reactions could be

$$A + B \longrightarrow R$$

$$A + B \longrightarrow S$$

If we assume that R is the desired product and S the undesired product, our aim in the choice and design of the reactor would be to maximize the production of R and to minimize the production of S. This can be expressed in terms of selectivity, which is defined as the ratio of the number of moles of reactant converted into desired product to the number of moles of reactant converted into undesired product. For two parallel reactions the selectivity reduces to the ratio of the moles of R produced to the moles of S produced. The rates of production of R and S can be expressed as

$$r_R = k_1 C_A^{\alpha_1} C_B^{\beta_1} C_R^{\rho} \qquad r_S = k_2 C_A^{\alpha_2} C_B^{\beta_2} C_S^{\sigma}$$

and the ratio of the rates of production of R and S will be

$$\frac{r_R}{r_S} = \frac{k_1 C_A^{\alpha_1} C_B^{\beta_1} C_R^{\rho}}{k_2 C_A^{\alpha_2} C_B^{\beta_2} C_S^{\sigma}} \tag{6-19}$$

The reaction conditions should be chosen so that this ratio will always be at its highest value. We shall now analyze several possible situations; in order to simplify matters, we shall assume that $\rho = \sigma = 0$; thus,

$$\frac{r_R}{r_S} = \frac{k_1}{k_2} C_A^{\alpha_1 - \alpha_2} C_B^{\beta_1 - \beta_2} \tag{6-20}$$

If $\alpha_1 > \alpha_2$ and/or $\beta_1 > \beta_2$, the desired reaction is of higher order than the undesired reaction. For this case the ratio of the desired reaction rate to the undesired reaction rate will be greater when the concentration of the reactant is high. The situation becomes less favorable as the reactant concentrations decrease. The batch or tubular-flow reactor would therefore be preferable because they operate at higher average concentrations of reactant than the CSTR. If, however, for some operational reasons it is necessary to use a CSTR, a number of them should be used in series to increase the yield and selectivity beyond that attainable with a single CSTR. The capacities of each reactor should also be progressively larger.

Where the undesired reaction is of the higher order, the concentration of reactants should be as low as possible and a CSTR would be the obvious choice. If a tubular-flow reactor or batch reactor is required for some reason other than kinetic considerations, a stepwise addition of reactants could be used in order to keep the reactant concentration as low as possible during the course of the reaction.

A situation that commonly occurs would be of the type $\alpha_1 > \alpha_2$ but $\beta_1 \approx \beta_2$. The selectivity for this case can be improved by using a feed in which the ratio of A to B is not equal to the stoichiometric requirement. The excess reactant would, of course, have to be recovered from the product stream and recycled to the reactor.

Example 6-5 The kinetics of the parallel reactions

$$A + B \longrightarrow R \quad \text{and} \quad A + B \longrightarrow S$$

have been studied, and expressions have been deduced for their kinetics when the reaction is carried out in an "indifferent" solvent. The equations are

$$\frac{dC_R}{dt} = k_1 C_A^{0.5} C_B^{0.5}$$

$$\frac{dC_S}{dt} = k_2 C_A C_B$$

where the concentrations are in moles per liter. The densities of pure A and pure B are 10 mol/L, but A and B are also available in solvent at concentrations of 5 mol/L of solution. What will be the ratio of R to S in the reactor exit stream for a CSTR and for the case of a plug-flow reactor when A and B are fed at the stoichiometric ratios? Is there any advantage in feeding at a ratio other than the stoichiometric ratio? For the study assume an 80 percent conversion of A to products.

SOLUTION The overall fractional yield (Denbigh, 1944) of R can be expressed as

$$\Phi = \frac{C_{R_f}}{C_{A_0} - C_{A_f}} \tag{6-21}$$

This form of the expression for Φ assumes that no R was present in the feed to the reactor. The overall fractional yield can be evaluated by integration of the expression for the instantaneous fractional yield

$$\phi = -\frac{dC_R}{dC_A}$$

which represents the moles of R formed per unit of A consumed. Thus,

$$\Phi = \frac{1}{C_{A_0} - C_{A_f}} \int_{C_{A_0}}^{C_{A_f}} \phi \, dC_A \qquad (6\text{-}22)$$

We shall examine three cases for each type of reactor:

1. A mixed feed of pure A and pure B
2. A mixed feed of solutions of A and B
3. A feed in which $C_B = 5C_A$

For cases 1 and 2, $C_A = C_B$ at all times. For 1, $C_{A_0} = \frac{10}{2}$ mol/L, and for 2, $C_{A_0} = \frac{5}{2}$ mol/L. For 3, we shall assume that $C_{B_0} = 2.5$ mol/L and $C_{A_0} = 0.5$ mol/L.

We can now write the expression for ϕ for the three cases

$$\phi = -\frac{dC_R}{dC_A} = \frac{dC_R}{dC_R + dC_S} = \frac{k_1 C_A^{0.5} C_B^{0.5}}{k_1 C_A^{0.5} C_B^{0.5} + k_2 C_A C_B}$$

For cases 1 and 2, where $C_A = C_B$ and κ is defined as k_2/k_1,

$$\phi = \frac{1}{1 + (k_2/k_1)C_A} = \frac{1}{1 + \kappa C_A}$$

and for case 3

$$\phi = \frac{1}{1 + 2.24\kappa C_A}$$

(a) For the CSTR C_A is equal to C_{A_f} everywhere in the reactor. Thus, $C_{A_f} = 1, 0.5$, and 0.1 mol/L for cases 1, 2 and 3, respectively.
Case 1

$$\phi = \frac{1}{1 + \kappa}$$

If we assume that $\kappa = 1$, then

$$\phi = \frac{1}{1 + 1} = 0.5$$

Therefore, the ratio of R to S will be 0.5: 0.5 = 1 : 1.

Case 2 Again, assuming $\kappa = 1$, we have

$$\phi = \frac{1}{1 + 0.5} = 0.667$$

Thus, for this case the ratio of R to S will be 2 : 1.

Case 3

$$\phi = \frac{1}{1 + 0.224} = 0.817$$

For this case the ratio of R to S will be 4.46.

(*b*) For plug flow

$$\Phi = -\frac{1}{C_{A_0} - C_{A_f}} \int_{C_{A_0}}^{C_{A_f}} \phi \, dC_A$$

Case 1

$$\Phi = -\frac{1}{5 - 1} \int_5^1 \frac{dC_A}{1 + \kappa C_A} = \frac{1}{4\kappa} \ln(1 + \kappa C_A) \Big|_1^5$$

If we again assume that $\kappa = 1$, then

$$\Phi = \frac{1}{4} \ln \frac{1 + 5}{1 + 1} = 0.275$$

and the ratio of R to S in the reactor exit stream will be 0.379 mol of R per mole of S produced.

Case 2

$$\Phi = -\frac{1}{2.5 - 0.5} \int_{2.5}^{0.5} \frac{dC_A}{1 + C_A} = 0.424$$

and the ratio of R to S is 0.736.

Case 3 Assuming that $\kappa = 1$, we have

$$\Phi = -\frac{1}{0.5 - 0.1} \int_{0.5}^{0.1} \frac{dC_A}{1 + 2.24C_A} = \frac{1}{0.4 + 2.24} \ln(1 + 2.24C_A) \Big|_{0.1}^{0.5} = 0.613$$

and the ratio of R to S is 1.58 : 1.

Discussion The numerical results are summarized in Table 6-4. It is apparent (not only from this example but also from inspection of the equations that model parallel reactions) that the control of yield and product distribution lies in the reactant concentration. In the example case and, as one

Table 6-4 Selectivity in parallel reactions

Feed concentration, mol/L		R/S ratio in product, mol/mol	
C_A	C_B	CSTR	Plug-flow reactor
5	5	1	0.379
2.5	2.5	2	0.736
0.5	2.5	4.46	1.58

would expect, in general, the high reactant concentration favored the reaction of higher order and the low reactant concentration favored the reaction of lower order.

The same general conclusion can also be reached when the selectivities for CSTR and plug flow cases are compared. The CSTR consistently gave a higher ratio of R to S than the equivalent feed composition to a plug-flow reactor. This is so because the CSTR operates always and everywhere at the final concentration of the reactants, whereas the entire spectrum of concentration from inlet to exit is obtained in the plug-flow reactor. There is an advantage with respect to selectivity in operating at a ratio other than the stoichiometric ratio in that one of the reactants is always at a lower concentration than it would be compared with operating at the stoichiometric ratio.

In this example we dealt only with the effect of reactant concentration on selectivity in two parallel reactions. We assumed for the purpose of this comparison that the conversion of reactant A into products was 80 percent for each case examined. We did not study the effect of changes of the parameter varied on reactor size, equipment, or the cost involved in the separation of R and S. It is obvious that the reactor size will increase as reactant concentration in the feed is decreased and that the cost and difficulties involved in separating R and S will increase as the concentration of these components in the indifferent solvent decreases. A complete study must consider all the factors involved in production cost and product specifications.

In this example we assumed that $k_1 = k_2$. The ratio of reaction rates of the two parallel reactions can also be affected by adjusting the temperature, thereby altering the ratio k_1/k_2 if the reactions have different energies of activation. It will be recalled that a reaction with a high activation energy is more temperature-sensitive than a reaction with a low activation energy. Therefore, if the activation energy of the desired reaction is higher than that of the undesired reaction, a high temperature will favor a high selectivity, and, conversely, if $E_1 < E_2$, a low temperature would be desirable. In the overall economic evaluation the effect of temperature on reaction rate and hence on reactor size must also be considered.

Series Reactions

To avoid obscuring the principles with unnecessary details we shall postulate the simplest possible case and consider first-order reactions in which the volume does not change as reaction proceeds. The unimolecular reactions we shall assume are

$$A \xrightarrow{k_1} R$$

$$R \xrightarrow{k_2} S$$

where R is the desired product. For a batch or tubular-flow reactor material balances written for the three species will give

$$-\frac{dC_A}{dt} = k_1 C_A \qquad \frac{dC_R}{dt} = k_1 C_A - k_2 C_R \qquad \frac{dC_S}{dt} = k_2 C_R$$

These three simultaneous equations can readily be solved. We shall assume that $C_R = C_S = 0$ and $C_A = C_{A_0}$ at $t = 0$. The concentration of R as a function of time will be

$$C_R = C_{A_0} \frac{k_1}{k_1 - k_2} (e^{-k_2 t} - e^{-k_1 t}) \qquad (6\text{-}23)$$

that of S, the undesired product,

$$C_S = C_{A_0} \left(1 + \frac{k_2}{k_1 - k_2} e^{-k_1 t} - \frac{k_1}{k_1 - k_2} e^{-k_2 t} \right) \qquad (6\text{-}24)$$

and the reactant will disappear according to

$$C_A = C_{A_0} e^{-k_1 t} \qquad (6\text{-}25)$$

Equations (6-23) to (6-25) are plotted in Fig. 6-7. The form of the curves is independent of the relative magnitude of k_1 and k_2, and the concentration of the desired product species R always goes through a maximum. The maximum value attainable for C_R and the time at which it will occur can easily be obtained by differentiating Eq. (6-23) with respect to time and setting the derivative equal to zero. The time at which the maximum concentration of R occurs is

$$t_{max} = \frac{\ln (k_2/k_1)}{k_2 - k_1} \qquad (6\text{-}26)$$

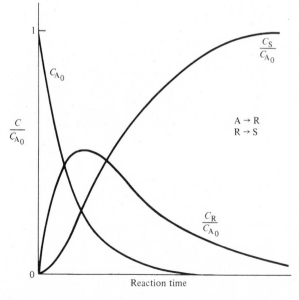

Figure 6-7 Reactant and product concentrations as function of time-series reactions in plug-flow or batch reactor.

Substituting this value into Eq. (6-23) yields the maximum concentration of R

$$C_{R_{max}} = C_{A0}\left(\frac{k_1}{k_2}\right)^{k_2/(k_2-k_1)} \tag{6-27}$$

Equation (6-27) is shown graphically in Fig. 6-8. It should be noted that high yields of R can be attained only if k_1/k_2 is large. This means that the reaction rate for the first reaction in the series should be high compared with that of the second, undesired reaction.

Similar equations can be derived for a CSTR by taking material balances on the three species at steady state. The expressions for the concentrations as a function of the average residence time in the reactor $\bar{t}$ will be

$$C_A = \frac{C_{A0}}{1 + k_1\bar{t}} \tag{6-28}$$

$$C_R = \frac{C_{A0}k_1\bar{t}}{(1 + k_1\bar{t})(1 + k_2\bar{t})} \tag{6-29}$$

$$C_S = \frac{C_{A0}k_1k_2\bar{t}^2}{(1 + k_1\bar{t})(1 + k_2\bar{t})} \tag{6-30}$$

These expressions are shown graphically in Fig. 6-9 for the case where k_1/k_2 equals 2. Again, it should be noted that the concentration of R goes through a

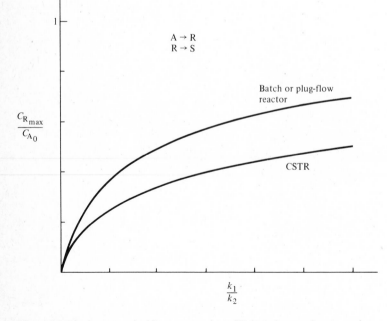

Figure 6-8 Maximum attainable concentration of R in series reaction.

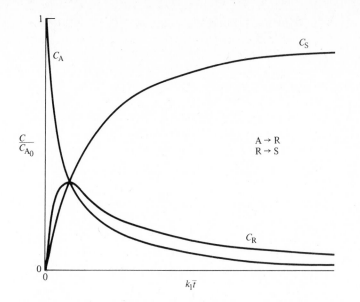

Figure 6-9 Reactant and product concentration as function of time-series reaction in CSTR.

maximum. By differentiation of Eq. (6-29) the average residence time required for this maximum is calculated to occur at

$$\bar{t}_{max} = \frac{1}{\sqrt{k_1 k_2}} \tag{6-31}$$

and the maximum concentration of R attainable is

$$C_{R_{max}} = \frac{C_{A_0}}{[(k_1/k_2)^{1/2} + 1]^2} \tag{6-32}$$

Equation (6-32) is shown graphically in Fig. 6-8 as the lower line. The maximum conversion to R in the CSTR is always less than that attainable in the batch or tubular-flow reactor, and the deleterious effect of ideal mixing is clearly evident in this figure.

We can now attempt to draw some conclusions concerning the optimum reactor type and operating conditions. To obtain high concentrations of R, k_1/k_2 must be as large as possible. Some control over this ratio can be exercised by appropriate choice of operating temperature, as discussed for parallel reactions. If $E_1 > E_2$, a high temperature should be chosen, and, conversely, if $E_1 < E_2$, a low temperature would favor a high conversion. The reactor type should be batch or tubular flow; if stirred tanks must be used, several should be used in series to approach plug flow.

As the concentration of desired product increases, so will the rate of the additional undesired reaction. The conclusion to be reached is therefore that the desired product should be removed as soon as it is formed. This could be done by distillation or extraction or some other separation operation. An alternative could be to operate with a low yield per pass through the reactor and to recycle the reactant after removal of the product. The final choice would be made in accordance with the economic optimum.

Example 6-6 Chlorobenzene has been produced commercially for a long time, and the technology of its production has changed very little in recent years. It is produced by the chlorination of benzene in either a batch or a continuous chlorination at atmospheric pressure and at about 55 to 60°C. The reactions that can occur are

$$\text{Benzene} + \text{Cl}_2 \longrightarrow \text{chlorobenzene} + \text{HCl} \qquad (1)$$

$$\text{Chlorobenzene} + \text{Cl}_2 \longrightarrow \text{dichlorobenzene} + \text{HCl} \qquad (2)$$

$$\text{Dichlorobenzene} + \text{Cl}_2 \longrightarrow \text{trichlorobenzene} + \text{HCl} \qquad (3)$$

Ortho-, and paradichlorobenzene are produced, but we shall lump them together and refer to them merely as dichlorobenzene. According to Stephenson (1966), k_1 is about 10 times larger than k_2. Since only negligible quantities of trichlorobenzene are produced, k_3 is evidently much smaller than k_1 and k_2 and so we shall neglect reaction (3).

Compare the composition of the products to be expected for chlorination performed in batch reactor and CSTR. Assume constant volume.

SOLUTION We are interested in composition in this problem and not in the residence time or volume requirements for a reactor. We therefore use an approach that is similar in principle to the approach taken in Example 6-5, where time was eliminated from consideration.

Let us rewrite the reactions; if A refers to benzene, B to chlorine, C to chlorobenzene, D to the dichlorobenzenes, and E to HCl the reactions now are

$$\text{A} + \text{B} \longrightarrow \text{C} + \text{E} \qquad (1)$$

$$\text{C} + \text{B} \longrightarrow \text{D} + \text{E} \qquad (2)$$

In accordance with the kinetics described by McMullin (1948), the rates of change of composition will be

$$\frac{dC_A}{dt} = -k_1 C_A C_B \qquad (3)$$

$$\frac{dC_C}{dt} = k_1 C_A C_B - k_2 C_C C_B \qquad (4)$$

$$\frac{dC_D}{dt} = k_2 C_C C_B \qquad (5)$$

and $k_1 \approx 10k_2$.

(a) *Batch or plug-flow reactor* We now eliminate time as a variable Dividing Eq. (4) by (3) leads to

$$\frac{dC_C}{dC_A} = -1 + \frac{k_2}{k_1}\frac{C_C}{C_A}$$

Defining $\kappa \equiv k_2/k_1 \neq 1$ and rewriting, we get

$$\frac{dC_C}{dC_A} - \kappa \frac{C_C}{C_A} = -1 \qquad (6)$$

This equation can be integrated; applying the condition that $C_C = 0$ when $C_A = C_{A_0}$ yields

$$\frac{C_C}{C_{A_0}} = \frac{1}{\kappa - 1}\left[\frac{C_A}{C_{A_0}} - \left(\frac{C_A}{C_{A_0}}\right)^{\kappa}\right] \qquad (7)$$

Similarly, the expression involving D can be obtained by dividing Eq. (5) by (3) and eliminating C_C

$$\frac{dC_D}{dC_A} = -\kappa \frac{C_C}{C_A}$$

$$\frac{dC_D}{dC_A} = \frac{\kappa}{1 - \kappa}\left[1 - \left(\frac{C_A}{C_{A_0}}\right)^{\kappa - 1}\right] \qquad (8)$$

Integrating Eq. (8) and applying the condition that $C_D = 0$ when $C_A = C_{A_0}$ yields

$$\frac{C_D}{C_{A_0}} = \frac{\kappa}{1 - \kappa}\left[\frac{C_A}{C_{A_0}} - \frac{1}{\kappa}\left(\frac{C_A}{C_{A_0}}\right)^{\kappa} + \frac{1 - \kappa}{\kappa}\right] \qquad (9)$$

Equations (7) and (9) can now be used to examine the ratios that will develop in a batch or plug-flow reactor.

We shall assume that chlorine is sparged continuously into the reactor so that its concentration is constant and that the density and volume of the solution remain constant. A simple calculation for 80 percent conversion of benzene yields

$$\frac{C_C}{C_{A_0}} = \frac{1}{0.1 - 1}(0.2 - 0.2^{0.1}) = 0.7237$$

$$\frac{C_D}{C_{A_0}} = \frac{0.1}{1 - 0.1}\left(0.2 - \frac{1}{0.1}0.2^{0.1} + \frac{1 - 0.1}{0.1}\right) = 0.0763$$

To put the results on a convenient, common basis we can relate the production of mono- and dichlorobenzene to the moles of chlorine consumed per mole of benzene charged.

By material balance

$$\frac{C_{B_0}}{C_{A_0}} = \frac{C_B}{C_{A_0}} + \frac{C_C}{C_{A_0}} + \frac{2C_D}{C_{A_0}}$$

or

$$\frac{C_{B_0} - C_B}{C_{A_0}} = \frac{C_C}{C_{A_0}} + \frac{2C_D}{C_{A_0}} \tag{10}$$

For 80 percent conversion

$$\frac{C_{B_0} - C_B}{C_{A_0}} = 0.7237 + (2)(0.0763) = 0.8763$$

The complete results are presented in Table 6-5 and in graphical form in Fig. 6-10.

(b) *CSTR* For the CSTR the concentrations and fractional yields can be obtained from material-balance considerations. This is the result, of course, of the fact that the reactant and product concentrations in the CSTR are

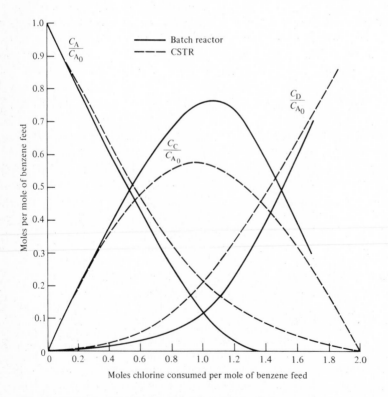

Figure 6-10 Product distribution in benzene chlorination.

Table 6-5 Composition of chlorobenzenes and chlorine consumption for batch reactor†

Unconverted benzene, mol	Monochlorobenzene produced, mol	Dichlorobenzene produced, mol	Chlorine consumed, mol
1	0	0	0
0.9	0.0995	0.0005	0.1005
0.8	0.1977	0.0023	0.2023
0.5	0.4811	0.0189	0.5186
0.2	0.7237	0.0763	0.8763
0.1	0.7715	0.1285	1.0285
0.0744‡	0.7743	0.1483	1.071
10^{-2}	0.6900	0.3000	1.290
10^{-3}	0.5558	0.4432	1.442
10^{-4}	0.4422	0.5577	1.558
10^{-6}	0.2791	0.7209	1.721

† Values are moles per mole of benzene fed.
‡ Maximum production of monochlorobenzene.

always at the exit concentrations, C_A, C_B, C_C, etc. If we assume an average residence time of $\bar{t}$, then

$$C_A = C_{A_0} + (-k_1 C_A C_B)\bar{t} \tag{11}$$

$$C_C = C_{C_0} + (k_1 C_A C_B - k_2 C_C C_B)\bar{t} \tag{12}$$

$$C_D = C_{D_0} + k_2 C_C C_B \bar{t} \tag{13}$$

Eliminating C_B from Eqs. (12) and (13) by use of Eq. (11) and assuming that $C_{C_0} = C_{D_0} = 0$, we obtain

$$\frac{C_C}{C_{A_0}} = \frac{(1 - C_A/C_{A_0})(C_A/C_{A_0})}{(C_A/C_{A_0}) + \kappa(1 - C_A/C_{A_0})} \tag{14}$$

and

$$\frac{C_D}{C_{A_0}} = \kappa\left(\frac{C_C}{C_{A_0}}\right)\left(\frac{C_{A_0}}{C_A} - 1\right) \tag{15}$$

We now calculate, as an example calculation, the product compositions for 80 percent conversion of benzene

$$\frac{C_C}{C_{A_0}} = \frac{(1 - 0.2)(0.2)}{0.2 + 0.1(1 - 0.2)} = 0.5714$$

$$\frac{C_D}{C_{A_0}} = 0.1(0.5714)\left(\frac{1}{0.2} - 1\right) = 0.2286$$

$$\frac{C_B}{C_{A_0}} = 0.5714 + (2)(0.2286) = 1.029$$

Table 6-6 Composition of chlorobenzenes and chlorine consumption for CSTR†

Unconsumed benzene, mol	Chlorobenzene produced, mol	Dichlorobenzene produced, mol	Chlorine consumed, mol
1	0	0	0
0.9	0.0989	0.0011	0.1011
0.8	0.1951	0.0049	0.2049
0.5	0.4545	0.0455	0.5455
0.240‡	0.5722‡	0.1825	0.9423
0.2	0.5714	0.2286	1.0286
0.1	0.4737	0.4263	1.3263
10^{-2}	0.0908	0.8992	1.8892
10^{-3}	0.0099	0.9891	1.9881
10^{-4}	0.0010	0.9989	1.9988
15^{-5}	0.0001	0.9999	1.9999

† Values are moles per mole of benzene fed.
‡ Maximum production of monochlorobenzene.

Complete results are presented in Table 6-6 and in graphical form in Fig. 6-10.

Discussion It is apparent by inspection of Tables 6-5 and 6-6 and Fig. 6-10 that for a given amount of chlorine consumption the conversion of benzene is greater in the plug-flow or batch reactor than in the CSTR. Also, the selectivity for monochlorobenzene is better in the batch reactor. If monochlorobenzene is the desired product and dichlorobenzene production is to be minimized, the consumption of chlorine must be kept low and the conversion of benzene should be kept low. If, however, the dichlorobenzene is the desired product, chlorine consumption should be high and the conversion of benzene can also be high.

These conclusions can be compared with commercial practice as described by Stephenson (1966). In commercial practice the chlorine is sparged into the bottom of cast-iron or mild-steel chlorinators of up to about 10,000 gal capacity. The reaction is allowed to proceed until about 75 percent of the benzene is chlorinated. The yield of monochlorobenzene is about 75 percent based on benzene. For the kinetic model developed in this example, if $C_A/C_{A_0} = 0.25$ (75 percent of benzene is reacted), the value of C_C/C_{A_0} would be 0.690, that is, 69 percent yield.

In the continuous chlorination of benzene, liquid from the chlorinator is distilled to remove the chlorinated benzene, and the unreacted benzene is recycled to the reactor. In other words, an excess of benzene is used compared with the batch case. Thus the results and conclusions obtained in this example are confirmed by commercial practice.

6-11 THERMAL BEHAVIOR OF A CSTR

Interesting and important temperature effects can occur in a continuous reactor system with exothermic reactions. The rate of heat generation increases as the rate of reaction increases, and the reaction rate increases with the temperature in accordance with Arrhenius's law. The coupling of these factors and their effect on the rate of heat removal from the reactor can result in the existence of two steady-state operating conditions.

A plot of heat generation rate for a CSTR as a function of temperature is shown in Fig. 6-11. The sigmoid shape is typical and is the result of two opposing effects. The reaction rate increases as the temperature increases, which, of course, results in an increase in the rate of heat generation. At high temperatures, however, the reaction rate is so great that the reaction is essentially complete and virtually no reactant remains. Therefore since a further increase in temperature cannot produce any additional reaction, the curve flattens out.

The equation for this curve can easily be obtained. We shall assume an irreversible first-order reaction and no change in density as a result of the reaction. The material-balance equation based on reactant A can be written as

$$vC_{A_0} = r_A V + C_A v = kC_A V + C_A v$$

The concentration of A at any time will therefore be

$$C_A = \frac{C_{A_0}}{1 + kV/v} \tag{6-33}$$

The rate at which heat will be generated due to reaction will be

$$Q_R = -kC_A V \, \Delta H \tag{6-34}$$

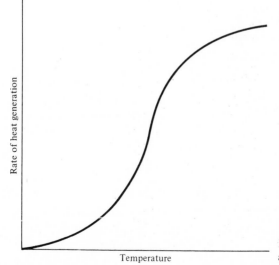

Figure 6-11 Rate of heat generation in a CSTR as a function of temperature.

Rate of heat generation

Temperature

Combining Eqs. (6-33) and (6-34) and remembering that the reaction rate constant can be expressed as $k = k_0 e^{-E/RT}$ yields the equation for the heat-generation curve

$$Q_R = - \frac{C_{A_0} v \, \Delta H}{1 + (v/Vk_0)e^{E/RT}} \tag{6-35}$$

If we assume that the CSTR is operating at steady-state conditions, the rate of heat generation must equal the rate of heat removal from the reactor. The energy-balance equation will be

$$Q_R = v\rho C_p(T - T_0) + UA(T - T_c) \tag{6-36}$$

The first term on the right-hand side of Eq. (6-36) is the heat removal by the product stream relative to T_0, the inlet temperature of the feedstream. The second term is the rate of heat removal by heat transfer. We note that the right-hand side of Eq. (6-36) is linear in T and that the rate of heat removal can be written as

$$Q_r = (v\rho C_p + UA)T - v\rho C_p T_0 - UAT_c \tag{6-37}$$

Equation (6-37) is shown in Fig. 6-12 for several temperatures of operation along with Eq. (6-35) for the rate of heat generation. In order for a steady state to be attained, Q_R must equal Q_r. This condition would occur at the intersection of the heat-generation and heat-removal lines. At a high rate of heat removal, as represented by line 1, one steady-state point occurs, and this at a low reactor operating temperature. Also, at low heat-removal rate as represented by line 4 only one steady state would occur, this time at a high reactor temperature. Line 2 intersects the heat-generation curve at three points. The upper and lower

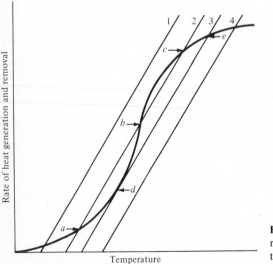

Figure 6-12 Rate of heat generation and removal in a CSTR as a function of temperature.

points are stable in that if the reactor were started from a cold condition, it would settle down at the lower operating point a; if it were started from a hot condition, or if it were started up without cooling until point b had been passed, it would settle down at c. Point b is an unstable point. If, for example, a small upward fluctuation of reactor temperature should occur, the heat-generation rate would be greater than the heat-removal rate and the temperature would continue to rise until point c was reached. The converse would be true if a slight downward fluctuation of temperature would occur, the system finally settling down at a. Point d on line 3 also represents an unstable situation, and any small upward fluctuation in temperature would result in the temperature rising to the steady state represented by e.

Usually the upper stable or stationary state is the desired one. Inspection of Eqs. (6-35) and (6-37) shows that the design or operating factors that favor the upper stationary state are high average residence times, high values for T_0 and T_c, and small values of UA. If the reaction is sufficiently exothermic and can provide enough heat to raise the reactants to the desired temperature and to overcome any heat losses in the system, it is said to be *autothermal*, i.e., self-supporting in its thermal requirements.

6-12 THERMAL BEHAVIOR OF A TUBULAR-FLOW REACTOR

The thermal behavior of a tubular-flow reactor will differ from that of a CSTR, whose contents are everywhere identical. The temperature in a tubular-flow reactor will rise along the reactor length for an exothermic reaction unless effective cooling is provided. For multiple steady states to appear it is necessary that a feedback mechanism be provided so that heat generated at one section of the reactor can pass back to an earlier section. In the CSTR the feedback was provided by the mixing. In the tubular-flow reactor the feedback will be provided if axial heat conduction is significant or if there is a transfer of heat from the product stream to the feed stream. One scheme for providing autothermal operation in the tubular-flow reactor is shown in Fig. 6-13a, in which the hot product and cold feed streams exchange heat in an external exchanger. The temperature profiles are shown in Fig. 6-13b.

If no appropriate thermal feedback mechanism is provided, the reaction will take place at the lower stationary state, where the reaction rate may be negligible. For the scheme shown in Fig. 6-13 the reaction could be extinguished if the temperature of the feed entering the reactor dropped below some critical value due, for example, to fouling of the heat-exchange surface.

One of the important effects that cannot be overlooked in the design of a tubular-flow reactor is the development of a radial temperature gradient in a highly exothermic reaction with wall cooling. The temperature profile, with temperatures near the tube axis much higher than those near the tube wall, can have an important effect on the reactor design because of the sensitivity of the reaction rate to temperature. The actual reaction rate can be much higher than the

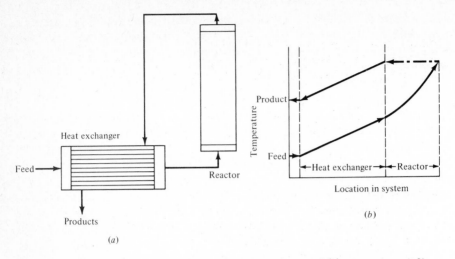

Figure 6-13 Autothermal operation: (a) heat-interchange scheme and (b) temperature profile.

reaction rate as calculated on the basis of a constant temperature across the tube. This effect is usually of more importance in the case of packed beds, and further discussion will wait until heterogeneous reactions have been discussed briefly.

6-13 SOME STABILITY AND START-UP CONSIDERATIONS

Aris and Amundson (1958) made a thorough study of a CSTR reacting system of the type we have been discussing. They solved the equations for this nonlinear system by linearization and present their results in the form of phase-plane trajectories. Their study included not only the situation to be expected for a reactor that is not under temperature control but also the trajectories and stability results for a reactor under temperature control for several values of a control parameter that is proportional to the feedback gain.

The parameters used by Aris and Amundson are summarized below:

$$\frac{\Delta H}{\rho C_p} = 200 \text{ K/(g mol/L)} \qquad \frac{UA}{\rho V C_p} = 1 \text{ min}^{-1} \qquad \frac{v}{V} = 1 \text{ min}^{-1}$$

$$C_{A_0} = 1 \text{ g mol/L} \qquad T_0 = 350 \text{ K}$$

$$\frac{E}{R} = 10 \text{ kcal/g mol} \qquad k_0 = e^{25} \text{ min}^{-1}$$

The reaction is assumed to be first order, and the reaction is

$$A \longrightarrow P$$

The trajectories for an uncontrolled reactor are presented in Fig. 6-14 with concentration and temperature as the state variables. Three steady-state solutions in which the rate of heat removal equals the rate of heat generation are noted. Two stable steady states, A and C, occur at $T = 354$ K and $C_A = 0.964$ g mol/L and at 441 K and 0.0885 g mol/L, respectively. An unstable steady state is observed at B, $T = 400$ K and 0.5 g mol/L. If the phase plane is divided along the line GBF, all transients that start from initial conditions to the right of the line will end at steady-state C, which is probably the desired reactor exit concentration, whereas those which start on the left will end up at A.

The danger inherent in a noncautious start-up can also be deduced from Fig. 6-14. A noncautious start-up could be one, for example, in which the reactor would initially be filled with reactant at the feed concentration and heat-up begun. By inspection of the trajectories it is apparent that such a start-up would move far out to the right before returning to the desired steady state at C. Such a trajectory could result in excessively high and dangerous temperatures.

In the part of their study which included reactor temperature control a closed loop was postulated in which the coolant flow rate is manipulated in response to deviations in reactor temperature. By proper choice of the control parameter the reactor can be stabilized so that the intermediate steady state becomes the only steady state. At somewhat smaller values of the control parameter a limit cycle developed about the unstable steady state, and the trajectories converged on this closed contour.

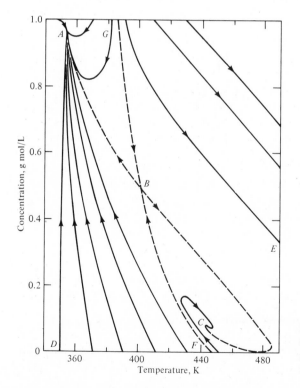

Figure 6-14 Concentration-temperature trajectories for uncontrolled reactor. [*From Aris and Amundson (1958), by permission.*]

6-14 HETEROGENEOUS REACTIONS

A heterogeneous reaction requires that at least two phases be present. The phases may be gas, liquid, or solid, and examples of large-scale commercial reactions can be cited for all possible phase combinations. They include the combustion of solid fuels and the roasting of ores as examples of gas-solid reactions, pickling steel as a liquid-solid reaction, chlorination as a gas-liquid reaction, and the precipitation of sodium bicarbonate in the Solvay soda process in which gas, liquid, and solid phases are involved. These are all examples of high-tonnage noncatalytic heterogeneous reactions. Some high-tonnage examples of catalytic heterogeneous reactions include ammonia synthesis as well as oxidation to produce nitric acid, SO_2 to SO_3 oxidation to produce sulfuric acid, petroleum cracking, and a host of petrochemical operations.

The field of heterogeneous reactions is extremely broad. In the discussion to follow we shall restrict ourselves to the gas-solid case. Since even this case is too broad for a complete discussion within the scope of this chapter, we shall further limit ourselves to a discussion and analysis of some of the problems that arise in the design of isothermal and nonisothermal packed tubular reactors. The matter of instability will also be touched upon.

Tubular-Reactor Design Problems

The problems involved in the design of a chemical reactor packed with a solid, which may be catalyst or a reactant, are similar in principle to those already discussed. The design revolves around energy and material balances coupled with the appropriate kinetic relationships.

The solid packing in the tubular reactor can have significant effects on the flow conditions, including departure from plug-flow conditions. Eddy and molecular diffusion can take place in the radial as well as in the axial directions. Radial or transverse diffusion would tend to bring reactor performance closer to what one would expect for perfect plug flow, whereas longitudinal diffusion or dispersion would tend to invalidate the plug-flow assumption. Although the longitudinal diffusion coefficient generally is larger than the radial diffusion coefficient, longitudinal flux is usually much smaller because of the small axial concentration gradients when the ratio of tube length to diameter is large. If the ratio of the length to particle diameter is large, we can usually ignore longitudinal dispersion effects compared with the effect of bulk flow.

Isothermal Tubular Reactor

The isothermal tubular, or fixed-bed, reactor is the simplest gas-solid reactor to analyze, especially if axial dispersion effects are negligible. There will be no radial exchange of mass or energy, and if the ratio of reactor length to particle diameter is large, axial dispersion can also be safely neglected.

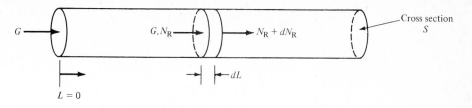

Figure 6-15 Differential element of tubular reactor.

We can write the conservation equations over an element of length dL of a reactor with constant cross section S (Fig. 6-15). A flat fluid-velocity profile is assumed, i.e., plug flow, and G is the mass flow rate per unit cross-sectional area. Let N_R denote the concentration of product species R entering the volume element expressed as moles R per unit mass of fluid, and $N_R + dN_R$ will be the exit concentration. The catalyst has a bulk density ρ_b and a global reaction rate at the concentration N_R of $\mathscr{R}_{N_R}$, where $\mathscr{R}$ is in units of moles per unit time per mass of catalyst. A balance on the product species R will yield

$$G \, dN_R = \mathscr{R}_{N_R} \rho_b \, dL \qquad (6\text{-}38)$$

This expression can be integrated to give the required length (and therefore volume) of tubular reactor

$$L = \frac{G}{\rho_b} \int_{N_{R_0}}^{N_{R_L}} \frac{dN_R}{\mathscr{R}_{N_R}} \qquad (6\text{-}39)$$

The mass of catalyst contained within the reactor will be

$$W = \rho_b LS \qquad (6\text{-}40)$$

The global reaction rate will generally be a function of reactant and product species, temperature, and flow rate. An a priori calculation is not possible at present, and the relationship between $\mathscr{R}$ and the various parameters that determine its value must be determined experimentally. Equation (6-39) can then be evaluated by graphical or numerical integration. If the reaction rate is a function of pressure, the pressure drop along the reactor must also be evaluated and considered in the reactor design.

Isothermal packed reactors are rare in practice, and the only case in which true isothermal operation is possible would occur if the heat of reaction were zero. A reaction of this type will serve as the basis for the following example.

Example 6-7 Most of the carbon disulfide produced is based on methane as the source of carbon. The reaction is

$$CH_4(g) + 2S_2(g) \;\rightleftharpoons\; CS_2(g) + 2H_2S(g)$$

The sulfur vapor at the condition of the reaction, about 650°C, is a mixture of diatomic, hexatomic, and octatomic molecules. The kinetic data, however, are correlated satisfactorily by assuming a second-order reaction between methane and diatomic sulfur, which is probably the true reactant. Forney and Smith (1951) report that at 600°C

$$\mathcal{R} = 0.26 P_{CH_4} P_{S_2} \tag{1}$$

where the units of partial pressures are atmospheres and $\mathcal{R}$ is in gram moles of CS_2 per gram of catalyst per hour.

CS_2 reactors are operated at about 650°C and isothermally. Isothermal operation is possible because at a temperature of about 650°C the heat of reaction is zero. The reaction is exothermic at lower temperatures and endothermic at higher temperatures. Therefore, adiabatic reaction at 650°C is also equivalent to isothermal operation.

Estimate the mass of catalyst required to produce 3 t/h CS_2. A 10 percent excess of sulfur is used, and 90 percent conversion of CH_4 to CS_2 is desired. The reactor operates at atmospheric pressure.

SOLUTION The global reaction rate, expressed in terms of partial pressures of the reactant species, can easily be related to the fractional conversion of CH_4. We shall consider 100 mol of inlet gas to the reactor and calculate its composition at fractional conversion X. The compositions will be:

Component	Composition	
	Inlet, %	Conversion X
CH_4	31.25	$31.25(1 - X)$
S_2	68.75	$68.75 - 62.5(1 - X)$
H_2S	—	$62.5X$
CS_2	—	$31.25X$
Total	100	100

The partial pressures of the reactant species will be

$$P_{CH_4} = 0.3125(1 - X)$$

$$P_{S_2} = 0.6875 - 0.625(1 - X)$$

We can now write the global-reaction rate equation at 600°C,

$$\mathcal{R}_{N_R} = (0.26)(0.3125)(1 - X)[0.6875 - 0.625(1 - X)]$$

The reaction will be carried out, however, at 650°C, and so we must correct for this temperature. Stephenson (1966) quotes an activation energy of 34,400 cal/g mol for this reaction. With this value the global reaction rate at 650°C will be

$$\mathcal{R}_{N_R} = (0.76)(0.3125)(1 - X)[0.6875 - 0.625(1 - X)] \tag{2}$$

We can now calculate N_R, the moles of product present per unit mass of material flowing. The mass of material flowing is equal to M_F per mole of feed, where M_F is the feed molecular weight, and $31.25X$ mol of product is present per 100 mol of feed. N_R can therefore be expressed as

$$N_R = \frac{0.3125X}{M_F} \qquad \text{and} \qquad dN_R = \frac{0.3125}{M_F} dX \tag{3}$$

The expression to be integrated, upon combining Eqs. (6-39) and (6-40), is

$$W = GS \int_0^{N_R} \frac{dN_R}{\mathcal{R}_{N_R}} = \frac{GS}{M_F} \int_0^X \frac{0.3125 \, dX}{R_{N_R}} \tag{4}$$

The term GS/M_F is the moles of feed to the reactor per hour. For a CS_2 production rate of 3000 kg/h at a methane to CS_2 conversion of 90 percent the required feed rate will be

$$\frac{GS}{M_F} = \frac{(3000)(1000)}{(76)(0.3125)(0.9)} = 1.4 \times 10^5 \text{ mol/h}$$

Appropriate substitutions into Eq. (4) yields the expression to be integrated

$$W = 1.4 \times 10^5 \int_0^{0.9} \frac{dX}{(0.76)(1 - X)[0.6875 - 0.625(1 - X)]} \tag{5}$$

The integral can be evaluated numerically or analytically

$$W = (1.4 \times 10^5)(8.81) = 1.234 \times 10^6 \text{ g catalyst required}$$

or *1234 kg catalyst*. The sensitivity to temperature is easily shown. If the reactor had been operated at 600°C, catalyst requirement would have been 3605 kg.

Adiabatic Operation

As mentioned earlier, the possibility that true isothermal operation can be attained in a gas-solid tubular reactor is remote, and nonisothermal operation would normally obtain. Adiabatic operation or operation closely approaching adiabatic would most likely occur in practice. Since in adiabatic operation no exchange of heat with the surroundings would take place, no radial temperature gradients would be present. The heat due to reaction would result in a change of enthalpy of the reacting fluid, and if we can assume that no temperature gradient exists within the individual solid particles and that axial heat conduction can be neglected, a simple energy balance can be written on the fluid entering and leaving a differential volume of length dL (Fig. 6-15)

$$GC_p \, dT = \rho_0(-\Delta H)\mathcal{R}_{N_R} \, dL \tag{6-41}$$

Combining Eq. (6-41) with (6-38), the material balance, yields

$$C_p \frac{dT}{dN_R} = -\Delta H$$

which upon integration with the additional assumption that $\Delta H / C_p$ is constant yields

$$T = T_0 + \frac{-\Delta H}{\bar{C}_p} N_R \tag{6-42}$$

where T_0 is the entering temperature and $\bar{C}_p$ is the average heat capacity of the flowing fluid. The adiabatic reaction path would therefore appear as shown in Fig. 6-16, as a straight line of slope $\bar{C}_p / (-\Delta H)$ on the TN_R plane. The line shown is for an exothermic reaction.

The global reaction rate in Eq. (6-40) is, of course, a function of reactant concentrations and reaction temperature. If data are available for $\mathcal{R}_{N_R}$, they can also be presented on the TN_R plane as isorate contours. Such a presentation is shown in Fig. 6-16 for a reversible exothermic reaction. The dotted line is the locus of the maxima of each contour. This line therefore represents the locus of points for which $\partial \mathcal{R}_{N_R} / \partial T$ is zero and is the locus of temperatures at which the reaction rate is a maximum for a given conversion. Reactor size can be evaluated from Fig. 6-16 by integration of the mass balance, Eq. (6-39). $\mathcal{R}_{N_R}$ is obtained as a function of N_R from the points at which the adiabatic reaction path intersects the isorate contours.

When we observe the reaction rates as we proceed along the adiabatic reaction path, we note that they increase at first and then decrease at more elevated conversions and temperatures. This behavior can easily be explained. The effect of an increase in temperature is twofold: it increases the rate of reaction but also reduces the value of the equilibrium constant, i.e., the maxi-

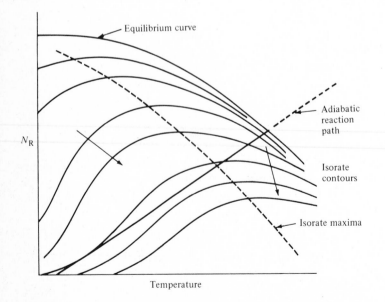

Figure 6-16 Adiabatic tubular-reactor performance.

mum conversion. At low conversions the reactants are far from equilibrium, and the rate of the forward reaction will be relatively high while the rate of the reverse reaction is low because product concentration is low. At higher conversions the rate of the reverse reaction will increase. In addition, for a reversible exothermic reaction the activation energy of the reverse reaction will be greater than that of the forward reaction. Thus, an increase in temperature will favor the reverse-reaction velocity constant relative to the forward-reaction velocity constant.

To minimize reactor size it would be desirable to operate along the maximum-rate curve. This could be achieved, in principle, by preheating the feed to the temperature at which the maximum rate intersects the feed composition. The reaction could then proceed but only to the extent of a differential conversion. The fluid could then be removed from the reactor, cooled to the maximum-reaction-rate temperature, and reintroduced into the reactor and the reaction again allowed to proceed to a differential conversion, this time along the new adiabatic reaction path, etc.

Obviously, such a scheme would require an infinite number of reactors and heat exchangers if the reactor temperature profile is to coincide with the maximum-reaction-rate curve. A more normal scheme is shown in Fig. 6-17, in which a sequence of four reactors and heat exchangers is required. The feed enters at T_i and is preheated to A. Reaction occurs from A to B, C to D, E to F, and G to H and cooling from B to C, D to E, and F to G. The optimum number of reactors, conversions in each reactor, temperature, etc., must be decided on the basis of appropriate economic evaluations, and possible alternative operational schemes must also be analyzed.

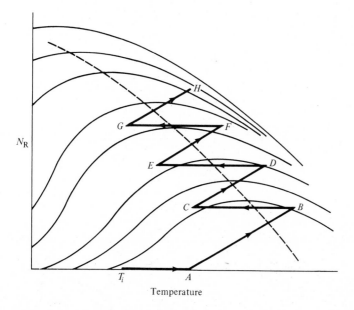

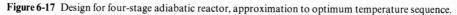

Figure 6-17 Design for four-stage adiabatic reactor, approximation to optimum temperature sequence.

Example 6-8 The oxidation of sulfur dioxide is an outstanding example of a reversible exothermic reaction in which a temperature profile is maintained in the adiabatic reactor to obtain optimum utilization of catalyst and high conversion. Estimate the weight of catalyst required in the first two reactors of a sulfuric acid plant in which the feed gas that enters the first reactor has a composition of 8% SO_2, 13% O_2, and 79% N_2. The feed enters at 410°C, and a conversion of 70 percent is attained in the first reactor. The exit gas is cooled to 450°C and enters the second reactor, where a conversion to 90 percent of the entering SO_2 is reached. Estimate catalyst weight per 1000 kg H_2SO_4 (100%) of daily capacity.

SOLUTION The data generated by Calderbank (1953) will be used in the solution. The expression developed for the global reaction rate with a vanadium catalyst is

$$\mathscr{R}_{N_R} = P_{O_2} P_{SO_2}^{1/2} \exp\left(-\frac{31,000}{RT} + 12.07\right) - \frac{P_{SO_3} P_{O_2}^{1/2}}{P_{SO_2}^{1/2}} \exp\left(-\frac{53,600}{RT} + 22.75\right)$$

For a feed gas of the same composition as in this example Calderbank presented the equation for the adiabatic reaction path

$$T = 235X + T_0$$

where T_0 is the entering temperature and X is the fractional conversion. The equilibrium constant as a function of temperature is

$$\ln K_p = \frac{22,600}{RT} - 10.68$$

Expressions for K_p and $\mathscr{R}_{N_R}$ as functions of fractional conversion can be developed easily. Taking as a basis 100 mol of gas entering the reactor, we can calculate the composition at any fractional conversion and the partial pressure of the various components:

Component	Moles in	Moles at fractional conversion X	Partial pressure, atm
SO_2	8	$8(1-X)$	$\dfrac{8(1-X)}{100-4X}$
O_2	13	$13 - \dfrac{8X}{2}$	$\dfrac{13-4X}{100-4X}$
N_2	79	79	$\dfrac{79}{100-4X}$
SO_3	$\cdots$	$8X$	$\dfrac{8X}{100-4X}$
Total	100	$100-4X$	1

K_p as a function of fractional conversion becomes

$$K_p = \frac{P_{SO_3}}{P_{SO_2} P_{O_2}^{1/2}} = \frac{X(100 - 4X)^{1/2}}{(1 - X)(13 - 4X)^{1/2}} = \exp\left(\frac{22{,}600}{RT} - 10.68\right)$$

The equilibrium fractional conversions are shown as a function of temperature in Fig. 6-18. The adiabatic reaction path $ABCD$ is also shown, where the reaction occurs from A to B and from C to D. It should be noted that 90 percent conversion cannot be achieved in one reactor. The maximum conversion possible, as can be noted by the intersection of line AB with the equilibrium curve, is about 78 percent.

The global rate equation expressed in terms of fractional conversion and temperature becomes

$$\mathscr{R}_{N_R} = \frac{(13 - 4X)(8 - 8X)^{1/2}}{(100 - 4X)^{3/2}} \exp\left(-\frac{31{,}000}{RT} + 12.07\right)$$
$$- \frac{(8X)(13 - 4X)^{1/2}}{(100 - 4X)(8 - 8X)^{1/2}} \exp\left(-\frac{53{,}600}{RT} + 22.75\right)$$

We are now in a position to calculate the catalyst requirements for the prescribed duty. By combining Eqs. (6-39) and (6-40) it can easily be shown that

$$W = F_A \int_{X_0}^{X_A} \frac{dX}{\mathscr{R}_{N_R}}$$

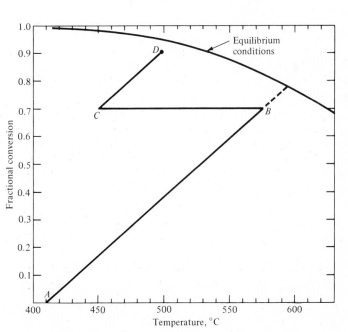

Figure 6-18 Adiabatic operation lines and equilibrium conditions for two-stage SO_2 oxidation.

where W is the weight of catalyst and F_A is the number of moles of reactant feed per second. This equation can be solved by graphical or numerical integration. The reaction rate as a function of conversion and temperature is shown in Table 6-7.

The catalyst weight will therefore be

$$W = \int_0^{0.7} \frac{dX}{\mathscr{R}_{N_R}} + \int_{0.7}^{0.9} \frac{dX}{\mathscr{R}_{N_R}}$$

Numerical integration yields 207.7×10^3 and 133.2×10^3 g catalyst-second per mole for the first and second reactors, respectively. The catalyst weight per daily 1000 kg of 100% H_2SO_4 will be

$$W = \frac{(1000)(207.7 + 133.2) \times 10^3}{(24)(3600)(98.1)} = 40.22 \text{ kg}$$

It can be pointed out that commercial sulfuric acid plants usually conduct the oxidation in four stages with an ultimate conversion of 98 percent or greater. In areas where pollution standards exist the concentration of SO_2

Table 6-7

X	T, K	$R_{N_R} \times 10^6$, (g mol product)/(g cat · s)	$\frac{1}{R_{N_R}} \times 10^{-3}$
		First reactor	
0	683	0.771	1297
0.05	694.7	1.092	916.1
0.10	706.5	1.523	656.8
0.15	718.2	2.093	477.8
0.20	730	2.836	352.2
0.25	741.7	3.786	264.2
0.30	753.5	4.978	200.9
0.35	765.2	6.441	155.3
0.40	777	8.188	122.1
0.45	788.7	10.20	98.0
0.50	800.5	12.4	80.6
0.55	812.2	14.6	68.5
0.60	824	16.4	60.8
0.65	835.7	17.19	58.2
0.70	847.5	15.6	64.2
		Second reactor	
0.70	723	1.167	856.8
0.75	734.7	1.432	698.4
0.80	746.5	1.654	604.7
0.85	758.2	1.705	586.3
0.90	770	1.275	784.3

discharged to the atmosphere is limited to 250 ppm. This concentration would correspond to a conversion of 99.7 percent, a conversion that cannot be reached by conventional procedures due to the thermodynamic equilibrium. In double absorption the gas leaving the third converter stage is sent to an absorption tower, in which the SO_3 is absorbed in sulfuric acid. The lean gas, now containing only SO_2, O_2, and N_2, is reheated and introduced into the final converter stage. An overall conversion of 99.7 percent and the concomitant reduction in exit SO_2 concentration are easily achieved by this procedure.

Nonisothermal Tubular Reactor

We shall consider that the reaction $A \rightarrow R$ is taking place in a reactor with heat being exchanged between the reactor and the surroundings. If the reaction is exothermic and heat is being removed at the wall, a radial temperature gradient will be established and the reacting fluid will be hotter at the tube axis than at the wall. The reaction rate will therefore be highest along the axis, as will the consumption rate of reactant. This, in turn, will result in transverse concentration gradients, reactant will diffuse toward the tube axis, and an outward flow of product will take place from tube axis toward the tube wall. The transverse temperature and concentration gradients will invalidate the simple one-dimensional analysis that was made in the isothermal case, and the balance equations must be written for the two dimensions L and r.

The analysis will be made for an elementary annulus (Fig. 6-19) of length dL and thickness dr. We neglect the effect of longitudinal dispersion and heat conduction and assume equimolar counterdiffusion. The terms in the mass-balance equation will be

Entering axially by mass flow $= 2\pi r\, dr\, GN_R$

Entering radially by diffusion $= -2\pi r\, dL\, D_r \dfrac{\partial N_R}{\partial r}$

Leaving axially by mass flow $= 2\pi r\, dr\, G\left(N_R + \dfrac{\partial N_R}{\partial L}\, dL\right)$

Leaving radially by diffusion $= -2\pi(r + dr)\, dL\, D_r\left(\dfrac{\partial N_R}{\partial r} + \dfrac{\partial^2 N_R}{\partial r^2}\, dr\right)$

Produced by reaction $= 2\pi r\, dr\, dL\, \rho_b \mathscr{R}_{N_R}$

and the mass-balance equation becomes

$$G\frac{\partial N_R}{\partial L} - D_r\left(\frac{\partial^2 N_R}{\partial r^2} + \frac{1}{r}\frac{\partial N_R}{\partial r}\right) = \rho_b \mathscr{R}_{N_R} \qquad (6\text{-}43)$$

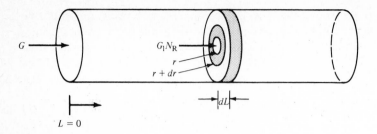

Figure 6-19 Differential section of two-dimensional tubular reactor.

The energy-balance equation can be developed by similar considerations and, with the additional assumption that the bed packing temperature is everywhere at the temperature of the gas stream, will be

$$C_p G \frac{\partial T}{\partial L} - k_r \left(\frac{\partial^2 T}{\partial r^2} + \frac{1}{r} \frac{\partial T}{\partial r} \right) = \rho_b \, \Delta H \, \mathscr{R}_{N_R} \tag{6-44}$$

Even this highly simplified analysis results in a pair of coupled equations that are also highly nonlinear because of the effect of temperature on reaction rate. The equation can be solved by numerical methods, and the typical form of the solutions for fractional conversion and temperature as a function of reactor length and tube radius is shown in Fig. 6-20.

By inspection of Fig. 6-20 it can be easily appreciated that extremely high temperatures can be reached at the tube axis. These temperatures can be

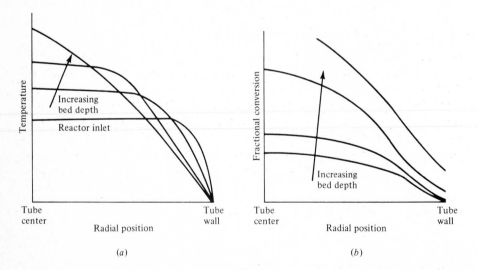

Figure 6-20 Typical profiles for a two-dimensional tubular reactor: (a) temperature profile and (b) conversion profile.

sufficiently high to damage the catalyst by overheating. Although it is not apparent by inspection of the equations, a maximum temperature is also obtained along the reactor length, and the location of this "hot" zone can move with changes in catalyst activity.

6-15 TUBULAR-REACTOR STABILITY

A detailed presentation of reactor stability and multiple steady-state considerations is well beyond the scope of this chapter. An extremely simple one-dimensional model can be used, however, to demonstrate that multiple steady states can be obtained in a tubular reactor. For this presentation we consider the situation obtained in a thin slice of a tubular reactor packed with a solid catalyst.

The entering concentration of reactant C_b is assumed constant, and the concentration of reactant at the catalyst surface is C_i. The rate of reaction in the slice can be written as

$$\text{Reaction rate} = k'C_i$$

and the rate of reactant transport to the catalyst surface will be

$$\text{Transport rate} = k''A(C_b - C_i)$$

where A is the area for mass or heat transfer in the slice. At steady state

$$k'C_i = k''A(C_b - C_i) \tag{6-45}$$

Solving Eq. (6-45) for C_i and substituting into the expression for reaction rate yields

$$\text{Reaction rate} = k'\frac{k''AC_b}{k' + k''A}$$

and the rate of heat generation due to reaction will be

$$\text{Rate of heat generation} = \frac{k'k''AC_b}{k' + k''A}(-\Delta H) \tag{6-46}$$

The parameter k' is an exponential function of temperature in accordance with Eq. (6-8), while k'' will have a weak dependence on temperature. At steady state the rate of heat generation by reaction must equal the rate of heat removal due to transport, and

$$\text{Rate of heat removal} = h'A(T_i - T_{b_0}) \tag{6-47}$$

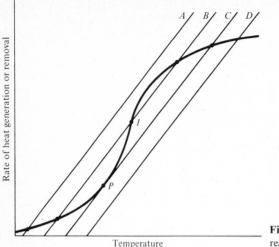

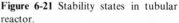

Figure 6-21 Stability states in tubular reactor.

Equations (6-46) and (6-47) are plotted in Fig. 6-21. The straight lines represent the rate of heat removal for several values of T_0, and the sigmoid curve represents the rate of heat generation due to reaction in accordance with Eq. (6-46). The intersections of the straight line with the reaction curve represent the simultaneous solutions of Eqs. (6-46) and (6-47) and are possible operating points.

A reactant stream that entered at a temperature corresponding to line A would encounter one stable but undesirable operating point at low temperature and conversion. Two possible stable operating points are observed on line B, one at a low conversion and low temperature and one at a high conversion and high temperature. The intermediate intersection point I can be referred to as an ignition temperature. The minimum ignition temperature P is at the point where line C is tangent to the reaction curve, and a slight perturbation could result in a low-temperature stable operating point being reached. If the entrance temperature corresponds to line D, only one stable operating point is possible and this at a high temperature and high conversion.

It should be apparent that an infinite family of steady-state profiles can be obtained in a packed-bed reactor and that the steady state achieved will depend upon the start-up and entrance condition. They should be carefully studied and analyzed during the design stages to prevent unpleasant surprises.

An important class of reactors is that in which the incoming feed stream is heated by heat interchange with the outlet stream or within the reactor itself. In such reactors relatively small perturbations in temperature can quench the reaction or result in trajectories to a new steady state that might subject the catalyst and system to undesirable temperatures. These effects must be considered especially during considerations of start-up and shutdown. The interested reader is referred to works that deal with reactor design for detailed expositions.

6-16 FINAL COMMENT

As has been hinted at several points, the theory of the design of heterogeneous tubular reactors is much more advanced than the discussion presented here. As in many other cases of chemical engineering, however, the experimental information in support of the theory is still insufficient to enable a complete tubular reactor design to be carried out with full confidence. Knowledge of effective thermal conductivities and of heat and mass transfer between gas and solid are especially important in view of the exponential effect of temperature on reaction rate.

PROBLEMS

6-1 Species A in a solvent medium at a concentration of 2 g mol/L is converted into R in a batch reactor according to first-order kinetics. Conversion of 95 percent is achieved during a 4-h cycle that includes 3 h of reaction plus 1 h for charging, discharging, and reactor cleaning between batches. Since the market for R has been growing, it has been proposed that production capacity be doubled by adding an identical reactor in parallel with the present one. If operations could be made continuous, however, downstream operations could be significantly simplified. You are asked to consider the possibility of operating the new reactor as a CSTR in series with the present one, which could also be operated as a CSTR. What should the size of the new reactor be compared with that of the present reactor, and should it be located upstream or downstream of the existing one?

6-2 Operation of a batch reactor requires that part of the total time be devoted to filling, discharging, and cleaning the reactor. Develop an expression relating the fractional conversion of A to t_s, the reactor shutdown time per batch for the maximum production rate for the isomerization reaction $A \rightarrow R$. The reaction is governed by first-order kinetics. Note that as t_s decreases, the conversion of A for maximum production rate also decreases. Discuss the effects of working at the optimum X_A on the costs involved in operations upstream and downstream of the reactor.

6-3 One of our products contains a small amount of an undesirable by-product A. The concentration of A in the untreated reactor product stream is 1 percent, but the product specifications require that no more than 0.04% A be present. The laboratory has developed a process that converts A into a volatile material which can easily be stripped. The A-decomposition reaction is first order with a reaction velocity constant k equal to 5.1 h^{-1}.

(a) What volume of CSTR would be required to produce specification-grade material from a feed stream of 2 m^3/h?

(b) For start-up it has been suggested that the reactor be filled rapidly and a feed rate of 1.8 m^3/h maintained until specification-grade material is obtained. For how long would off-specification product be made?

(c) If a batch start-up policy were to be adopted, how much time would be required before the design feed rate could be initiated?

6-4 A CSTR is to be used for the homogeneous polymerization of the monomer M to yield the marketable dimer and trimer products. An unmarketable tetramer is also produced. The reaction sequence involves an initiation step to an activated complex M*, followed by a series of propagation reactions as follows:

$$M \longrightarrow M^* \qquad R_i = k_i C_M \qquad k_i = 0.06 \ min^{-1}$$

$$M^* + M \longrightarrow M_2 \qquad R_1 = k_1 C_{M^*} C_M \qquad k_1 = 0.3 \ L/(g \ mol \cdot min)$$

$$M^* + M_2 \longrightarrow M_3 \qquad R_2 = k_2 C_{M^*} C_{M_2} \qquad k_2 = 0.12 \ L/(g \ mol \cdot min)$$

$$M^* + M_3 \longrightarrow M_4 \qquad R_3 = k_3 C_{M^*} C_{M_3} \qquad k_3 = 0.03 \ L/(g \ mol \cdot min)$$

There is no change in specific volume as a result of reaction. The reactor effluent undergoes a rapid chemical quench immediately after leaving the reactor to kill the activated monomer according to $M^* \to M$. The reactor is designed for a 10-min average residence time for feeds with monomer concentration in the range of 1 to 4 mol/L. An operating manual for use by the plant operators is being prepared. As part of the operating manual you are to prepare:

(a) A graph of the weight fraction of products after quench on a solvent- and monomer-free basis vs. monomer feed concentration

(b) A graph of fractional monomer conversion vs. monomer feed concentration

(c) A graph of concentration of products in the reactor effluent after quench vs. monomer feed concentration

6-5 Our research department has come up with a process to produce R and S. Although the actual reaction path is complicated, the irreversible kinetics can be represented quite simply by

$$A \xrightarrow{\ k_1\ } R \xrightarrow{\ k_2\ } S$$

Without catalyst, $k_1 = 0.4 \text{ h}^{-1}$ and $k_2 = 0.04 \text{ h}^{-1}$. When a catalyst is used, k_1 is not affected but $k_2 = 0.6 \text{ h}^{-1}$. There is no volume change due to reaction. Although A can easily be separated from R and S by distillation, no scheme for separating R and S has been developed. It is believed, however, that the market will accept R if it contains no more than 4 mol % S and S if it contains no more than 4 mol % R. It is proposed that this hypothesis be tested with some marketable R and S produced in a 1-m^3-batch pilot-plant reactor. Reactant A is available as a solution containing 5 g mol/L.

(a) How much reaction time will be required to produce a batch of R? What will be the concentration of A in the solution to be distilled?

(b) How much reaction time will be required to produce a batch of S? What will be the concentration of A in the solution to be distilled?

6-6 A process to produce marketable R has been developed. The homogeneous reaction is

$$A + B \xrightarrow{\hspace{1.5cm}} R \tag{1}$$

At the reaction conditions the reactant species B undergoes the disproportionation reaction

$$2B \xrightarrow{\hspace{1.5cm}} S + T \tag{2}$$

but no market exists for S and T. The laboratory work has shown that reaction (1) is first order with respect to each of the reactants and that reaction (2) is second order with respect to B. The ratio k_2/k_1 is approximately equal to 1.2 under the conditions at which the reaction will be carried out in the production plant. There is no volume change as a result of the reaction. To minimize the extent of the undesired disproportionation reaction it has been suggested that a CSTR be used. A mixed feed in which $C_{A_0} = C_{B_0} = 4$ g mol/L would be used, and a fractional conversion of 90 percent of B is planned.

(a) What fraction of the B would be converted to the desired product R?

(b) Suggest other schemes that would result in an increase in the fractional yield of R. Consider also the possibility (1) that A could easily be recovered from the product stream for recycle and (2) that A would be difficult to recover for recycle.

6-7 In general, overall reactant conversion can be increased by specifying a larger reactor or by recovering unreacted reactant in a separation step and recycling the recovered reactant to the reactor. Consider reactant A being fed to a CSTR that is followed by a separation operation. All the A in the reactant effluent is recovered in the separator as pure A and is recycled to the reactor. Develop expressions relating overall conversion of A to the recycle ratio expressed as moles A recycled per mole of makeup A and to reactor volume. Assume negligible volume change during reaction and consider the cases where the kinetics are irreversible and are zero, first, and second order with respect to A. Discuss the effects of recycle operation on equipment size and operating costs.

6-8 A dilute aqueous solution of acetic acid is used in one stage of a chemical process. The solution is produced by the hydrolysis of acetic anhydride in two isothermal equal-volume CSTRs. The solution is pumped to the acid consumer through a long 50-mm-ID pipeline. A bright and alert young chemical engineer observed that the acid received by the consumer corresponds to 95 percent hydrolysis of anhydride although the reactor effluent composition corresponds to 90 percent hydrolysis, the minimum conversion that the consumer can accept for its process. The engineer suggests that the production rate can be increased by operating at a throughput such that the conversion obtained by the time the reactor effluent reaches the consumer would be 90 percent. By how much could the throughput rate be increased? Additional data unearthed by the engineer are as follows:

Individual reactor working volume, 3 m^3
Acetic anhydride concentration in reactor feed, 0.4 g mol/L
Reaction kinetics, pseudo first order with respect to the anhydride

6-9 In aqueous medium A reacts according to the following irreversible, homogeneous first-order reactions

$$A \xrightarrow{\;k_1\;} R \xrightarrow{\;k_2\;} S$$

The reactions are carried out in an isothermal CSTR, and the desired species R is separated from the reactor effluent by solvent extraction with a highly selective organic solvent. A and S are completely insoluble in the solvent, which can be assumed to be immiscible with water. The partition coefficient with respect to R is

$$K = \frac{\text{molar concentration of R in solvent phase}}{\text{molar concentration of R in aqueous phase}} = 5$$

It has been suggested that the yield of R could be increased if the selective solvent were fed to the reactor along with the aqueous feed solution of A. Assuming that the aqueous and organic phases would be in equilibrium in the reactor, calculate the expected yield and relative production rates of R for several ratios of volume of solvent to volume of aqueous feed. Compare with the yield obtained without solvent. The concentration of A in the aqueous feed is 1.5 g mol/L, and 3 m^3/h is processed. The reactor was originally designed to produce the maximum attainable concentration of R.

6-10 The gas leaving the second reactor bed of Example 6-8 is cooled to 450°C, enters a third adiabatic bed, and leaves at 96 percent SO_2 conversion. This gas is cooled, and essentially all the SO_3 is absorbed in sulfuric acid. The gas then passes through a demister to remove droplets of acid, is heated to 430°C, and enters a fourth reactor bed. The composition of the gas leaving this bed is equivalent to a conversion of 99.7 percent of the original entering SO_2. Calculate the temperatures of the gases leaving the third and fourth reactors and the weight of catalyst required in each reactor.

6-11 Plot a curve for instantaneous rate of heat generation vs. reactor temperature for an exothermic reaction performed in a batch reactor. On this same figure plot, as in Fig. 6-12, the rate of heat removal with coolant temperature as the parameter. Show that a runaway reaction is possible under certain conditions of coolant temperature.

6-12 A process for producing P is being developed in the research department. The reaction is $A + B \rightarrow P$. Component B is a solvent which also participates in the reaction. The kinetics are pseudo first order with respect to A and k at 80°C, the temperature at which most of the measurements were made, is 0.0022 s^{-1}. From careful rate measurements the value of the activation energy was estimated to be 21 kcal/g mol. The reaction is exothermic, and approximately 50 kcal is liberated per gram mole of A consumed.

To produce P in sufficient quantity to permit market evaluation and development, some batches of P are to be prepared in a jacketed vessel equipped with an internal cooling coil. The working volume of the vessel is 1 m^3, the total heat-transfer surface is 7 m^2, and the overall heat-transfer coefficient is expected to be about 500 cal/($m^2 \cdot s \cdot$ °C). Two methods have been suggested for operating the reactor: (1) rapidly fill reactor with cold feed solution at 25°C containing 2 g mol of A per liter and (2) heat solvent to 80°C, begin cooling, and simultaneously add reactant A to bring its concentration to 2 mol/L. This was the procedure used in the laboratory. In view of the highly exothermic nature of the reaction the cooling water at 25°C will initially be pumped at a high rate through the jacket and coil for both methods.

(a) Estimate the initial rate of disappearance of A for both methods. Would you expect any operating difficulties to develop?

(b) How would you recommend operating the reactor?

SEVEN

ECONOMIC ANALYSIS AND EVALUATION

> If the years are many you shall increase the price,
> and if the years are few you shall diminish the
> price.
>
> *Leviticus, 25 : 6*

Economic analysis and evaluation are often needed by the chemical engineer as a guide to planning new capacity and in studying how changes in plant size, selling price, and demand and alternative processes or equipment might affect profitability. In an economic analysis factors are assessed that are relevant to a process or project decision involving the expenditure of money or other resources. Such a decision requires the consideration of alternatives, and one of the alternatives that can usually be considered is to do nothing. The economic evaluation involves those factors that can be quantified in terms of money. The final decision must also include considerations of those factors that are difficult or impossible to reduce or to quantify in monetary terms. Some of these irreducible factors include, for example, the expected effects of a decision on safety and on corporate image as well as such factors as management judgment and estimates of the future activities of a major competitor.

In a large project it may be necessary to carry out a number of economic analyses and evaluations at different stages of the project. Preliminary economic evaluations are made before initial research is begun, and the evaluations become more comprehensive as development proceeds. A thorough economic analysis and evaluation would be carried out before company resources are committed to a full-scale plant and would be one of the major elements considered in such a decision.

7-1 PROJECT EVALUATION

Every project requires the commitment of resources in the form of money and in other forms that can also be expressed in monetary terms, e.g., engineering, research, development, and management talent. The commitment of this money must be justified in terms of making a profit or receiving a return on the funds or resources committed. The economic evaluation of a project requires that the costs of the project be estimated as well as the financial gain to be expected. These figures are then analyzed to obtain a quantified measure of the value or desirability of the project. Thus three stages or steps are involved in the economic evaluation of a project:

1. Estimates of the factors that influence profitability such as equipment costs, raw-material costs, selling price, and sales
2. A forecast of how these factors will vary throughout the life of the project
3. The treatment of these factors to yield some understandable measure or index of profitability

7-2 CASH-FLOW DIAGRAM

A diagram of the integrated or cumulative cash flow to be expected during the project life is useful and instructive. This diagram shows the predicted cash position of the project at any instant in time. A typical cumulative-cash-flow diagram for a project such as a new chemical plant is shown in Fig. 7-1. In such a diagram an expenditure is considered to be a negative cash flow and income is a positive cash flow. The project begins at time zero with a zero cash position. Cash flow is negative as the project begins and funds are expended on development and as design work progresses. The negative slope of the curve increases sharply at point *A* as large expenditures are made for equipment, buildings, and construction. As construction is completed, working capital is committed and start-up operations begin. After start-up is completed, production and sales begin at *B*. As income from sales exceeds production and operating costs, the curve begins to move upward, and at *C*, where the curve crosses the zero value for the cumulative cash flow, the income received has just balanced the previous expenditure on the project. This point can be termed the *breakeven point*. As the project continues, the cumulative positive-cash-flow position continues to increase. At the end of the project there is a final positive cash flow when the working capital is recovered and the salvage value of the plant, if any, is credited.

The maximum cumulative negative cash flow can be considered to represent the debt incurred by the plant. This debt is repaid out of the positive cash flow as the project continues. These cash flows take place over the life of the project, all of which is in the future in so far as the analysis and evaluation is concerned. It is therefore necessary to be able to bring these cash flows to some standard and common base in order to compare cash flows that take place at different times in the future.

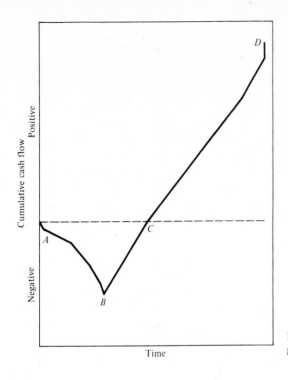

Figure 7-1 Cumulative-cash-flow diagram.

7-3 TIME VALUE OF MONEY

Interest is the compensation or rental paid for the use of borrowed capital. Capital borrowed for a long time would cost the borrower more than capital borrowed for a short time. The amount of capital on which interest is paid is termed the *principal*, and the *rate of interest* is the amount of interest earned per unit of capital per unit of time.

Interest rate i is usually based on 1 year as the unit of time. Thus, if the interest rate is 8 percent and is payable at the end of 1 year, we would expect to pay interest of $8 on a principal of $100. If the debt were retired at that time, the lender would receive $108. If the interest were calculated on a quarterly basis, the interest rate would be 2 percent per period, and the sum payable to the lender when the debt is retired can be calculated from

$$S = P(1 + i)^n \tag{7-1}$$

which is the equation for the future value S of a principal P after n interest periods at i fractional interest per period. This assumes that the interest rate is compounded; i.e., as the interest is earned, it is added to the capital. P is also termed the *present value* of a future payment S. The discount factor which discounts this future payment to its present value is $1/(1 + i)^n$.

This discount factor and Eq. (7-1) apply when money flows into or out of the company on a discrete time basis. In real life, however, money flows into and out of a company on an almost continuous basis. Funds are received as customers pay their accounts, funds are disbursed as bills come in, payrolls are met, and cash is invested. A more realistic concept is that of continuous compounding of interest and continuous cash flow. The assumption of continuous cash flow reflects the actual flow of cash in a company more nearly than does the assumption of cash flows on a discrete time basis. It will also be seen that discounting continuous cash flows is computationally simple. For continuous interest compounding, the number of interest periods per year can be considered to be infinite. The appropriate relationship between the present value and a future payment can easily be developed.

We shall call m the number of interest periods per year and i the nominal annual rate of interest. The future value of a sum of money P after t years will then be

$$S = P\left(1 + \frac{i}{m}\right)^{mt}$$

For continuous compounding m approaches infinity, hence

$$S = P\left[\lim_{m \to \infty} \left(1 + \frac{i}{m}\right)^{mt}\right]$$

The value for the limit term within the brackets is e^{it} so that

$$S = Pe^{it} \tag{7-2}$$

or

$$P = Se^{-it} \tag{7-3}$$

The discount factor which will give us the present value of a payment S received at an instant in time t years in the future is e^{-it}.

Present Value of a Uniform Cash Flow

Some cash flows will be at a uniform rate through at least part of the project life. The present value of such a uniform cash flow can be easily calculated. If the rate of cash flow is $\dot{R}$, the flow during an increment of time dt will be $\dot{R}\,dt$ and the present value of that flow will be $\dot{R}e^{-it}\,dt$. The present worth of the total uniform cash flow from the present, time zero, until time t can be obtained by integration, so that

$$P = \int_0^t \dot{R}e^{-it}\,dt = \frac{\dot{R}(1 - e^{-it})}{i}$$

The total uniform cash flow was $\dot{R}t$, so that the present value can be rewritten as

$$P = \frac{R(1 - e^{it})}{it} \tag{7-4}$$

where R is the total cash flow that occurred over t years at the uniform rate of $\dot{R}$ dollars per year. The term $(1 - e^{-it})/it$ is the discount factor for cash effects that occur uniformly over a period of years. The present value of a uniform cash flow during a period of 1 year beginning t years in the future will be

$$P = \frac{Re^{-it}(1 - e^{-i})}{i} \qquad (7\text{-}5)$$

Discount factors for continuous interest compounding are presented in Tables 7-1 to 7-3. Although both the interest rate i and the time period t appear in Tables 7-1 and 7-2, they appear as the product it. Each cash-flow discount function has the same value for each combination of interest rate and time period with the same product.

Table 7-1 Discount factors for cash flows which occur at an instant in time after the reference point

$100it$ = percentage rate of return × number of years after the reference point

$100it$	0	1	2	3	4	5	6	7	8	9
0	1.0000	.9901	.9802	.9704	.9608	.9512	.9418	.9324	.9231	.9139
10	.9048	.8958	.8869	.8781	.8694	.8607	.8521	.8437	.8353	.8270
20	.8187	.8106	.8025	.7945	.7866	.7788	.7711	.7634	.7558	.7483
30	.7408	.7334	.7261	.7189	.7118	.7047	.6977	.6907	.6839	.6771
40	.6703	.6637	.6570	.6505	.6440	.6376	.6313	.6250	.6188	.6126
50	.6065	.6005	.5945	.5886	.5827	.5770	.5712	.5655	.5599	.5543
60	.5488	.5434	.5379	.5326	.5273	.5220	.5169	.5117	.5066	.5016
70	.4966	.4916	.4868	.4819	.4771	.4724	.4677	.4630	.4584	.4538
80	.4493	.4449	.4404	.4360	.4317	.4274	.4232	.4190	.4148	.4107
90	.4066	.4025	.3985	.3946	.3906	.3867	.3829	.3791	.3753	.3716
100	.3679	.3642	.3606	.3570	.3535	.3499	.3465	.3430	.3396	.3362
110	.3329	.3296	.3263	.3230	.3198	.3166	.3135	.3104	.3073	.3042
120	.3012	.2982	.2952	.2923	.2894	.2865	.2837	.2808	.2780	.2753
130	.2725	.2698	.2671	.2645	.2618	.2592	.2567	.2541	.2516	.2491
140	.2466	.2441	.2417	.2393	.2369	.2346	.2322	.2299	.2276	.2254
150	.2231	.2209	.2187	.2165	.2144	.2122	.2101	.2080	.2060	.2039
160	.2019	.1999	.1979	.1959	.1940	.1921	.1901	.1882	.1864	.1845
170	.1827	.1809	.1791	.1773	.1755	.1738	.1720	.1703	.1686	.1670
180	.1653	.1637	.1620	.1604	.1588	.1572	.1557	.1541	.1526	.1511
190	.1496	.1481	.1466	.1451	.1437	.1423	.1409	.1395	.1381	.1367
200	.1353	.1340	.1327	.1313	.1300	.1287	.1275	.1262	.1249	.1237
210	.1225	.1212	.1200	.1188	.1177	.1165	.1153	.1142	.1130	.1119
220	.1108	.1097	.1086	.1075	.1065	.1054	.1044	.1033	.1023	.1013
230	.1003	.0993	.0983	.0973	.0963	.0954	.0944	.0935	.0926	.0916
240	.0907	.0898	.0889	.0880	.0872	.0863	.0854	.0846	.0837	.0829
250	.0821	.0813	.0805	.0797	.0789	.0781	.0773	.0765	.0758	.0750
260	.0743	.0735	.0728	.0721	.0714	.0707	.0699	.0693	.0686	.0679
270	.0672	.0665	.0659	.0652	.0646	.0639	.0633	.0627	.0620	.0614
280	.0608	.0602	.0596	.0590	.0584	.0578	.0573	.0567	.0561	.0556
290	.0550	.0545	.0539	.0534	.0529	.0523	.0518	.0513	.0508	.0503

(continued)

Table 7-1—*Continued*

100it	0	1	2	3	4	5	6	7	8	9
300	.0498	.0493	.0488	.0483	.0478	.0474	.0469	.0464	.0460	.0455
310	.0450	.0446	.0442	.0437	.0433	.0429	.0424	.0420	.0416	.0412
320	.0408	.0404	.0400	.0396	.0392	.0388	.0384	.0380	.0376	.0373
330	.0369	.0365	.0362	.0358	.0354	.0351	.0347	.0344	.0340	.0337
340	.0334	.0330	.0327	.0324	.0321	.0317	.0314	.0311	.0308	.0305
350	.0302	.0299	.0296	.0293	.0290	.0287	.0284	.0282	.0279	.0276
360	.0273	.0271	.0268	.0265	.0263	.0260	.0257	.0255	.0252	.0250
370	.0247	.0245	.0242	.0240	.0238	.0235	.0233	.0231	.0228	.0226
380	.0224	.0221	.0219	.0217	.0215	.0213	.0211	.0209	.0207	.0204
390	.0202	.0200	.0198	.0196	.0194	.0193	.0191	.0189	.0187	.0185
400	.0183	.0181	.0180	.0178	.0176	.0174	.0172	.0171	.0169	.0167
410	.0166	.0164	.0162	.0161	.0159	.0158	.0156	.0155	.0153	.0151
420	.0150	.0148	.0147	.0146	.0144	.0143	.0141	.0140	.0138	.0137
430	.0136	.0134	.0133	.0132	.0130	.0129	.0128	.0127	.0125	.0124
440	.0123	.0122	.0120	.0119	.0118	.0117	.0116	.0114	.0113	.0112
450	.0111	.0110	.0109	.0108	.0107	.0106	.0105	.0104	.0103	.0102
460	.0101	.0100	.0099	.0098	.0097	.0096	.0095	.0094	.0093	.0092
470	.0091	.0090	.0089	.0088	.0087	.0087	.0086	.0085	.0084	.0083
480	.0082	.0081	.0081	.0080	.0079	.0078	.0078	.0077	.0076	.0075
490	.0074	.0074	.0073	.0072	.0072	.0071	.0070	.0069	.0069	.0068

100it	0	10	20	30	40	50	60	70	80	90
500	.0067	.0061	.0055	.0050	.0045	.0041	.0037	.0033	.0030	.0027
600	.0025	.0022	.0020	.0018	.0017	.0015	.0014	.0012	.0011	.0010
700	.0009	.0008	.0007	.0007	.0006	.0006	.0005	.0005	.0004	.0004
800	.0003	.0003	.0003	.0002	.0002	.0002	.0002	.0002	.0002	.0001
900	.0001	.0001	.0001	.0001	.0001	.0001	.0001	.0001	.0001	.0001
1000	.0000									

Table 7-2 Discount factors for cash flows which occur uniformly over a period of years starting with the reference point
$100it$ = percentage rate of return × number of years in period starting with the reference point

100it	0	1	2	3	4	5	6	7	8	9
0	1.0000	.9950	.9901	.9851	.9803	.9754	.9706	.9658	.9610	.9563
10	.9516	.9470	.9423	.9377	.9332	.9286	.9241	.9196	.9152	.9107
20	.9063	.9020	.8976	.8933	.8891	.8848	.8806	.8764	.8722	.8681
30	.8639	.8598	.8558	.8517	.8477	.8438	.8398	.8359	.8319	.8281
40	.8242	.8204	.8166	.8128	.8090	.8053	.8016	.7979	.7942	.7906
50	.7869	.7833	.7798	.7762	.7727	.7692	.7657	.7622	.7588	.7554
60	.7520	.7486	.7452	.7419	.7386	.7353	.7320	.7288	.7256	.7224
70	.7192	.7160	.7128	.7097	.7066	.7035	.7004	.6974	.6944	.6913
80	.6883	.6854	.6824	.6795	.6765	.6736	.6707	.6679	.6650	.6622
90	.6594	.6566	.6537	.6510	.6483	.6455	.6428	.6401	.6374	.6348

Table 7-2—*Continued*

100*it*	0	1	2	3	4	5	6	7	8	9
100	.6321	.6295	.6269	.6243	.6217	.6191	.6166	.6140	.6115	.6090
110	.6065	.6040	.6016	.5991	.5967	.5942	.5918	.5894	.5871	.5847
120	.5823	.5800	.5777	.5754	.5731	.5708	.5685	.5663	.5641	.5618
130	.5596	.5574	.5552	.5530	.5509	.5487	.5466	.5444	.5424	.5402
140	.5381	.5361	.5340	.5320	.5299	.5279	.5259	.5239	.5219	.5199
150	.5179	.5160	.5140	.5121	.5102	.5082	.5064	.5044	.5026	.5007
160	.4988	.4970	.4952	.4933	.4915	.4897	.4879	.4861	.4843	.4825
170	.4808	.4790	.4773	.4756	.4739	.4721	.4704	.4687	.4671	.4654
180	.4637	.4621	.4605	.4588	.4571	.4555	.4540	.4523	.4508	.4491
190	.4476	.4460	.4445	.4429	.4414	.4399	.4383	.4368	.4354	.4338
200	.4323	.4308	.4294	.4279	.4265	.4250	.4236	.4221	.4207	.4193
210	.4179	.4165	.4151	.4137	.4123	.4109	.4096	.4082	.4069	.4055
220	.4042	.4029	.4015	.4002	.3989	.3976	.3963	.3950	.3937	.3925
230	.3912	.3899	.3887	.3874	.3862	.3849	.3837	.3825	.3813	.3801
240	.3789	.3777	.3765	.3753	.3741	.3729	.3718	.3706	.3695	.3683
250	.3672	.3660	.3649	.3638	.3627	.3615	.3604	.3593	.3582	.3571
260	.3560	.3550	.3539	.3528	.3517	.3507	.3496	.3486	.3476	.3465
270	.3455	.3445	.3434	.3424	.3414	.3404	.3393	.3384	.3374	.3364
280	.3354	.3344	.3335	.3325	.3315	.3306	.3296	.3287	.3277	.3268
290	.3259	.3249	.3240	.3231	.3221	.3212	.3203	.3194	.3185	.3176
300	.3167	.3158	.3150	.3141	.3132	.3123	.3115	.3106	.3098	.3089
310	.3080	.3072	.3064	.3055	.3047	.3039	.3030	.3022	.3014	.3006
320	.2998	.2990	.2982	.2974	.2966	.2958	.2950	.2942	.2934	.2926
330	.2919	.2911	.2903	.2896	.2888	.2880	.2873	.2865	.2858	.2850
340	.2843	.2836	.2828	.2821	.2814	.2807	.2799	.2792	.2785	.2778
350	.2771	.2764	.2757	.2750	.2743	.2736	.2729	.2722	.2715	.2709
360	.2702	.2695	.2688	.2682	.2675	.2669	.2662	.2655	.2649	.2642
370	.2636	.2629	.2623	.2617	.2610	.2604	.2598	.2591	.2585	.2579
380	.2573	.2567	.2560	.2554	.2548	.2542	.2536	.2530	.2524	.2518
390	.2512	.2506	.2500	.2495	.2489	.2483	.2477	.2471	.2466	.2460
400	.2454	.2449	.2443	.2437	.2432	.2426	.2421	.2415	.2410	.2404
410	.2399	.2393	.2388	.2382	.2377	.2372	.2366	.2361	.2356	.2350
420	.2345	.2340	.2335	.2330	.2325	.2319	.2314	.2309	.2304	.2299
430	.2294	.2289	.2284	.2279	.2274	.2269	.2264	.2259	.2255	.2250
440	.2245	.2240	.2235	.2230	.2226	.2221	.2216	.2212	.2207	.2202
450	.2198	.2193	.2188	.2184	.2179	.2175	.2170	.2166	.2161	.2157
460	.2152	.2148	.2143	.2139	.2134	.2130	.2126	.2121	.2117	.2113
470	.2108	.2104	.2100	.2096	.2091	.2087	.2083	.2079	.2074	.2070
480	.2066	.2062	.2058	.2054	.2050	.2046	.2042	.2038	.2034	.2030
490	.2026	.2022	.2018	.2014	.2010	.2006	.2002	.1998	.1994	.1990

100*it*	0	10	20	30	40	50	60	70	80	90
500	.1987	.1949	.1912	.1877	1843	.1811	.1779	.1749	.1719	.1690
600	.1663	.1636	.1610	.1584	.1560	.1536	.1513	.1491	.1469	.1448
700	.1427	.1407	.1388	.1369	.1351	.1333	.1315	.1298	.1282	.1265
800	.1250	.1234	.1219	.1206	.1190	.1176	.1163	.1149	.1136	.1123
900	.1111	.1099	.1087	.1075	.1064	.1053	.1043	.1031	.1020	.1010

(*continued*)

Table 7-2—*Continued*

100*it*	0	10	20	30	40	50	60	70	80	90
1000	.1000	.0990	.0980	.0971	.0962	.0952	.0943	.0935	.0926	.0917
1100	.0909	.0901	.0893	.0885	.0877	.0869	.0862	.0855	.0847	.0840
1200	.0833	.0826	.0820	.0813	.0806	.0800	.0794	.0787	.0781	.0775
1300	.0769	.0763	.0758	.0752	.0746	.0741	.0735	.0730	.0725	.0719
1400	.0714	.0709	.0704	.0699	.0694	.0690	.0685	.0680	.0676	.0671
1500	.0667	.0662	.0658	.0654	.0649	.0645	.0641	.0637	.0633	.0629
1600	.0625	.0621	.0617	.0613	.0610	.0606	.0602	.0599	.0595	.0592
1700	.0588	.0585	.0581	.0578	.0575	.0571	.0568	.0565	.0562	.0559
1800	.0556	.0552	.0549	.0546	.0543	.0541	.0538	.0535	.0532	.0529
1900	.0526	.0524	.0521	.0518	.0515	.0513	.0510	.0508	.0505	.0502
2000	.0500									

Example 7-1 (*a*) You are offered a payment of $100 to be given to you after 1 year. If you are willing to accept a nominal 12 percent interest rate continuously compounded, how much is this offer worth now? (*b*) How much should you be willing to pay now to receive $100 per year for 20 years? (*c*) How much should you be willing to pay now for you and your heirs to receive $100 per year in perpetuity?

SOLUTION (*a*) The question being asked is: What is the present worth of an instantaneous payment of $100 in 1 year's time? The appropriate discount factor can be read from Table 7-1. For $100it = 12$, the discount factor is 0.8869 and

$$P_1 = (100)(0.8869) = 88.69$$

You should therefore be willing to pay $88.69 now to receive $100 after 1 year.

(*b*) If we assume that the $100 is to be received at the end of each year, we need to sum up the present worths of $100 after 1, 2, ... years until the final payment of $100 after 20 years. The calculation can be made by summing the appropriate discount factors from Table 7-1

$$P_{20} = \$100(0.8869 + 0.7866 + \cdots + 0.1003) = \$713.18$$

To obtain this result we summed $e^{-i} + e^{-ei} + \cdots + e^{-20i}$. We can reduce the computational effort if we recall that

$$\sum (1 + x + x^2 + \cdots + x^m) = \frac{1 - x^m}{1 - x}$$

Thus, we can obtain the appropriate discount factor from

$$\frac{1 - e^{-im}}{1 - e^{-i}} - 1$$

Table 7-3 Discount factors for cash flows which occur uniformly over 1-year periods after the reference point

Year	1%	2%	3%	4%	5%	6%	7%	8%	9%	10%	11%	12%	13%	14%	15%	16%	17%	18%	19%	20%
0-1	.9950	.9901	.9851	.9803	.9754	.9706	.9658	.9610	.9563	.9516	.9470	.9423	.9377	.9332	.9286	.9241	.9196	.9152	.9107	.9063
1-2	.9851	.9705	.9560	.9418	.9278	.9141	.9005	.8872	.8740	.8611	.8483	.8358	.8234	.8112	.7993	.7875	.7759	.7644	.7531	.7421
2-3	.9753	.9512	.9278	.9049	.8826	.8608	.8396	.8189	.7988	.7791	.7600	.7413	.7230	.7053	.6879	.6710	.6546	.6385	.6228	.6075
3-4	.9656	.9324	.9004	.8694	.8395	.8107	.7829	.7560	.7300	.7050	.6808	.6574	.6349	.6131	.5921	.5718	.5522	.5333	.5150	.4974
4-5	.9560	.9140	.8737	.8353	.7986	.7635	.7299	.6979	.6672	.6379	.6099	.5831	.5575	.5330	.5096	.4873	.4659	.4455	.4259	.4072
5-6	.9465	.8959	.8479	.8026	.7596	.7190	.6806	.6442	.6098	.5772	.5463	.5172	.4895	.4634	.4386	.4152	.3931	.3721	.3522	.3334
6-7	.9371	.8781	.8229	.7711	.7226	.6772	.6346	.5947	.5573	.5223	.4894	.4588	.4299	.4029	.3775	.3538	.3316	.3108	.2913	.2730
7-8	.9278	.8607	.7985	.7409	.6874	.6377	.5917	.5490	.5093	.4726	.4385	.4069	.3775	.3502	.3250	.3015	.2798	.2596	.2409	.2235
8-9	.9185	.8437	.7749	.7118	.6538	6006	.5517	.5068	.4655	.4276	.3928	.3609	.3314	.3045	.2797	.2569	.2360	.2168	.1992	.1830
9-10	.9094	.8270	.7520	.6839	.6219	.5656	.5144	.4678	.4254	.3869	.3519	.3201	.2910	.2647	.2407	.2189	.1991	.1811	.1647	.1498
10-11	.9003	.8106	.7298	.6571	.5916	.5327	.4796	.4318	.3888	.3501	.3152	.2839	.2556	.2301	.2072	.1866	.1680	.1513	.1362	.1227
11-12	.8914	.7946	.7082	.6312	.5628	.5016	.4472	.3986	.3553	.3168	.2824	.2518	.2244	.2000	.1783	.1590	.1417	.1264	.1126	.1004
12-13	.8825	.7788	.6873	.6065	.5353	.4724	.4169	.3680	.3248	.2866	.2530	.2233	.1970	.1739	.1535	.1355	.1196	.1055	.0932	.0822
13-14	.8737	.7634	.6670	.5827	.5092	.4449	.3888	.3397	.2968	.2593	.2266	.1981	.1730	.1512	.1321	.1154	.1009	.0882	.0770	.0673
14-15	.8650	.7483	.6473	.5599	.4844	.4190	.3625	.3136	.2713	.2347	.2030	.1757	.1519	.1314	.1137	.0984	.0851	.0736	.0637	.0551
15-16	.8564	.7335	.6282	.5380	.4608	.3946	.3380	.2895	.2479	.2123	.1819	.1558	.1334	.1143	.0979	.0838	.0718	.0615	.0527	.0451
16-17	.8479	.7189	.6096	.5169	.4383	.3716	.3151	.2672	.2266	.1921	.1629	.1382	.1172	.0993	.0842	.0714	.0606	.0514	.0436	.0369
17-18	.8395	.7047	.5916	.4966	.4169	.3500	.2938	.2467	.2071	.1739	.1460	.1225	.1029	.0864	.0725	.0609	.0511	.0429	.0360	.0303
18-19	.8311	.6908	.5741	.4772	.3966	.3296	.2740	.2277	.1893	.1573	.1308	.1087	.0903	.0751	.0624	.0519	.0431	.0358	.0298	.0248
19-20	.8228	.6771	.5571	.4584	.3772	.3104	.2554	.2102	.1730	.1423	.1171	.0964	.0793	.0653	.0537	.0442	.0364	.0299	.0246	.0203
20-21	.8147	.6637	.5407	.4405	.3588	.2923	.2382	.1940	.1581	.1288	.1049	.0855	.0697	.0568	.0462	.0377	.0307	.0250	.0204	.0166
21-22	.8065	.6505	.5247	.4232	.3413	.2753	.2221	.1791	.1445	.1165	.0940	.0758	.0612	.0493	.0398	.0321	.0259	.0209	.0169	.0136
22-23	.7985	.6376	.5092	.4066	.3247	.2593	.2071	.1653	.1320	.1054	.0842	.0673	.0537	.0429	.0343	.0274	.0218	.0175	.0139	.0111
23-24	.7906	.6250	.4941	.3907	.3089	.2442	.1931	.1526	.1207	.0954	.0754	.0596	.0472	.0373	.0295	.0233	.0184	.0146	.0115	.0091
24-25	.7827	.6126	.4795	.3753	.2938	.2300	.1800	.1409	.1103	.0863	.0676	.0529	.0414	.0324	.0254	.0199	.0156	.0122	.0095	.0075
25-30	.7596	.5772	.4386	.3334	.2535	.1928	.1466	.1115	.0849	.0646	.0492	.0374	.0285	.0217	.0165	.0126	.0096	.0073	.0056	.0043
30-35	.7226	.5223	.3775	.2730	.1974	.1428	.1033	.0748	.0541	.0392	.0284	.0205	.0149	.0108	.0078	.0057	.0041	.0030	.0022	.0016
35-40	.6874	.4726	.3250	.2235	.1538	.1058	.0728	.0501	.0345	.0238	.0164	.0113	.0078	.0054	.0037	.0025	.0018	.0012	.0008	.0006
40-45	.6538	.4276	.2797	.1830	.1197	.0784	.0513	.0336	.0220	.0144	.0094	.0062	.0041	.0027	.0017	.0011	.0008	.0005	.0003	.0002
45-50	.6219	.3869	.2407	.1498	.0933	.0581	.0362	.0225	.0140	.0087	.0054	.0034	.0021	.0013	.0008	.0005	.0003	.0002	.0001	.0001

(continued)

163

Table 7-3—*Continued*

Year	21%	22%	23%	24%	25%	26%	27%	28%	29%	30%	31%	32%	33%	34%	35%	36%	37%	38%	39%	40%
0-1	.9020	.8976	.8933	.8890	.8848	.8806	.8764	.8722	.8681	.8640	.8598	.8558	.8517	.8477	.8438	.8398	.8359	.8319	.8281	.8242
1-2	.7311	.7204	.7098	.6993	.6891	.6790	.6690	.6592	.6495	.6400	.6307	.6214	.6123	.6034	.5946	.5859	.5774	.5689	.5606	.5525
2-3	.5926	.5781	.5639	.5501	.5367	.5235	.5107	.4982	.4860	.4741	.4626	.4512	.4402	.4295	.4190	.4088	.3988	.3891	.3796	.3703
3-4	.4804	.4639	.4481	.4327	.4179	.4037	.3899	.3765	.3637	.3513	.3393	.3277	.3165	.3057	.2953	.2852	.2755	.2661	.2570	.2482
4-5	.3894	.3723	.3560	.3404	.3255	.3112	.2976	.2846	.2721	.2602	.2488	.2379	.2275	.2176	.2081	.1990	.1903	.1820	.1740	.1664
5-6	.3156	.2988	.2829	.2678	.2535	.2400	.2272	.2151	.2036	.1928	.1825	.1728	.1636	.1549	.1466	.1388	.1314	.1244	.1178	.1115
6-7	.2558	.2398	.2247	.2106	.1974	.1850	.1734	.1626	.1524	.1428	.1339	.1255	.1176	.1102	.1033	.0968	.0908	.0851	.0798	.0748
7-8	.2074	.1924	.1786	.1657	.1538	.1427	.1324	.1229	.1140	.1058	.0982	.0911	.0845	.0785	.0728	.0676	.0627	.0582	.0540	.0501
8-9	.1681	.1544	.1419	.1303	.1197	.1100	.1011	.0929	.0853	.0784	.0720	.0662	.0608	.0558	.0513	.0471	.0433	.0398	.0366	.0336
9-10	.1363	.1239	.1127	.1025	.0933	.0848	.0772	.0702	.0638	.0581	.0528	.0480	.0437	.0397	.0362	.0329	.0299	.0272	.0248	.0225
10-11	.1105	.0995	.0896	.0807	.0726	.0654	.0589	.0530	.0478	.0430	.0387	.0349	.0314	.0283	.0255	.0229	.0207	.0186	.0168	.0151
11-12	.0895	.0798	.0711	.0634	.0566	.0504	.0450	.0401	.0357	.0319	.0284	.0253	.0226	.0201	.0180	.0160	.0143	.0127	.0113	.0101
12-13	.0726	.0641	.0565	.0499	.0441	.0389	.0343	.0303	.0267	.0236	.0208	.0184	.0162	.0143	.0127	.0112	.0099	.0087	.0077	.0068
13-14	.0588	.0514	.0449	.0393	.0343	.0300	.0262	.0229	.0200	.0175	.0153	.0134	.0117	.0102	.0089	.0078	.0068	.0060	.0052	.0045
14-15	.0477	.0413	.0357	.0309	.0267	.0231	.0200	.0173	.0150	.0130	.0112	.0097	.0084	.0073	.0063	.0054	.0047	.0041	.0035	.0030
15-16	.0387	.0331	.0284	.0243	.0208	.0178	.0153	.0131	.0112	.0096	.0082	.0071	.0060	.0052	.0044	.0038	.0033	.0028	.0024	.0020
16-17	.0313	.0266	.0225	.0191	.0162	.0137	.0117	.0099	.0084	.0071	.0060	.0051	.0043	.0037	.0031	.0027	.0022	.0019	.0016	.0014
17-18	.0254	.0213	.0179	.0150	.0126	.0106	.0089	.0075	.0063	.0053	.0044	.0037	.0031	.0026	.0022	.0018	.0015	.0013	.0011	.0009
18-19	.0206	.0171	.0142	.0118	.0098	.0082	.0068	.0057	.0047	.0039	.0032	.0027	.0023	.0019	.0016	.0013	.0011	.0009	.0007	.0006
19-20	.0167	.0137	.0113	.0093	.0077	.0063	.0052	.0043	.0035	.0029	.0024	.0020	.0016	.0013	.0011	.0009	.0007	.0006	.0005	.0004
20-21	.0135	.0110	.0090	.0073	.0060	.0049	.0040	.0032	.0026	.0021	.0017	.0014	.0012	.0009	.0008	.0006	.0005	.0004	.0003	.0003
21-22	.0110	.0088	.0071	.0058	.0046	.0038	.0030	.0024	.0020	.0016	.0013	.0010	.0008	.0007	.0005	.0004	.0004	.0003	.0002	.0002
22-23	.0089	.0071	.0057	.0045	.0036	.0029	.0023	.0018	.0015	.0012	.0009	.0007	.0006	.0005	.0004	.0003	.0002	.0002	.0002	.0001
23-24	.0072	.0057	.0045	.0036	.0028	.0022	.0018	.0014	.0011	.0009	.0007	.0005	.0004	.0003	.0003	.0002	.0002	.0001	.0001	.0001
24-25	.0058	.0046	.0036	.0028	.0022	.0017	.0013	.0011	.0008	.0007	.0005	.0004	.0003	.0002	.0002	.0002	.0001	.0001	.0001	.0001
25-30	.0032	.0025	.0019	.0014	.0011	.0008	.0006	.0005	.0004	.0003	.0002	.0002	.0001	.0001	.0001	.0001				
30-35	.0011	.0008	.0006	.0004	.0003	.0002	.0002	.0001	.0001	.0001										
35-40	.0004	.0003	.0002	.0001	.0001	.0001														
40-45	.0001	.0001	.0001																	

where $m = n + 1$. By minor algebraic manipulation this expression becomes

$$\frac{1 - e^{-in}}{e^{i} - 1}$$

Therefore, for our example the discount factor becomes

$$\frac{1 - e^{-(0.12)(20)}}{e^{0.12} - 1} = 7.1318$$

and
$$P_{20} = (\$100)(7.1318) = \$713.18$$

Note: If the $100 per year were to be received as a uniform cash flow, we would use the appropriate discount factor from Table 7-2. For $100it = (100)(0.12)(20) = 240$, the discount factor is 0.3789, and

$$P_{20} = (100)(20)(0.3789) = \$757.80$$

(c) We shall assume that the money is received at the end of each year. Our problem is to calculate what sum of money now will provide $100 at the end of 1 year plus sufficient seed money to continue the cycle

$$S - 100 = P_{\infty}$$

But for 1 year

$$S = P_{\infty} e^{i}$$

therefore

$$P_{\infty}(e^{i} - 1) = 100 \quad \text{and} \quad P_{\infty} = \frac{\$100}{e^{0.12} - 1} = \$784.33$$

7-4 EVALUATION CRITERIA

The purpose of an economic evaluation is to assess the desirability of a proposed investment. The goal is to combine all the factors that contribute to the cost and to the return on the project into a single index of profitability or financial attractiveness. Unfortunately, no single quantitative measure or index has been universally accepted as the "best" index. Each of the indexes in common use has advantages, and it is necessary to understand them all.

The commonly used indexes can be divided into three general categories depending on whether the index has units of time, money, or rate of return. Within each category there are indexes which are relatively crude and also quick and easy to calculate as well as more sophisticated and realistic indexes which require more time and effort to obtain.

Time Indexes

The *payback time* (also referred to as payout time, cash recovery period, payoff time, payout period) is a crude, rough-and-ready but commonly used index. It is the time necessary for all the depreciable investment to be recovered by the income received from the project and can be read directly from the cumulative-cash-flow diagram.

An idealized cumulative-cash-flow diagram is shown in Fig. 7-2. The negative cash flow at time zero from A to B would represent investment in land I_l, the flow from B to C represents investment during plant construction I_D, and that from C to D is the provision of working capital I_W. Profitable operation begins at D. The payback time is shown on this diagram as P, which represents the time required after the beginning of profitable operation for the depreciable capital I_D to be recovered.

An alternative definition for the payback time is the time needed for the cumulative expenditure to be exactly balanced by the net income. This would be represented by point E in Fig. 7-2, the point where the cumulative-cash-flow line crosses the zero cash position.

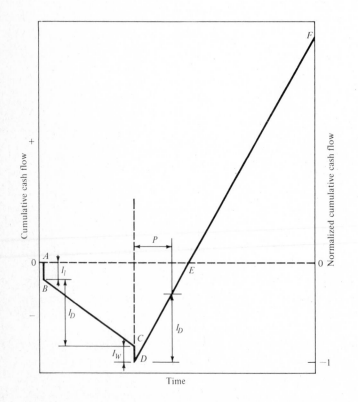

Figure 7-2 Idealized cumulative-cash-flow diagram.

The *equivalent maximum investment period* (EMIP) is a more sophisticated version of the payback time. Proposed by Allen (1967), it is defined as the negative area enclosed by the cumulative-cash-flow curve up to the breakeven point E divided by the maximum cumulative expenditure, represented by D in Fig. 7-2

$$\text{EMIP} = \frac{\text{area } (ABCDEA)}{\overline{0D}}$$

The units of EMIP are (dollars × years)/dollars = years. The EMIP is equivalent to the period that the maximum cumulative project debt would be outstanding if all of the debt were incurred simultaneously and then all repaid instantaneously at a later time.

The cumulative-cash-flow curve can be normalized by dividing the cumulative cash position by the maximum cumulative expenditure. The normalized cumulative cash flow is shown as the right-hand ordinate of Fig. 7-2. Under these normalized conditions the maximum cumulative debt becomes -1, and the EMIP is merely the area between the zero cash position at zero time and the zero cash position at the breakeven point E.

For these time indexes the desirable project is that with the shortest time. The evaluation and selection of a project based on a time index ignores the effects of cash flow that occurs after the payback time or the EMIP. EMIP is somewhat more realistic than the payback time because it takes into account the pattern of expenditure and of income, and it can be expected to distinguish between projects with the same payback time. A project with a small value for EMIP usually means that the cash-flow rate at the breakeven point is high, so that the project usually will also be attractive in terms of the other indexes that will be considered.

Values for the payout time that would be considered attractive are in the range of 2 years or less for a high-risk project and less than 5 years for a medium-risk project. A value of 3 years for EMIP would be considered attractive for a medium-risk project.

Money or Cash Indexes

The value for the *cumulative cash position* at the end of the project's estimated life or at the end of a specified number of years represents a crude evaluation procedure. The net cash position can be read directly off the cumulative-cash-flow diagram. This method can be used to compare alternative projects but takes no account of the size or the pattern of the investment necessary to generate the net cash position.

The *present-value* or *present-worth method* computes the present values of each component of the cash flow, taking into account their magnitude and timing, and sums them up to give the present value for the entire project from beginning to end. The method requires that an acceptable interest rate or rate of return on invested capital be established, and this rate is then used in the calculation of the present value of the various cash flows.

For the cash indexes the project with the higher value for the cash index will be the more attractive income-producing project. If an investment is being considered, e.g., the choice between competing equipment items, the alternative with the lowest value for the cash index is the more attractive one. Only the present-worth method, however, takes proper account not only of the time value of money but also of the magnitude of the various cash flows and their timings. Although the present-value method requires more computational effort than the evaluation of cumulative-cash-flow position does, the greatest effort involved in an economic evaluation is that necessary to obtain a reasonably reliable cumulative-cash-flow diagram. The additional effort necessary to convert the information presented in such a diagram into a present value is marginal and yields an index that takes proper account of the cash-flow timing and magnitude and of the interest rate.

Rate-of-Return Indexes

The *rate of return on investment* is a simple procedure that measures the ratio of average yearly net profit to the total cumulative investment and expresses it as a percentage. The more attractive alternative would be that which gives the higher rate of return. This simple index ignores the pattern of cash flow and is crude. It is useful, nevertheless, in a number of applications, especially where a choice has to be made between competing alternative investments that satisfy the same objective.

The *discounted-cash-flow* (DCF) *rate of return* is also known by a number of other names, such as internal rate of return, investor's method, profitability index, and interest rate of return. It is similar to the present-value method in that it deals properly with all cash flows, their magnitude, and timing. It differs, however, in its treatment of the interest rate. In the present-value method the interest rate was specified in advance, and this rate was then used in the evaluation of the present worth of the entire project. In the DCF rate of return the interest rate is calculated as that interest rate which makes the present value of the project equal to zero. The DCF rate of return therefore represents the maximum rate of interest at which money could be borrowed to finance the project, the net cash return being just adequate to pay all the principal and interest due over the life of the loan, i.e., the life of the project. When using this criterion, the alternative yielding the largest DCF rate of return is the desirable alternative provided that the DCF of return is equal to or greater than the cost of capital.

7-5 COMPARISON BETWEEN PRESENT WORTH AND DCF RETURN

Both the present-value and the DCF-return methods take proper account of all cash flows, their magnitude, and timing and the time value of money. The DCF rate of return can be obtained only by a trial-and-error calculation, but computer programs are available or can easily be generated for both DCF-return and for present-value methods. There has been much discussion of the relative merits of these two methods, and situations have been found where they lead to different conclusions.

The present-value method yields a direct cash measure of the attractiveness of a project, and the values are conveniently additive. The same present value can be obtained, however, for projects with widely different initial investments. Present value at the cost of capital is the economic profit because if the net present worth or value is zero when it is calculated at an interest or discount rate equal to the cost of capital, the positive cash flow has been just sufficient to pay the cost of capital used in the project. Therefore present value measures profit. If the goal of the company is to maximize profit, it should use the present value at the cost of capital as its decision criterion for investment.

The DCF rate of a return does not give any indication of the cash value of the project, and the returns are not additive. It measures how efficiently the capital is being used but gives no indication of how large the profits will be. Therefore it is an appropriate criterion when the supply of capital is restricted and the capital must be rationed to those projects which can use it most efficiently.

Any method of economic evaluation should be used with discretion and with due regard for its merits and demerits. Each index provides information that is helpful in making project decisions. No major investment decision should be on the basis of a single criterion. If different criteria lead to different conclusions, this is a signal that a closer study is required to identify the reasons for the discrepancy.

The several methods described above for economic evaluation will now be illustrated and amplified with the help of examples.

Example 7-2 Our company has been attempting to reduce energy losses by recovering heat wherever possible. Heat presently being lost can be recovered from a hot stream by addition of a heat exchanger and pump. Three alternatives have survived from those proposed by the design group, and because the alternatives are mutually exclusive, only one can be selected. Company policy for investments of this type, which are low-risk, based on well-known and tested technology, and good for the company image, is to require a minimum rate of return on investment of 10 percent with tax effects neglected. The three alternatives are summarized in Table 7-4. Which

Table 7-4 Heat-saving alternatives

Alternative	Total installed cost	Annual value of heat saved	Annual operating costs
1	$25,500	$7200	$150
2	31,000	8600	250
3	33,500	9500	300

alternative would you recommend? Maintenance plus other fixed charges are 15 percent of the initial investment per year.

SOLUTION The first step is to calculate the money saved per year and the rate of return for each alternative:

Alternative 1:

$$\text{Annual savings} = 7200 - 150 - (0.15)(25,500) = \$3225$$

$$\text{Rate of return} = \frac{3225}{25,500} \, 100 = 12.6\%$$

Alternative 2:

$$\text{Annual savings} = 8600 - 250 - (0.15)(31,000) = \$3700$$

$$\text{Rate of return} = \frac{3700}{31,000} \, 100 = 11.9\%$$

Alternative 3:

$$\text{Annual savings} = 9500 - 300 - (0.15)(33,500) = \$4175$$

$$\text{Rate of return} = \frac{4175}{33,500} \, 100 = 12.5\%$$

Since the rate of return for each of the alternatives is above the 10 percent minimum, all the alternatives remain as candidates for selection. Because the alternatives are mutually exclusive, the selection can be based on the rate of return of incremental savings for incremental investment. The selection is based on a comparison between pairs. The survivor of the pair is then compared with the next remaining alternative. The final survivor of this step-by-step examination will be the recommended alternative.

Compare alternatives 1 and 2:

$$\Delta(\text{saving}) = 3700 - 3225 = \$475$$

$$\Delta(\text{investment}) = 31,000 - 25,500 = \$5500$$

$$\text{Rate of return} = \frac{475}{5500} \, 100 = 8.6\%$$

The rate of return for the incremental investment is less than 10 percent, so alternative 2 is rejected compared with alternative 1.

Compare alternatives 1 and 3:

$$\Delta(\text{saving}) = 4175 - 3225 = \$950$$

$$\Delta(\text{investment}) = 33,500 - 25,500 = \$8000$$

$$\text{Rate of return} = \frac{950}{8000} \, 100 = 11.9\%$$

This rate of return is greater than the minimum required, so alternative 1 is eliminated by alternative 3, which would be the investment to recommend.

Example 7-3 Our company has developed a process to produce isopentaloyl and is now considering construction of a plant to produce 5×10^6 lb/yr. Plant construction at a total depreciable investment of $2 million will require 1 year. At the beginning of construction $300,000 will be disbursed, an additional $1 million paid after 6 months, and the remaining $1 million will be paid after 12 months. Working capital of $400,000 is to be provided at the beginning of production, and an additional $100,000 will be provided at the beginning of the fourth year of production. The working capital will be recovered after 10 years of production when the project terminates. Sales are expected to be 2×10^6 lb during the first year of production, 3.5×10^6 lb during the second year, 4.5×10^6 lb during the third year, and 5×10^6 lb/yr thereafter until the project terminates. The product will sell for 50 cents per pound, variable costs are 15 cents per pound, and fixed costs (not including depreciation) are $200,000 per year. Profits are taxed at 45 percent, and there is no tax credit for investment. Depreciation will be over 10 years on a straight-line basis.

Calculate investment criteria as follows: (*a*) payback time, (*b*) EMIP, (*c*) net present worth at a rate of return of 15 percent, and (*d*) DCF rate of return. For parts (*c*) and (*d*) use continuous-interest compounding.

SOLUTION All the investment criteria will require knowledge of the cash flows, which will be calculated first for each year of sales. For the first year of production:

No.	Item	Calculation	Value
(1)	Sales income	$(2 \times 10^6)(0.5)$	$1,000,000
(2)	Expenses	$200,000 + (2 \times 10^6)(0.15)$	500,000
(3)	Operating Income	(2) − (1)	500,000
(4)	Depreciation	$\dfrac{2,300,000}{10}$	230,000
(5)	Gross profit	(3) − (4)	270,000
(6)	Taxes	(5) × 0.45	121,500
(7)	Net profit	(5) − (6)	148,500
(8)	Cash flow	(4) + (7)	378,500

The cash flows due to sales are calculated in a similar manner for the remaining years. The cash flows over the entire life of the project are summarized in Table 7-5.

The cumulative cash flows are plotted in Fig. 7-3.

(*a*) Payback time can be read from Figure 7-3 as 3.4 years. At that time the cumulative cash flow is − $500,000, and all the depreciable capital would have been recovered.

Table 7-5 Annual and cumulative cash flows

Year	Cash flow	Cumulative cash flow
-1	$ - 300,000	$ - 300,000
-0.5	- 1,000,000	- 1,300,000
0	- 1,000,000	- 2,300,000
0	- 400,000	- 2,700,000
0-1	378,500	- 2,321,500
1-2	667,250	- 1,654,250
2-3	859,750	- 794,500
3	- 100,000	- 894,500
3-4	956,000	61,500
4-5	956,000	1,017,500
5-6	956,000	1,973,500
6-7	956,000	2,929,500
7-8	956,000	3,885,500
8-9	956,000	4,841,500
9-10	956,000	5,797,500
10	500,000	6,297,500

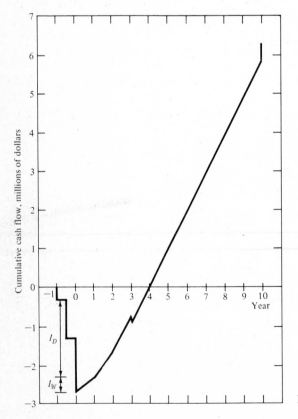

Figure 7-3 Cumulative cash flows.

(b) To determine the EMIP the total negative area of the cumulative-cash-flow diagram is measured, and this is divided by the maximum project debt. The area calculated from Fig. 7-3 is 6,968,000 (dollars times years), and the maximum cumulative project debt is $2,700,000.

$$\text{EMIP} = \frac{6,968,000}{2,700,000} = 2.58 \text{ years}$$

Note: Let us assume that the depreciable investment had been made under a different schedule, for example $1 million at time -1, $1 million at time -0.5, and the remaining $300,000 at time 0. This would have had no effect on the payback time, which would have remained at 3.4 years. The EMIP would have increased, however, from the 2.58 years at the original schedule of expenditures to 2.84 years for this revised schedule. This demonstrates that the EMIP is, indeed, somewhat sensitive to the pattern of expenditure.

(c) The calculation of the net present worth (NPW) is summarized in Table 7-6, columns 3 and 4. An entry like 0–1 in the time column means that the cash flow occurs uniformly during that year. The summation of the elements of column 4 gives the net present worth of the venture as $1,338,800.

Note: The present worth of the uniform cash flow from beginning of year 3 to the end of the tenth year could have been calculated using Eq. (7-4)

Table 7-6 Net-present-worth calculations

Time, yr (1)	Cash flow, thousands (2)	Discount factor $i = 0.15$ (3)	Present worth, thousands (4)	Discount factor $i = 0.25$ (5)	Present worth, thousands (6)	Discount factor $i = 0.23$ (7)	Present worth, thousands (8)
-1	$-300	1.1618	$-349	1.2840	$-385.2	1.2587	$-377.6
-0.5	$-1000	1.0778	$-1078	1.1331	$-1133.1	1.1219	$-1121.9
0	$-1400	1.000	$-1400	1.0000	$-1400	1.0000	$-1400
0–1	378.5	0.9286	351.5	0.8848	334.9	0.8933	338.1
1–2	667.3	0.7993	533.4	0.6891	459.8	0.7098	473.6
2–3	859.8	0.6879	591.5	0.5367	461.5	0.5639	484.8
3	$-100	0.6376	$-63.8	0.4724	$-47.2	0.5016	$-50.2
3–4	956	0.5921	566.0	0.4179	399.5	0.4481	428.4
4–5	956	0.5096	487.2	0.3255	311.2	0.3560	340.3
5–6	956	0.4386	419.3	0.2535	242.3	0.2829	270.5
6–7	956	0.3775	360.9	0.1974	188.7	0.2247	214.8
7–8	956	0.3250	310.7	0.1538	147.0	0.1786	170.7
8–9	956	0.2797	267.4	0.1197	114.4	0.1419	135.7
9–10	956	0.2047	230.1	0.0933	89.2	0.1127	107.7
10	500	0.2231	111.6	0.0821	41.1	0.1003	50.1
			$\sum = 1338.8$		$\sum = -175.9$		$\sum = 65$

for the uniform cash flow during 7 years and discounting from year 3 with Eq. (7-3). The calculation would have been

$$\frac{[(956)(7)](1 - e^{(-0.15)(7)})}{(0.15)(7)} e^{(-0.15)(3)} = 2641.7$$

(*d*) The calculation of the DCF rate of return requires a trial-and-error procedure. A value for the rate of return is selected, and the corresponding NPW is calculated. If the NPW is not zero, a new value for *i* is selected and the cycle repeated. Graphical interpolation can speed up the operation.

We already have one pair of values for *i* and NPW from part (*c*). A higher value of $i = 0.25$ will be chosen for the next trial. The calculations are summarized in columns 5 and 6 of Table 7-6, and the NPW at this rate of return is $-\$175,900$.

The next trial will be with a smaller value for *i*, $i = 0.23$. The calculations are summarized in columns 7 and 8, and the net present worth is now positive, \$65,000. A further trial could be made, but we shall accept the value of $i = 23.5$ percent obtained by interpolation.

7-6 REALISTIC ECONOMIC EVALUATION

The data upon which the evaluation in Example 7-3 was performed were characterized by a certainty and finality that are unrealistic. A chemical plant operates in a dynamic environment, and the factors involved in profitability considerations change with time; markets grow or decline, prices rise and fall, technological obsolescence occurs. This dynamism and its effects must be reflected in any realistic analysis, but none of these factors were even hinted at in Example 7-3. How allowances for change can be incorporated into economic analyses will be considered in later chapters that deal with forecasting, projections, and uncertainty. In this chapter we will deal with inflation, another aspect of a modern and dynamic environment, and how to incorporate its effects into economic evaluations.

7-7 CHARACTERISTICS OF INFLATION

In the preceding discussion, in which the time value of money was presented, it was tacitly assumed that the currency had a constant real value. It was assumed that a dollar obtained at some time in the future would have the same purchasing power as a dollar today. Historically, however, the purchasing power of the monetary unit, no matter what the currency is, erodes and depreciates with time, and there is little if any reason to expect that this trend will change in the future. Wilsher (1970) presents an entertaining and witty study of the changes in purchasing power of one of the world's major currencies over a 100-year period.

This depreciation in the value of money, termed *inflation*, is distinct and separate from discounting, which is applied to allow for the time value or rental price of money. In spite of the effect that inflation can have on economic analyses for investment decisions, it has largely been ignored in most engineering economics textbooks commonly used in chemical engineering studies. If it does appear, it is usually dealt with briefly and in passing.

Inflation is a rise in the price level or a fall in the purchasing power of the monetary unit. Inflation is usually expressed as a rate, e.g., the rate of decrease in purchasing power or the rate of increase of some national price level (Burford et al., 1977). The official rate of inflation is based on a specified mixture of goods required to maintain the standard of living of the average citizen. The annual rate of increase in cash outlay required to purchase this "basket" of goods is the official rate of inflation. It is reported monthly by the Bureau of Labor Statistics and depends on the goods or basket chosen.

The situation can be very different insofar as a company or firm is concerned. Table 7-7 presents the values for two indexes from 1973 to 1977 and the corresponding rates of inflation. A basket of industrial chemicals that would have cost $103 in 1973 would cost $224 in 1977, whereas a basket of process-industries equipment that costs $500 in 1977 could have been purchased for $344 in 1973. In other words, the purchasing power of the dollar had eroded by more than 50 percent with respect to industrial chemicals, but it had eroded by only 31 percent as far as the purchase of process-plant equipment is concerned. It is highly unlikely, therefore, that the actual inflation rate experienced by a particular project or company will be identical to the official rate of inflation, although this official rate is the best available measure of the general rate at which the purchasing power of the currency erodes.

Table 7-7 Cost indexes and inflation rates

	Index		Annual inflation rate, %	
Year	M & S equipment cost†	Industrial chemical prices‡	M & S equipment cost†	Industrial chemical prices‡
1973	344	103		
			15	48
1974	398	152		
			12	36
1975	444	207		
			6	6
1976	472	219		
			6	2
1977	500	224		

† Marshall and Stevens, Process Industries Installed-Equipment Cost Index, 1926 = 100.

‡ Bureau of Labor Statistics, 1967 = 100.

Burford et al. (1977) discuss the overall impact of inflation on the typical business investment that involves future revenues and expenses. The components of the investment do not respond to inflation in exactly the same way.

With respect to the individual firm the impact of inflation will depend primarily upon the financial, pricing, and inventory policy of the firm and upon depreciation. The effects stem primarily from differences in the historical value of a good compared with the real value of the good when payment is made for it. In inventory-valuation procedures, for example, the first-in first-out (FIFO) method means that inventoried material is charged to product cost at historical levels even though, as a result of inflation, they must be replaced at higher prices. The last-in first-out (LIFO) method reduces the extent of inventory undercosting and also decreases taxable income.

Depreciation policies are restricted by law, and annual depreciation charges are based on acquisition cost. No matter what depreciation policy is permitted, in no case can the total depreciation exceed the initial cost of acquisition. This historical-cost depreciation basis means that the portion of the revenue designated for the recovery of depreciable capital has less purchasing power than when the capital was originally invested. There also is an increase in the relative tax liability because the deduction from income for depreciation purposes is measured in historical dollars rather than in current (or future) dollars.

A realistic assessment of project profitability requires that the money values used in the analysis be corrected for the erosion in purchasing power of the currency. In allowing for inflation an attempt should be made to predict future costs and prices.

7-8 ALLOWING FOR INFLATION

Let us assume that the annual rate of inflation is f. A present sum of money P can purchase a certain amount of goods. The same amount of goods purchased 1 year from now would cost Pe^f. If f is constant, then, in general, Pe^{ft} would have the same purchasing power at the end of t years as P dollars would have today. Similarly, a cash flow S received t years in the future will have a current real value of Se^{-ft}.

The calculations involving the time value of money can now be combined with those concerned with inflation. A cash flow of S that occurs t years in the future with its current real value of Se^{-ft} would now have to be discounted at the financial interest rate i, so that

$$P = Se^{-ft}e^{-it}$$

or
$$P = Se^{-(f+i)t} \tag{7-6}$$

If the inflation rate is not constant, so that different rates apply each year, we can write

$$P = Se^{-it} \prod_{j=1}^{t} e^{-f_j} \tag{7-7}$$

Watson and Holland (1977) in their analysis prefer to use discrete cash flows and discrete annual compound interest rather than continuous compounding in order to make use of published annual accounts. Discounting by annual or by continuous-interest compounding methods can be made to be equivalent, and slightly different nominal interest rates would have to be used to give equal results. In the example to follow continuous-interest compounding will be used although cost and price levels for each year will be calculated on the basis of annual interest compounding at the inflation rate.

Example 7-4 Calculate the net present worth at $i = 0.15$ for the investment described in Example 7-3 with the effects of inflation included. Assume that prices inflate at the rate of 5 percent per year. Prices and costs can be assumed to be constant during each year at the levels reached at the beginning of the year. Working capital is to be maintained by the provision of $100,000 at the beginning of each year if the working capital at the end of the preceding year dropped below 20 percent of the annual sales for that year.

SOLUTION The summary of the calculations is presented in Table 7-8. Cash-flow calculations for the first year of operations (0–1) are as shown in Example 7-3. The calculations for the third year of operations (3–4) will be presented.

Table 7-8 Net-present-worth calculations with inflation

Time, yr	Cash flow, thousands	Discount factor $i = 0.15, f = 0.05$	Present worth, thousands
−1	$−300	1.2214	$−366.4
−0.5	−1000	1.1052	−1105.2
0	−1400	1.0000	−1400
0–1	378.5	0.9063	343
1–2	695.4	0.7421	516.1
2–3	936.6	0.6075	569
3	−100	0.5488	−54.9
3–4	1090.4	0.4974	542.4
4	−100	0.4493	−44.9
4–5	1139.7	0.4072	464.1
5	−100	0.3679	−36.8
5–6	1191.5	0.3334	397.2
6–7	1245.9	0.2730	340.1
7–8	1303	0.2235	291.2
8–9	1363	0.1830	249.4
9	−100	0.1653	−16.5
9–10	1426	0.1498	213.6
10	800	0.1353	108.2
			$\Sigma = 1009.6$

Income from sales The price level at the beginning of the year will be $(\$0.50)(1.05^3) = \$0.5788$ per pound of product, and this price will be maintained throughout the year

$$\text{Income} = (5.0 \times 10^6)(0.5788) = \$2894 \times 10^3$$

Production expenses Production expenses, variable as well as fixed, will also inflate at 5 percent per year. Fixed costs are fixed only with respect to production rate but not with respect to a rise in the cost of purchased goods and services

$$\text{Production expenses} = [200,000 + (5 \times 10^6)(0.15)](1.05^3)$$

$$= \$1099.7 \times 10^3$$

$$\text{Operating Income} = \$(2894 - 1099.7) \times 10^3 = \$1794.3 \times 10^3$$

Gross profit Gross profit is obtained by deducting the annual depreciation from the operating income. The depreciation is $230,000, as before, and is based on the historical cost of the depreciable investment

$$\text{Gross profit} = (1794.3 - 230) \times 10^3 = \$1564.3 \times 10^3$$

$$\text{Taxes} = (1.564.3 \times 10^3)(0.45) = \$703.9 \times 10^3$$

$$\text{Net profit} = (1564.3 - 703.9) \times 10^3 = \$860.4 \times 10^3$$

$$\text{Annual cash flow} = \$(1794.3 - 703.9) \times 10^3 = \$1090.4 \times 10^3$$

This is discounted to year 0 by 0.4974, the discount factor at 20 percent for a uniform cash flow during year 3–4. The 20 percent is the rate of return of 15 percent plus the 5 percent rate at which the currency erodes in purchasing power. The present worth in terms of purchasing power at time zero is therefore $542,400.

We must now check to see if the working capital is sufficient for next year based on this year's performance. The sales income was 2894×10^3, and 20 percent of this is $579,000. The total working capital that had been provided to date was $500,000, so an additional $100,000 is to be provided at the beginning of the following year.

The net present worth of the venture is $1,009,600 as compared to $1,338,800 for the same venture when inflation was not considered (Table 7-6). Thus, inflation resulted in reduced profitability.

7-9 EFFECT OF INFLATION ON FUTURE EQUIPMENT NEEDS

The engineer is frequently called upon to decide or help decide whether to select equipment size in accordance with today's needs or to oversize for an expected larger capacity requirement in the future. If the turndown ratio of oversize equipment for future capacity is such that it could operate efficiently and effectively at today's underdesign capacity, an economic analysis is required to reach a decision. The following alternatives could be considered:

I. Build today's capacity today, build a new facility with the required future capacity at a future date, and scrap the initial small-capacity unit
II. Build a unit today that has sufficient capacity for the expected future requirement
III. Build today's capacity today and add parallel capacity at a future date

A complete economic analysis of these alternatives would require, among other things, consideration of depreciation, depreciation policy, tax considerations, effect of abandonment on depreciation and taxes, and expected life of the project. Since our purpose in this section is to consider only the effect of inflation, the example that follows is simplified and will deal only with the present real worths of the capital expenditures for the three alternatives.

Example 7-5 A plant is presently being designed for a production capacity of $\bar{P}_1$ lb/yr. Management believes that in 5 years an increase of 100 percent in capacity will be needed. You are to examine whether higher-capacity equipment, where appropriate, should be purchased and installed now. The rate of return used is to be 22 percent, and the equipment cost index is expected to inflate at 10 percent per year. For the purposes of this example, neglect depreciation, taxes, dismantling costs, and salvage.

SOLUTION A general solution for selection between alternatives I and II will first be developed.

Let $\bar{P}_1$ = production capacity today
$\bar{P}_2$ = future production capacity in t yr
C_1 = installed equipment cost for P_1
C_2 = installed equipment cost for P_2

The relationship between cost and scale can be expressed by

$$C_2 = C_1 \left(\frac{\bar{P}_2}{\bar{P}_1}\right)^m$$

where the value of the scale-up exponent m is usually about 0.6. The present worth of alternative II is therefore

$$PW_{II} = C_1 \left(\frac{\bar{P}_2}{\bar{P}_1}\right)^m$$

The cost for $\bar{P}_2$ in future dollars at time t will be $C_2 e^{ft}$, and in order to supply these dollars we would need today

$$PW = C_2 e^{ft} e^{-it}$$

The total present worth of alternative I is therefore

$$PW_1 = C_1 + C_2 e^{-(i-f)t}$$

At the breakeven time t_e, both alternatives will have the same present worth, and

$$C_1 \left(\frac{\bar{P}_2}{\bar{P}_1} \right)^m = C_1 + C_2 e^{-(i-f)t_e}$$

Define $\bar{P}_2 / \bar{P}_1 \equiv R$. Substituting for C_2, we get

$$C_1 (R)^m = C_1 + C_1 R^m e^{-(i-f)t_e}$$

and both alternatives will be equal at

$$t_e = \frac{\ln [R^m/(R^m - 1)]}{i - f}$$

If the new capacity will be needed in less than t_e years, alternative II is preferable to alternative I.

For the example case and for $m = 0.6$

$$t_e = \frac{\ln [2.0^{0.6}/(2.0^{0.6} - 1)]}{0.22 - 0.10} = 8.98$$

The increased capacity will be needed only 5 years hence, and so in this case alternative II would be selected over alternative I.

Note: In the event of no inflation, $f = 0$, and $t_e = 4.9$ years. The decision would then be alternative I, i.e., build capacity as required now and build the large plant in accordance with the capacity requirement then. The effect of inflation was to encourage investment in equipment that would not be operating at its full potential.

We must now look at alternative III. The capacity to be added at t in accordance with alternative III is $P_2 - P_1$. The cost in future dollars at time t is

$$C_1 \left(\frac{\bar{P}_2 - \bar{P}_1}{\bar{P}_1} \right)^m e^{ft}$$

The total present worth of alternative III will be

$$PW_{III} = C_1 + C_1 \left(\frac{\bar{P}_2 - \bar{P}_1}{\bar{P}_1} \right)^m e^{-(i-f)t} = C_1 [1 + (R - 1)^m e^{-(i-f)t}]$$

$$PW_{II} = C_1 R^m$$

$$PW_1 = C_1 (1 + R^m e^{-(i-f)t})$$

In the present example we are comparing III with II, and because we are dealing with investments, we want the smaller PW. Therefore, if

$$R^m < 1 + (R - 1)^m e^{-(i-f)t}$$

we would pick alternative II in preference to III. In our case

$$R^m = 2.0^{0.6} = 1.52$$

$$1 + (R - 1)^m e^{-(i-f)t} = 1 + (2.0 - 1)^{0.6} e^{-(0.22-0.10)(5)} = 1.55$$

The right-hand-side of the inequality is larger than the left-hand-side so the preference remains alternative II.

The conclusion for this example is that those items of plant equipment which can perform and operate well and efficiently at conditions well under design should be included in the initial plant at sizes and capacities in accordance with the eventual expected increased production capacity.

7-10 CAPITAL-COST ESTIMATES

Thus far in our economic evaluation and analysis we have made the tacit assumption that we know the values of the various cost elements. We shall now briefly discuss the elements of process-plant costs and note several procedures for their estimation.

Capital cost represents the monies required to finance the construction of a manufacturing facility, to bring it into operation, and to keep it in operation. Capital costs are composed of two parts, fixed capital and working capital.

Fixed capital is money required to provide all the facilities necessary to perform the processing operation. Fixed capital includes the cost of process machinery and equipment directly involved in the manufacturing operation as well as the costs involved in engineering and design, land purchase, site preparation, buildings, storage facilities, provision of utilities and services, and the provision of employees' amenities, such as restaurant and recreation facilities. Fixed capital also includes start-up expenses.

Working capital is the money required to start the manufacturing operation and to allow the project to meet its obligations. Working capital will include cash for wages and salaries and other accounts payable, accounts receivable, raw material and product in storage, and material in process. Working capital will generally be about 10 to 20 percent of the capital investment, or about 15 percent of the annual product sales value. It is recoverable at the termination of the project. Typical working-capital requirements are shown in Table 7-9.

Table 7-9 Typical working-capital requirements

Item	Requirement
Raw-material inventory	1 month at raw-material cost
Product inventory	1 month at processing cost
Wages	1 month labor cost
Cash	1 month processing cost excluding raw material
Accounts receivable	1 month sales
Accounts payable	1 month variable cost
Taxes payable	6 months tax cost

7-11 COST INDEXES

Most of the estimating procedures we are about to discuss are based on historical cost data or information. This information can be updated by using appropriate indexes, several of which were introduced in our discussion of inflation. The present cost can be obtained by multiplying the historical cost by the ratio of the present value of the index to its value at the time the historical data were valid.

$$\text{Present cost} = \text{historical cost} \times \frac{\text{present index value}}{\text{historical index value}}$$

Many indexes are available, and they cover practically every area of interest. The most important for the chemical engineer are the *Chemical Engineering* plant-cost, the Marshall and Stevens installed-equipment-cost, and the Nelson refinery-construction indexes. Values for these indexes are shown in Table 7-10.

Chemical Engineering Plant-Cost Index

This index is based on four major components of plant cost. These components and their weights are equipment, machinery, and supports, 61 percent; construction labor, 22 percent; buildings, 7 percent; and engineering and supervision labor, 10 percent. The index is based on 1957–1959 = 100 and is published bimonthly in *Chemical Engineering*.

Table 7-10 Cost indexes

Year	*Chemical Engineering* plant-cost index 1957–1959 = 100	Marshall and Stevens process-industry installed-equipment cost, 1926 = 100	Nelson (inflation) refinery-construction index, 1946 = 100
1965	104	244	261
1966	107	252	273
1967	110	263	287
1968	114	273	304
1969	119	285	329
1970	126	303	365
1971	132	321	406
1972	137	332	439
1973	144	344	468
1974	165	398	523
1975	182	444	576
1976	192	472	616
1977	204	505	653
1978	219	545	701
1979 (mid)	222	593	740

Marshall and Stevens Installed-Equipment-Cost Index

Two Marshall and Stevens equipment-cost indexes are available, an all-industry index and a process-industry index. The all-industry index is the average of individual indexes for 47 different types of industrial, commercial, and housing equipment. The process-industry index, shown in Table 7-10, is based on eight process industries. The industries and their weights in the index are cement, 2 percent; chemicals, 48 percent; clay products, 2 percent; glass, 3 percent; paint, 5 percent; paper, 10 percent; petroleum, 22 percent; and rubber, 8 percent. This index, based on 1926 = 100, appears bimonthly in *Chemical Engineering*.

Nelson Refinery-Construction-Cost Index

Refinery-construction-cost elements and their weights as used in this index are skilled labor, 30 percent; common labor, 30 percent; iron and steel, 24 percent; building materials, 8 percent; and miscellaneous equipment, 8 percent. The index is based on 1946 = 100 and appears monthly in the *Oil and Gas Journal*.

Another index that can be useful to the chemical engineer is the construction-cost index appearing in the *Engineering News-Record*. Its elements include costs of steel, lumber, cement, and labor. It is a veteran index and was based on 1913 = 100. Since then additional bases have been used, 1926 = 100 and 1949 = 100.

7-12 FIXED-CAPITAL ESTIMATES

The accuracy of an estimate depends on how much is known about the project and how much effort is invested in the preparation of the estimate. The estimate can range from a quickie or ball-park estimate of low accuracy to an estimate with an accuracy in the range of ±5 percent. The cost of making a high-accuracy estimate is not inconsiderable and can be in the range of 0.5 to 1 percent of the cost of a multimillion dollar project.

The American Association of Cost Engineers (1958) proposed five classifications for estimate types:

1. Order-of-magnitude, or ratio, estimate based on previous cost data. It is suitable for preliminary screening and has a probable accuracy poorer than ±30 percent.
2. Study, or factored, estimate is based on knowledge of the major items of equipment. The probable accuracy is ±30 percent.
3. Preliminary, or budget-authorization, estimate is based on sufficient data to permit the project to be budgeted. The expected accuracy is ±20 percent.
4. Definitive, or project-control, estimate is based on almost complete data but before all drawings and specifications have been finished. The expected accuracy is ±10 percent.

5. A detailed firm or contractor's estimate would be based on complete engineering drawings, specifications, and site survey. The probable accuracy would be ± 5 percent.

Ratio Methods

Ratio methods do not require that the process be designed, but they do require previous data. The simplest and quickest but least accurate method is based on the observation that in the chemical industry the turnover ratio, the ratio of annual sales to fixed capital, approximates unity for many plants. Thus, to produce 10 million pounds per year of a product which sells for $1 per pound, i.e., annual sales of $10 million, a plant would require, according to this procedure, a fixed-capital investment of $10 million.

Estimates can be based on the capital cost per ton of annual production. Guthrie (1970) has developed capital costs for 54 chemical processes. His total investment costs assume a developed site and include 15 to 20 percent of the capital costs to represent working capital and start-up costs. Guthrie cites $7.5 million as the total investment cost for a 100,000 ton/yr acetone plant in 1968. The investment per annual ton is therefore $75. On the basis of constant cost per ton of annual capacity, a 150,000 ton/yr acetone plant would require a total investment of $(150,000)(75) = \$11.25$ million, based on 1968 dollars. This procedure should not be used if the capacity considered is more than twice or less than half of the capacity of the plant for which the cost data per ton of annual capacity were obtained. The cost on an undeveloped site might be 30 to 40 percent more than that on a developed site and enlargements of, or addition to, an existing plant could cost 20 to 30 percent less than the figures cited for a developed site.

Costs can be related to scale by the so-called *six-tenths-factor rule* (Williams, 1947). It is based on the observation that a straight line is obtained when capacity vs. equipment or process cost is plotted on log-log coordinates. For equipment costs the slope is usually about 0.6. In equation form this straight-line relationship can be expressed as

$$C_2 = C_1 \left(\frac{\bar{P}_2}{\bar{P}_1} \right)^m \tag{7-8}$$

C_1 is the cost for equipment or process with capacity $\bar{P}_1$, and C_2 is the cost of the same equipment or process with capacity $\bar{P}_2$. Values for the cost-capacity exponent m can vary from less than 0.2 to more than 1.0 for equipment items (Peters and Timmerhaus, 1968; Jelen, 1970). Some cost-capacity exponents for chemical plants are presented in Table 7-11, and the values for m range from 0.45 to 0.80. A value of 0.6 can be used for equipment costs and 0.7 for complete plant when no other information is available.

Table 7-11 Total capital investment for some chemical plants

Plant	Capacity, 10^3 tons/yr	Total capital investment,† millions	Cost-capacity exponent
Acetic acid	10	$ 2.1	0.68
Acetone	100	7.5	0.45
Butadiene	100	7.5	0.68
Ethylene oxide	100	29.0	0.78
Formaldehyde	100	4.5	0.55
Hydrogen peroxide	100	7.8	0.75
Isoprene	100	16.5	0.55
Methanol	100	6.8	0.60
Phosphoric acid	10	1.2	0.60
Polyethylene	100	8.5	0.65
Urea	100	3.8	0.70
Vinyl chloride	100	9.0	0.80

† Construction on a developed site. Investment includes fixed capital for battery-limits plant, working capital, and start-up expenses, 1968 dollars.
Source: From Guthrie (1970), by permission.

According to the six-tenths-factor rule and using 0.45 as the value for m, the total investment for a 150,000 ton/yr acetone plant would be

$$\$7,500,000 \left(\frac{150,000}{100,000}\right)^{0.45} = \$9 \times 10^6$$

compared with the estimate based on constant cost per annual ton of $11.25 million. This demonstrates the investment advantages of large-scale plants. A 150,000 ton/yr acetone plant would cost $60 per annual ton capacity, compared with $75 per ton for a 100,000 ton/yr plant.

Factor Methods

Factor methods are based on the cost of the purchased equipment. The simplest procedure proposed by Lang (1947, 1948) assumes that the fixed-capital investment can be estimated by multiplying the total equipment cost by an appropriate factor. The Lang factors are shown in Table 7-12. Fixed-capital investment or total capital investment is estimated by multiplying the total delivered-equipment cost by the appropriate factor in the table. These factors include costs for land and contractor's fees.

Although the Lang factors take the type of plant into account, they cannot account for process variations within the types. Greater accuracy can be achieved if factors are used in accordance with individual equipment types. Wroth (1960) proposed multiplying factors for a number of individual items of process equipment. The purchased cost multiplied by the factor will give an

Table 7-12 Lang multiplication factors

	Factor for	
Type of plant	Fixed capital investment	Total capital investment
Solid processing	3.9	4.6
Solid and fluid processing	4.1	4.9
Fluid processing	4.8	5.7

Source: From Peters and Timmerhaus (1968), by permission.

estimate of the installed cost including cost of site development, buildings, electrical installations, painting, carpentry, foundations, structures, piping, installation, engineering, overhead, supervision and contractor's fee, and rentals. Some of the factors produced by Wroth are blowers and fans, 2.5; centrifugal motor-driven compressors (less motor), 2.0; heat exchangers, 4.8; instruments, 4.1; electric motors, 8.5; centrifugal motor-driven pumps (less motor), 7.0; process tanks, 4.1; and towers and columns, 4.0.

Example 7-6 Estimate the cost for a chemical plant processing fluids and requiring the following purchased equipment items:

Blowers and fans	$ 8,000
Centrifugal pumps (less motor)	20,000
Compressor (less motor)	100,000
Heat exchangers	120,000
Motors	40,000
Process tanks	48,000
Distillation towers	160,000
	496,000

SOLUTION When the Lang factor of 4.8 is used, the fixed-capital cost will be

$$(4.8)(496,000) = \$2,380,800$$

With the Wroth factors the fixed-capital cost would be

$$(8000)(2.5) + (20,000)(7.0) + (100,000)(2.0) + (120,000)(4.8) +$$
$$(40,000)(8.5) + (48,000)(4.1) + (160,000)(4.0) = \$2,112,800$$

Hirsch and Glazier (1960) proposed an equation for cost-estimation purposes that is suitable for computer calculation. The notation for their equation, presented below, is specific to this equation only:

$$C = F_I[E(1 + F_F + F_P + F_M) + E_A + E_E] \tag{7-9}$$

where C = total battery-limits investment

E = total purchased equipment cost, FOB, less incremental cost for alloys

E_E = cost of erected equipment

E_A = incremental cost of alloy

F_I = indirect cost factor representing contractor's overhead and profit, engineering, supervision, and contingencies, usually taken as 1.4

F_F = cost factor for field labor

F_M = cost factor for miscellaneous items such as instruments, insulation, foundations, structural steel, buildings, painting, wiring, freight, and field supervision

F_P = cost factor for piping materials

The factors F_F, F_M, and F_P are defined by

$$\log F_F = 0.635 - 0.154 \log 0.001E - 0.992\frac{e}{E} + 0.506\frac{f}{E} \qquad (7\text{-}9a)$$

$$F_M = 0.344 + 0.033 \log 0.001E + 1.194\frac{t}{E} \qquad (7\text{-}9b)$$

$$\log F_P = -0.266 - 0.014 \log 0.001E - 0.15\frac{e}{E} + 0.556\frac{p}{E} \qquad (7\text{-}9c)$$

where e = total heat exchanger cost less incremental cost of alloy

f = total cost of field-fabricated vessels less incremental cost of alloy (vessels larger than 12 ft in diameter generally are field-fabricated)

p = total cost of pump plus pump driver less incremental cost of alloy

t = total cost of tower shells less incremental cost of alloy

Example 7-7 Using the Hirsch-Glazier equation, estimate the total battery-limits investment for a chemical plant whose estimated equipment costs are shown in Table 7-13.

SOLUTION The Hirsch-Glazier factors will be developed:

$$E = \$4,330,000$$

$$E_E = \$345,000$$

$$E_A = \$(110 + 200 + 200 + 350)(1000) = \$860,000$$

$$f = \$400,000$$

$$p = \$250,000$$

$$t = \$400,000$$

$$e = \$900,000$$

Table 7-13 Estimated major equipment costs (thousands) for Example 7-7

Equipment item	Cost, carbon-steel basis	Cost, alloy basis	Incremental cost of alloy
FOB equipment:			
Pumps	$ 250	$ 360	$110
Compressors	1380		
Towers, less than 12 ft			
diam, shells	400	600	200
Trays and internals	800	1000	200
Heat exchangers	900	1250	350
Drums and tanks, less			
than 12 ft diam	200		
Over 12 ft diam	400		
Total FOB equipment cost	$4330		
Erected equipment:			
Fired heater	285		
Tanks	60		
Total erected-equipment cost	$345		

F_F, F_M, and F_P are calculated from Eqs. (7-9)

$$\log F_F = 0.635 - 0.154 \log 4330 - 0.992 \frac{900,000}{4,330,000} + 0.506 \frac{400,000}{4,330,000}$$

$$= -0.08446$$

$$F_F = 0.823$$

$$F_M = 0.344 + 0.033 \log 4330 + 1.194 \frac{400,000}{4,330,000} = 0.574$$

$$\log F_P = -0.266 - 0.014 \log 4330 - 0.156 \frac{900,000}{4,330,000} + 0.556 \frac{250,000}{4,330,000}$$

$$= -0.3172$$

$$F_P = 0.482$$

Substituting into Eq. (7-9) and assuming $F_I = 1.4$ gives

$$C = 1.4[(4,330,000)(1 + 0.823 + 0.574 + 0.482) + 345,000 + 860,000]$$

$$= \$19,140,000$$

Detailed Estimates

More accurate capital cost estimates can be made if all of the individual cost components are considered. Peters and Timmerhaus (1968) developed the average values for these components as percentages of delivered-equipment costs for different types of typical plant. Their results are shown in Table 7-14. The actual percentages for individual process plant will depend on such factors as the type of process, its complexity and location, experience with the particular process, etc.

> **Example 7-8** Estimate the battery-limits cost for the plant described in Example 7-7. Use Table 7-14 and assume that the delivered-equipment cost will be 10 percent more than the FOB cost. The plant is classified as fluid-processing.

> SOLUTION To obtain the battery-limits cost and in order to be on the same basis as in Example 7-7 we shall omit the cost of service facilities and land. The percentage figures for buildings, yard improvements, engineering, and supervision and construction expenses will also be decreased compared with the values shown in Table 7-14 to account for the fact that part of these items are due to the service facilities. The total delivered-equipment cost will be taken as the FOB equipment cost multiplied by 1.1 plus the erected-equipment cost from Table 7-13.

Component	Cost, thousands	
Delivered equipment	$ 6,054	
Equipment installation, 47%	2,845	
Instrumentation, 18%	1,090	
Piping, 66%	3,996	
Electrical, 11%	666	
Buildings, 10%	605	
Yard improvements, 6%	363	
Total direct plant cost		$15,620
Engineering and supervision, 25%	1,514	
Construction expenses, 30%	1,816	
		$3,330
Total direct and indirect cost		$18,950
Contractor's fee, 5% of direct + indirect cost	948	
Contingency, 10% of direct + indirect cost	1,895	
		$2,843
Total battery-limits cost rounded off		$21,800

Table 7-14 Typical capital-investment components based on delivered-equipment cost

Item	Percent of delivered-equipment cost		
	Solid-processing plant	Solid-fluid processing plant	Fluid-processing plant
Direct costs:			
Purchased equipment, delivered (including fabricated equipment and process machinery)	100	100	100
Installation	45	39	47
Instrumentation and controls (installed)	9	13	18
Piping (installed)	16	31	66
Electrical (installed)	10	10	11
Building (including services)	25	29	18
Yard improvements	13	10	10
Service facilities (installed)	40	55	70
Land (if purchase is required)	6	6	6
Total direct plant cost	264	293	346
Indirect costs:			
Engineering and supervision	33	32	33
Construction expenses	39	34	41
Total direct and indirect costs	336	359	420
Contractor's fee (about 5% of direct and indirect plant costs)	17	18	21
Contingency (about 10% of direct and indirect plant costs)	34	36	42
Fixed-capital investment	387	413	483
Working capital (about 15% of total capital investment)	68	74	86
Total capital investment	455	487	569

Source: From Peters and Timmerhaus (1968), by permission.

This can be compared to the estimated battery-limits cost of $19,140,000 obtained in Example 7-7. The estimate obtained in the present example is probably high because of the important proportion of alloy equipment in the plant. The costs of items such as instrumentation, electrical, buildings, and yard improvements would be independent of the metal used in the process and its cost. Equipment installation would be only slightly dependent on the material of construction used in the equipment.

Accuracy of detailed estimates A maximum accuracy within approximately ± 5 percent can be obtained with detailed estimates if each individual cost component is carefully estimated. As much as possible of the estimate should be based on quotations from equipment vendors. Installation costs are then estimated from labor rates and efficiencies and worker-hour and material calculations for each type of equipment. Accurate estimates are made of other factors involved such as engineering, drafting, field supervision, construction equipment, procurement, profit management, and home-office costs. Estimates of this degree of accuracy are usually prepared only by contractors who are bidding on contracts from specifications and drawings. Their bids represent a sales price to them and the price must be sufficient to cover all the direct and indirect costs of the project and to leave a reasonable profit or fee for the contractor.

Detailed estimates of lesser accuracy that require less effort in their preparation can be based on unit costs. These, again, are based primarily on purchased-equipment cost. They are similar in principle to the procedure used in Example 7-8 but break the cost factors down to the individual equipment item; for each item cost factors for such elements as material, labor, piping, electrical, engineering, etc., would be used based on past experience and records. The estimate would then be completed by applying a factor for construction expense, contractor's fee and contingency.

A lengthier exposition on detailed estimating procedures is beyond the scope of this book. The interested reader is referred to Bauman (1964), Guthrie (1974), and similar works.

Offsite Facilities

The battery-limits plant requires that services and facilities be supplied so that it can function. The utilities that must be supplied include steam, electricity, cooling and process water, refrigeration, and compressed air. Fire protection and waste disposal must be supplied as well as, in most cases, outside-battery-limits raw-material and product storage.

The cost for service facilities, as can be seen from Table 7-14, is in the range of 40 to 70 percent of purchased-equipment cost. These percentages, however, can vary widely. A large, new plant at a new site would require a higher percentage of the total capital investment devoted to offsite facilities than a small process in an existing plant.

Start-Up Costs

During start-up the plant will generally not be operating at an optimal level and may not be producing specification-grade product. Changes and additions to equipment are usually necessary before the plant can operate satisfactorily. Start-up costs include money for equipment, materials, labor, and overheads. During early start-up operations raw materials and services may be consumed without any salable product being produced. The net start-up cost will include all the expenses less the value of any salable product that is made.

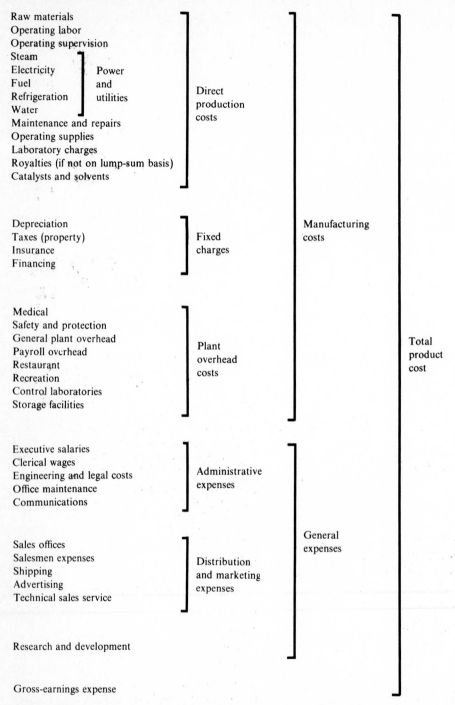

Raw materials
Operating labor
Operating supervision
Steam
Electricity ⎤ Power
Fuel ⎥ and
Refrigeration ⎥ utilities
Water ⎦
Maintenance and repairs
Operating supplies
Laboratory charges
Royalties (if not on lump-sum basis)
Catalysts and solvents
⎤ Direct
production
costs

Depreciation
Taxes (property)
Insurance
Financing
⎤ Fixed
charges

Medical
Safety and protection
General plant overhead
Payroll overhead
Restaurant
Recreation
Control laboratories
Storage facilities
⎤ Plant
overhead
costs

⎤ Manufacturing
costs

Executive salaries
Clerical wages
Engineering and legal costs
Office maintenance
Communications
⎤ Administrative
expenses

Sales offices
Salesmen expenses
Shipping
Advertising
Technical sales service
⎤ Distribution
and marketing
expenses

⎤ General
expenses

⎤ Total
product
cost

Research and development

Gross-earnings expense

Figure 7-4 Total product cost checklist for a typical chemical plant. [*From Peters, M. S. and K. D. Timmerhaus, 1968, Plant Design and Economics for Chemical Engineers, McGraw-Hill Book Company, New York.*]

Start-up expense may be as high as 12 percent of the fixed capital investment but can be much less for known and tested processes and technologies. For estimation purposes a value of 10 percent for start-up cost can be assumed.

7-13 TOTAL-PRODUCT-COST ESTIMATION

A checklist of the cost elements of the total product cost for a typical chemical plant is shown in Figure 7-4. The best source of data for product cost estimates is data from similar plants. If company records of such costs are available, quick and reliable estimates of manufacturing costs and general expenses can easily be made.

Manufacturing Costs

Manufacturing costs include all the expenses connected with the operation of the producing facility. These expenses are classified into the categories shown in Fig. 7-4: direct production costs, fixed charges, and plant overhead costs.

Direct production costs include expenses directly connected with the actual processing. These costs are a function of the production level and are termed *variable costs*. They vary almost linearly with production rate although some elements such as maintenance, for example, continue, albeit at a lower expenditure, even though the plant may be shut down.

Fixed charges are expenses which remain practically constant from year to year and do not vary with the plant production rate.

Plant costs include those expenses necessary to ensure that the manufacturing operation can proceed satisfactorily. They are concerned primarily with employee overhead expenses but also include control laboratories and storage facilities. They are relatively independent of production rate.

General Expenses

General expenses include other costs that are involved in the company operations, e.g., administration, distribution and marketing, and research and development. The charges for these operations are relatively constant and do not vary with the production rate. The gross-earnings expense in Fig. 7-4 refers to expenses due to income taxes.

7-14 DIRECT PRODUCTION COSTS

Raw Materials

Raw-materials costs usually constitute the major single item of costs in chemical manufacturing. The amounts of the raw materials consumed per unit of time or per unit of production can be determined from material-balance considerations.

In the event that salable by-products are obtained, a credit should be taken for them.

For preliminary cost estimations market prices can be used for estimating raw-material costs. Sources for price information are periodicals such as the *Chemical Marketing Reporter* and *European Chemical News*. Actual prices can differ from published prices. They are usually negotiated, and the price will depend on such factors as quantity, quality, duration of contract, and state of the market. Freight or transportation charges should be included in the costs. Information on shipping charges can be obtained from carriers or from company sources.

Operating Labor and Operating Supervision

The quantity of operating labor can be estimated from company experience or from literature on similar processes. If sufficient process information is available, the labor requirements can be estimated by an analysis of the work to be done.

When no information from other sources is available, Wessel (1952) has suggested that labor requirements can be related to plant capacity by the equation

$$\frac{\text{Operating worker-hours}}{\text{Ton of product}} = t \frac{\text{no. of process steps}}{(\text{capacity in tons/day})^{0.76}} \qquad (7\text{-}10)$$

where

$$t = \begin{cases} 23 & \text{for batch operations with a maximum of labor} \\ 17 & \text{for operations with average labor requirements} \\ 10 & \text{for well-instrumented continuous processes} \end{cases}$$

Equation (7-10) suggests that labor requirements can be scaled up or down by the 0.24 power of the capacity ratio. This value is in the range of 0.2 to 0.25, as suggested by O'Connell (1962).

Operating supervision can be estimated as a percentage of operating labor. The range of values is 10 to 25 percent, 15 percent being typical.

Utilities

Utility requirements can be estimated from the preliminary design, from company sources for similar operation, or from literature sources. Rates for the utilities can be obtained from company sources or by inquiry to outside sources. Rates can vary considerably with location and with demand.

Utilities may be purchased from a public utility at established rates or supplied by a central station supplying an entire plant. In the latter case the rates would depend upon the operating cost of the service installation. If the utility is self-generated but used by only one manufacturing unit, all its operating expenses would be charged to the consumer unit and the installation would be charged to the production unit as part of its fixed capital.

Other Elements

The additional elements of the direct production cost as shown in Fig. 7-4 are maintenance and repairs, operating supplies, laboratory charges, royalties, and catalysts and solvents.

To keep a plant in efficient operating condition expenses are incurred for maintenance and repairs. On an average, the costs for labor, materials, and supervision equal approximately 6 percent of the fixed-capital maintenance. The range will be from 2 to 20 percent, the higher figure prevailing when operating conditions are severe.

Operating supplies include such items as instrument charts, lubricants, control-laboratory chemicals, and similar expendables. The annual cost can be estimated as 15 percent of the total cost for maintenance and repairs or 0.5 to 1.0 percent of investment.

Laboratory charges are costs of laboratory tests for product-quality and operation control. They can be estimated by multiplying estimated worker-hours by a worker-hour rate for laboratory personnel. For quick estimates experience figures of 10 to 20 percent of operating labor can be used.

Royalties, if necessary, must be included in the direct manufacturing cost, and the cost charged should be based on the actual royalties charged by the licensor. If no information is available in the preliminary design stages, a figure of 1 to 5 percent of sales can be used.

Depending on the particular process, charges should also be included for catalysts and solvents that cannot be included as elements of the raw-material costs.

7-15 FIXED CHARGES

Fixed charges, or fixed costs, represent those costs which continue whether the plant is in production or not. Those shown in Fig. 7-4 are related to capital investment and include expenses for depreciation, local property taxes, and financing and interest charges.

Depreciation

The fixed capital invested in equipment and buildings in a new productive venture undergoes wear and tear, exhaustion, and obsolescence, resulting in a decrease in value. This decrease in value, known as *depreciation*, is part of the manufacturing expense.

Because income tax is based on income after all costs have been deducted, the depreciation rate is of importance in determining the income tax to be paid. The U.S. Bureau of Internal Revenue has established the probable useful life of various types of equipment and the alternative methods that may be employed in establishing the depreciation rate. For engineering projects the straight-line depreciation method is generally used. A useful-life period is assumed as well as a salvage value at the end of the useful life. The annual depreciation charges will be the difference between the initial cost and salvage value divided by the number of years of useful life. Depreciation methods are available which result in higher depreciation charges during the early years of the project and lower charges in the later years.

The annual depreciation rate for machinery and equipment is about 10 percent of the fixed-capital. Buildings are usually depreciated at an annual rate of about 3 percent of the initial cost.

Property Taxes

Property taxes depend upon the location of the plant and upon local regulations. Plants in cities generally pay higher property taxes than those in less populated areas. Some localities offer tax concessions in order to attract new enterprise. In the absence of specific data for local conditions an annual rate of 2 percent of investment may be assumed.

Insurance

Insurance rates can vary greatly depending upon the hazards involved and the amount of protection facilities provided. Annual insurance charges of 1 percent of the fixed-capital investment may be used for estimating purposes.

Financing

Financing charge or interest charges are paid for the use of borrowed capital. It is thus a fixed charge although it can be listed as a management cost rather than a manufacturing cost. Annual interest rates amount to 6 to 10 percent of the borrowed capital. If all the capital investment is supplied from company funds and no borrowed capital is employed, it is debatable whether interest should be charged as a cost. It cannot be charged as a cost for income tax purposes if owned money is used. Company policy should be consulted for inclusion of interest as a cost.

7-16 PLANT OVERHEAD COSTS

The costs of maintaining certain service functions that are required indirectly by the manufacturing unit are included in plant overhead costs. Some of the components of these overhead costs are presented in Table 7-15. Generally these charges are distributed over all the manufacturing units in direct proportion to the labor costs. For estimating plant-overhead costs a charge equivalent to 50 to 70 percent of the total expense for operating labor, supervision, and maintenance may be assumed. The higher value would be for plants involving more complex operations, and the lower figure would be for small manufacturing units or units with simple operations.

7-17 GENERAL EXPENSES

The general expenses comprise those overhead costs due to administration, distribution and marketing, and research and development. The administrative costs are connected with top-management activities and their supporting services. They may vary from company to company and even from plant to plant within a company. If no figures are available from company records or policy statements, a value of 10 to 20 percent of operating labor may be used for estimating purposes.

Distribution and marketing costs include charges due to selling the product. They include costs for salesmen, sales office, advertising, technical sales services, and containers. Shipping costs must be included if the product is sold on a delivered basis. Distribution and marketing costs range from 2 to 20 percent of the total product cost. The lower values would be for large-volume bulk chemical products while the larger values would apply to a new product or to a product sold in small amounts but to a large number of customers.

Research and development activities are necessary to develop new processes and products, to improve processes for existing products, to improve product quality, etc. The costs for personnel, equipment, supplies, and laboratories can be estimated as 2 to 5 percent of annual sales value.

7-18 GROSS-EARNINGS EXPENSES

The gross-earnings expenses refer to national and regional corporate income taxes. The gross earnings are the total income minus total production costs, and it is these earnings that are subject to income tax. Because income tax rates are based on total company gross earnings, the rate can vary from company to company. If income taxes are to be considered in the cost estimate, it is usual to assume the highest income tax rate. Net earnings will equal gross-earnings less income taxes.

7-19 PRELIMINARY TOTAL-PRODUCT-COST ESTIMATION

The average values for the various product cost elements cited above are presented in Table 7-15. Although such average values can be used for screening and for rapid and preliminary cost estimates, the estimating engineer will have to apply judgment in the choice of the specific values to be used within the ranges presented. The use of average values does not take into account such factors as differences between situations and process complexities or types and should be applied with due care and caution.

Table 7-15 Preliminary total-production-cost estimation

A. Direct production costs
   1. Materials
      a. Raw materials, estimate from price lists
      b. By-products, estimate from price lists
   2. Operating labor, estimate from literature or from similar operations
   3. Operating supervision, 10–25% of operating labor
   4. Utilities, from literature or similar operations
   5. Maintenance and repairs, 2–20% of fixed-capital investment
   6. Operating supplies, 15% of maintenance and repairs or 0.5–1% of fixed-capital investment
   7. Laboratory charges, 10–20% of operating labor
   8. Royalties, 1–5% of total product cost

B. Annual fixed charges
   1. Depreciation, 6–10% of fixed-capital investment
   2. Property taxes, 2% of fixed-capital investment
   3. Insurance, 1% of fixed-capital investment
   4. Financing, 6–10% of borrowed capital

C. Plant overhead costs, 50–70% of total expenses for operating labor, supervision, maintenance, and repairs

D. General expenses
   1. Administrative, 10–20% of operating labor
   2. Distribution and marketing, 2–20% of total product cost
   3. Research and development, 2–5% of sales

E. Gross-earnings expense, dependent upon total company earnings and income tax regulations; range of gross-earnings expense is 25–60% of gross earnings; gross earnings = total income less items A to D

PROBLEMS

7-1 A bond has a face value of $1000 and pays 5 percent per year semiannually, i.e., $25 every 6 months. It will reach maturity in $2\frac{1}{2}$ years. A purchaser wants to obtain a minimum yield of 8 percent (continuous interest compounding basis) on his investments. What is the maximum price he should be prepared to pay for this bond?

7-2 Mrs. Jones' son is approaching his sixth birthday, and Mrs. Jones decides to establish a fund that could assist in financing her son's education. She decides to invest a certain sum on each birthday from the sixth to the seventeenth, inclusive. The fund should exactly suffice to pay $3000 to her son on each birthday from the eighteenth to the twenty-first, inclusive. If the money can be invested at 6 percent interest (continuous interest compounding) how much will Mrs. Jones have to invest in the fund on each birthday?

7-3 Tank levels in a tank farm remote from the control room are read and logged twice daily by a plant operator. The labor costs involved are $10,000 per year. A proposal has been made that instrumentation to do this job be installed, thereby eliminating the labor cost. The installed cost of the instrumentation will be $50,000, its life should be 10 years, and annual maintenance costs will be $2000. What would be the rate of return of this investment? (Neglect the effects of this investment on depreciation, income tax, etc.)

7-4 Estimate the investment (straight-line depreciation over 10 years) that can be made to save $1000 per year of energy cost. Assume that annual costs due to maintenance, insurance, property taxes, etc., are equal to 8 percent of the investment. Income tax rate is 48 percent, and 10 percent rate of return after taxes is required.

7-5 A solution is to be stored and maintained at 150°F in a tank 20 ft in diameter and 25 ft tall. Heat losses for the uninsulated tank were calculated and are expected to be 900,000 Btu/h at the worst design conditions of surroundings at 0°F with a 20 mi/h wind. Heat losses would be reduced to 140,000 Btu/h if $1\frac{1}{2}$-in insulation were installed on sidewall and roof. Insulation will cost $6 per square foot installed, and steam costs $2 per million Btu. Other conditions are as in Prob. 7-4. The plant is in a Midwest location. Would you recommend the installation of thermal insulation?

7-6 You are to recommend which compressor is to be purchased for the supply of instrument air for a plant under construction. Two compressors have survived earlier screening and evaluation. Your company requires a return of at least 12 percent on all unnecessary investments. Neglecting depreciation, taxes, etc., which compressor would you recommend based on data in the table?

	Compressor A	Compressor B
Investment	$60,000	$40,000
Annual operating expenses	1,600	2,300
Annual maintenance, insurance, etc.	3,000	2,800
Salvage value	0	0
Expected life, years	20	10

7-7 The installation of insulation for a heat exchanger is being considered. The following information has been generated:

Insulation thickness, in	Btu/h saved	Cost for installed insulation
1	300,000	$18,000
2	350,000	24,000
3	370,000	27,000
4	380,000	28,000

Annual fixed charges are 10 percent of investment. The value of heat is $3 per million Btu. The exchanger operates 300 days/yr. Your company requires a 15 percent return on an investment of this type. What insulation thickness should be used?

7-8 Your company must purchase a reactor as part of a plant expansion program. Bids have been received for four reactors, each of which can supply the required service and all of which are estimated to have equal lives. The following additional data apply to the four bids:

Bid	Installed cost	Annual operating plus fixed costs
1	$100,000	$30,000
2	140,000	23,000
3	160,000	21,000
4	120,000	28,000

All other costs are equal for all of the bids. If your company demands a 16 percent return on any unnecessary investment, which of the bids should be accepted?

7-9 Calculate the net present worth of the following project on the basis of a 15 percent return. Capital investment will be $40 million, of which 20 percent is supplied at the beginning of the first year (time zero), 65 percent at the beginning of the second year, and 15 percent at the beginning of the third year, when production begins. Working capital of $1.5 million is also supplied at that time. The working capital will be recovered intact at project termination. Uniform positive cash flows due to sales less expenses other than depreciation over the 10-year productive life of the project are as follows:

Time, years	Flow, millions	Time, years	Flow, millions
2–3	$12.5	7–8	$17.5
3–4	15	8–9	17.5
4–5	17.5	9–10	15
5–6	20	10–11	15
6–7	20	11–12	15

Straight-line depreciation accounting is used, and income taxes of 35 percent are paid. Assume that these taxes are paid at the end of the year during which the profits were generated.

7-10 As a result of poor project management, delays in delivery, etc., completion of the project described in Prob. 7-9 was delayed by 1 year, and additional investment of $7.5 million was required. This sum was paid at the beginning of the fourth year. The other conditions of Prob. 7-9 will still be in effect, but the cash flows due to sales are displaced by 1 year. What is the net present worth of this project?

7-11 (*a*) What effect, if any, would a constant inflation rate of 5 percent have on the solution to Prob. 7-3?

(*b*) Would you recommend the proposal made in Prob. 7-3 if your company requires a return of 10 percent after taxes on any unnecessary investment, costs and prices inflate at 5 percent per year, and the tax rate is 48 percent? Assume that prices and costs are constant during each year. Straight-line depreciation over 10 years can be assumed.

7-12 Calculate the DCF rate of return for the project summarized in the following table. The tax rate is 48 percent, and project (economic) life is equal to the depreciable life. Straight-line depreciation is used.

Plant capacity	30×10^6 lb/yr
Fixed capital (depreciable)	\$10 million, $\frac{1}{2}$ at beginning and $\frac{1}{2}$ at end of construction
Working capital	\$1 million during first 3 years of production \$1.5 million until project termination
Construction	18 months
Start-up	3 months
Start-up costs	\$1.5 million
Project life	10 years
Annual variable costs	\$2.5 million based on 100% production
Annual fixed costs (not including depreciation)	\$2 million
Sales, annual	20% in first year 50% in second year 70% in third year 90% in fourth year 100% in fifth and following years
Sales price	45 cents per pound during first 3 years 35 cents per pound during next 3 years 30 cents per pound until project termination

EIGHT

FORECASTING THE FUTURE

> Said Rabbi Eudemus of Haifa: "Since the destruction of the Temple, the gift of prophecy has been denied to prophets and bestowed upon scholars."†
>
> *"Babylonian Talmud," Baba Bathra 12ª*

Forecasts and predictions provided the basis for the economic evaluations and analyses of the preceding chapter. We predicted product demand, raw material and labor costs, investment requirements, and even inflation rates and then incorporated these predictions into our evaluations. We therefore made the tacit assumption that it is possible to make projections into the future of factors that will affect revenue, costs and profitability. Planning for the future cannot proceed without forecasting even though it is difficult, if not impossible, to assess and predict values of all the factors that will influence profitability. In the next chapter we shall deal with the problems that arise from the fact that there will be uncertainties in our predicted values. In this chapter we present some of the techniques and procedures for making forecasts and projections.

---

† It should be noted that authorities differ on this point. For example: "Said Rabbi Johanan: 'Since the destruction of the Temple, the gift of prophecy has been denied to prophets and bestowed upon fools and children.'" Ibid., 12ᵇ.

8-1 HOW TO PREDICT

Schemes that can be used to make predictions have been presented and discussed by Bross (1963). *Persistence, trajectory,* and *cyclic predictions* are variations on the same theme, and all are based on historical data. These procedures maintain that what has happened in the past is likely to continue into the future. If, for example, a study of maintenance costs in your company shows that the annual cost of maintenance over the past 10 years was 4 percent of the invested capital, then *persistence* prediction would predict that the annual maintenance costs in the future will be 4 percent of the invested capital. *Trajectory* prediction assumes that the rate of change remains constant. If, for example, the average of your grades at the end of the first semester of chemical engineering studies was 70 percent, at the end of the second semester the average was 75 percent, and at the end of the third semester 80 percent, trajectory prediction would forecast an average for you of 85 percent at the end of the fourth semester. Obviously, long-term predictions by this scheme can be risky. If the analysis of past data indicates that some event shows a cyclic pattern of change with time, *cyclic* prediction assumes that this cyclic pattern will continue into the future.

Associative prediction is based on causality where logic or experience shows that a given event seems to follow or be associated with some other event. Economic analysts who predict business and economic conditions by following the behaviors of *leading indicators* are using associative prediction. An increase in the number of building starts should be followed after a reasonable time by an increase in the purchase of durable consumer goods such as refrigerators and washing machines, as an example.

Analog prediction is the common engineering and scientific prediction technique. Behavior is predicted from a model (usually mathematical in the case of economic factors) that describes the behavior of the system and extracts from it the predicted values for the elements of interest. The prediction will be good *if* the model represents the real situation in all relevant and significant aspects.

An additional scheme listed by Siddall (1972) and not to be overlooked is *intuitive* prediction, prediction based on experience, intuition, judgment, and knowledge of a field.

In the paragraphs to follow it will be apparent that each specific forecasting technique discussed is a prediction scheme included in the above listing even though the name attached to the specific technique may not be the same as that of its parent.

8-2 MARKET FORECASTING

The economic environment in which a chemical plant operates is a dynamic and not a static environment, and it undergoes continuous change. During the life of the plant the demand for its product will change, as will all the factors that determine its profitability, e.g., labor, raw materials, and utilities costs. Many of these factors can be included in a complete economic evaluation of a proposed plant, but one of the most important pieces of information required for design calculations and economic evaluations is the estimate of market demand for the product. An error of 15 to 20 percent in the estimate of the size or cost of a heat exchanger or of a distillation column would have a relatively small effect on the results of an economic evaluation and could easily be corrected in practice. The same percentage error in a market forecast could have a profound effect on the profitability of a proposed venture.

A chemical product will generally proceed through several stages during its life. The period of laboratory research and development can be compared to the period of gestation and infancy. During its childhood the growth in demand is usually at a high rate. Demand then grows but at a diminishing rate, as in adolescence, and as the product reaches maturity, the demand continues to grow but at a low, relatively constant rate. In some cases this growth rate decays into old age and death, as has happened to some chemical products, e.g., celluloid.

We have already seen in Chapter One that the products of the chemical industry can be roughly classified into three categories: long-established products whose growth rate is approximately equal to the growth rate of the total industrial production, newer products which have relatively recently moved into the bulk commodity class and whose demand rate exceeds that of the total industrial production, and, finally, new products exhibiting a high growth in demand as they enter into new applications and markets.

The methods used to forecast market demands must consider these factors. A forecasting procedure that may give satisfactory results for a long-established chemical can be woefully inadequate to deal with an innovative product. The methods used to forecast the market for a chemical product can be grouped into two categories, qualitative and quantitative. The qualitative, or subjective, methods are those which rely primarily on the judgment and experience of individuals, whereas the quantitative methods are based on numerical data, usually historical in nature.

8-3 QUALITATIVE METHODS

The qualitative methods are, to a great extent, intuitive predictions and are based on opinions gathered by personal interview, by the market-research interviewers, or by the use of questionnaires. The method involves selecting persons who are considered to be experts in the specific field and obtaining their forecasts of market growth and demand. The results obtained must be analyzed and the conflicting opinions weighed and judged in order to arrive at a consensus opinion. This technique, the sample-of-industry opinion, is a valuable one although a great deal depends on the judgement and skill of the market-research interviewer.

Another technique that can be used within the company obtains the consensus of management opinion. The various management groups are supplied with as much basic information as possible on markets, trends, competition, etc., and at a management conference all participants are expected to express their reasoned personal opinions and forecasts of the future. Although this method is simple and relatively cheap, it has limitations. It is highly subjective, as each member of the management group would have different factors to consider and there is no way of evaluating and weighting individual opinions. Although group action of this type can result in more factors and points of view being considered, it also suffers from the disadvantages of committee action.

Delphi Procedure

The Delphi procedure was designed to retain the advantages of a committee or panel of experts while minimizing the disadvantages. Rather than having face-to-face committee encounters, the group interaction is based on questionnaires presented to each member of the committee, who remains anonymous and unknown to the other members of the panel.

The technique as used to forecast the growth in sales of a given product would be somewhat as follows. The first questionnaire submitted to the "oracles" would contain questions like: By what years would you estimate that sales would reach certain given levels? The Delphi director, who receives these estimates, would present them to the panel in the second-round questionnaire in the form of median and upper and lower quartile results. Each panel member would then be asked to make new estimates, taking into account the estimates of the other members, and also to explain why he thinks other members of the panel are erring in their forecasts. These second-round results would be incorporated into the third-round questionnaire, which would now include the arguments and discussions of why the median and upper- and lower-quartile predictions are either too optimistic or pessimistic. The panel members are then asked to revise their forecasts if they wish to revise them and to present additional arguments or counterarguments.

The procedure would be terminated when there no longer is any significant shift in opinion, usually by the end of the third or fourth round. The median forecast then becomes the predicted forecast, and the uncertainty involved in the forecast is indicated by the range between the upper- and lower-quartile forecasts. The comments presented by the panel members in support of their forecasts and their arguments concerning other panelists' estimates can also help to identify factors that should be considered in market and product development.

Chinn and Cuddy (1971) have used the Delphi method, even though it was developed for long-range forecasting, for short-range evaluations. They found it to be an excellent tool for gaining consensus in a management group when diverging opinions exist.

8-4 QUANTITATIVE METHODS

Trend curves

In this method, historical production or demand data are plotted against time in such a way that a trend curve can be fitted to the data and then extrapolated into the future. The basic forecasting assumption is persistence or trajectory prediction; a growth pattern that has been established in the past is quite likely to continue into the future. Demand data for as long a period as possible should be obtained so that the appropriate trend curve can be fitted. The longer a particular form of the growth or trend curve has existed, the more likely it is that this growth trend will continue and represent the probable future development.

Trend curves have been used for a considerable time and for various purposes, and some of the more commonly used ones will be described. When the data for demand are plotted against time, it is possible to draw a smooth curve through the points and then to extrapolate to later years. Attempting to fit a mathematical equation to the data is preferable to using a free-hand method to smooth the data. The three types of commonly used equations are polynomials, exponentials, and modified exponentials (Gregg et al., 1964).

Polynomials Polynomial trend curves

$$D = a + bt + ct^2 \tag{8-1}$$

represent one of the possibilities for fitting trend data. If the coefficient c is equal to zero, the data for demand against time would plot as a straight line, indicating that demand is increasing by a constant amount each year. This is a highly unlikely situation for a chemical product. If c does not equal zero, the demand data would be represented by a parabola. If c has a positive value, this means that the demand would eventually skyrocket without limit, again a highly unlikely situation for a chemical product. If, on the other hand, c has a negative value, this means that there is a ceiling value for the demand which will then decline without limit. Such a situation should be viewed with suspicion unless there are market, legal, or technical factors that would support the trend-curve prediction that demand will decrease.

Exponentials Exponential trend equations are of the form

$$\log D = a + bt + ct^2 \tag{8-2}$$

The equation is a simple exponential or compound-interest trend if the constant c is zero and the demand increases by a constant proportion each year. This type of trend curve might well represent the growth of demand in a chemical product that is in the rapid-growth stage. The logarithmic parabolic curve obtained with a nonzero c is subject to the same comments made earlier for the parabolic polynomial trend curve.

If D or log D plots as a straight line vs. time, the coefficient c is equal to zero and the coefficients a and b can easily be evaluated by a least-squares linear regression. In the case of the polynomial, which will be a linear trend equation for this case,

$$a = \frac{\sum D \sum t^2 - \sum t \sum Dt}{n \sum t^2 - (\sum t)^2} \qquad b = \frac{n \sum Dt - \sum D \sum t}{n \sum t^2 - (\sum t)^2} \qquad (8\text{-}3)$$

where n is the number of years. The expression for a and b in the linear exponential trend equation will be identical if D is replaced by log D.

Modified exponentials The three commonly used modified-exponential equations are the simple modified exponential

$$D = a - bc^t \qquad (8\text{-}4)$$

logistic

$$D = \frac{1}{a + bc^t} \qquad (8\text{-}5)$$

and the Gompertz

$$\log D = a - bc^t \qquad (8\text{-}6)$$

In these three equations a, b, and c are constants, and $c < 1$. All three equations imply that the demand has a ceiling value that is approached asymptotically. Demand curves represented by the modified-exponential equations are shown in Fig. 8-1.

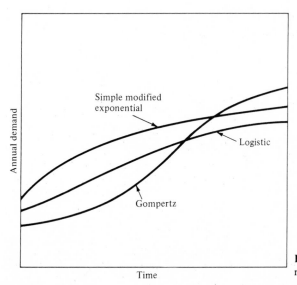

Figure 8-1 Modified-exponential demand curves.

In practice, the typical growth of demand for a chemical product, as outlined earlier, can be divided into three major phases of growth, i.e., the introductory consumption with a relatively low rate of growth in demand, an intermediate period of rapid growth in demand, and the final approach to maturity. When plotted as demand vs. time on arithmetic coordinates, this pattern yields an S-shaped curve. An examination of Fig. 8-1 shows that only the Gompertz equation yields such an S-shaped curve. As a result, the Gompertz equation is the trend equation that is generally most useful for extrapolation purposes.

The values for the three constants a, b and c in the Gompertz equation can be obtained from three values based on average demands over the time period for which data are available. The method involves dividing the period into three equal subperiods. If the number of time periods is not exactly divisible by 3, the first 1 or 2 years at the beginning of the period can be ignored. The average of the logarithm of the demands is then calculated for each period, where n is the number of years in the total period.

$$(\log D)_\text{I} = \frac{3}{n} \sum_{i=1}^{n/3} \log D_i \tag{8-7a}$$

$$(\log D)_\text{II} = \frac{3}{n} \sum_{i=(n/3)+1}^{2n/3} \log D_i \tag{8-7b}$$

$$(\log D)_\text{III} = \frac{3}{n} \sum_{i=(2n/3)+1}^{n} \log D_i \tag{8-7c}$$

The constants a, b, and c can then be calculated from

$$c^{n/3} = \frac{(\log D)_\text{III} - (\log D)_\text{II}}{(\log D)_\text{II} - (\log D)_\text{I}} \tag{8-8a}$$

$$a = \frac{(\log D)_\text{I}(\log D)_\text{III} - (\log D)_\text{II}^2}{(\log D)_\text{I} + (\log D)_\text{III} - 2(\log D)_\text{II}} \tag{8-8b}$$

$$b = \frac{n}{3} \frac{[(\log D)_\text{I} - (\log D)_\text{II}]^2}{(c + c^2 + \cdots + c^{n/3})[2(\log D)_\text{II} - (\log D)_\text{I} - (\log D)_\text{III}]} \tag{8-8c}$$

The use of these equations can be best illustrated by an example.

Example 8-1 Synthetic detergents were introduced during the 1940s as an almost direct substitute for soap. Since their introduction detergents have been taking over an increasing fraction of the market originally held by soap. Historical data, presented in Table 8-1, are available for soap and detergent sales. These data are to be examined for trend characteristics.

DISCUSSION AND SOLUTION The fraction of the total market for soap plus synthetic detergent that can be captured or taken over by the detergent cannot exceed unity. The data will therefore be examined first to see whether

Table 8-1 Annual soap and synthetic detergent sales in billions of pounds

Year	Soap	Detergent	Year	Soap	Detergent
1951	2.43	1.59	1961	1.25	3.85
1952	2.18	1.73	1962	1.15	3.86
1953	1.97	2.15	1963	1.08	3.90
1954	1.67	2.30	1964	0.99	4.10
1955	1.56	2.64	1965	0.97	4.16
1956	1.51	3.08	1966	0.97	4.30
1957	1.48	3.29	1967	0.97	4.58
1958	1.30	3.59	1968	0.98	4.79
1959	1.24	3.64	1969	0.94	5.01
1960	1.25	3.75	1970	0.93	5.19

Source: Facts and Figures for the Chemical Process Industries, *Chemical and Engineering News*, various years.

there appears to be a recognizable trend for the market fraction captured by synthetic detergents.

The fraction of market captured by the synthetic detergents is shown in Table 8-2 and plotted in Fig. 8-2. The shape of the curve indicates that either a simple modified-exponential trend curve or the Gompertz equation might fit the data. The coefficients will be estimated by the procedure outlined above. The total period is 20 years, of which the first 18 will be used for estimating the coefficients of the two trend-curve equations. These equations

Table 8-2 Fraction of detergent plus soap market due to detergent

Year	$D + S$, 10^9 lb	Detergent market fraction f	Year	$D + S$, 10^9 lb	Detergent market fraction f
1951	4.02	0.396	1961	5.10	0.755
1952	3.91	0.442	1962	5.01	0.770
1953	4.12	0.522	1963	4.98	0.783
1954	3.97	0.579	1964	5.09	0.806
1955	4.20	0.629	1965	5.13	0.811
1956	4.59	0.671	1966	5.27	0.816
1957	4.77	0.690	1967	5.55	0.825
1958	4.89	0.734	1968	5.77	0.830
1959	4.88	0.746	1969	5.95	0.842
1960	5.00	0.750	1970	6.12	0.848

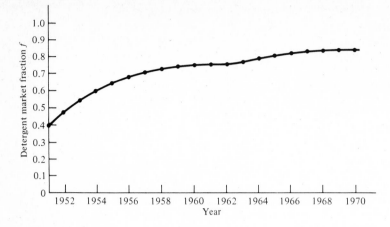

Figure 8-2 Fraction of soap plus detergent market due to detergent.

will then be used to forecast the market fraction due to detergent that would be expected for the last 2 years for which data are available, 1969 and 1970.

Year	Demand year (t)	Market fraction (f)	$\log f$
1951	1	0.396	−0.402
1952	2	0.442	−0.355
1953	3	0.522	−0.282
1954	4	0.579	−0.237
1955	5	0.629	−0.201
1956	6	0.671	−0.173
$f_{\text{I}} = 0.5398$		$(\log f)_{\text{I}} = -0.275$	
1957	7	0.690	−0.161
1958	8	0.734	−0.134
1959	9	0.746	−0.127
1960	10	0.750	−0.125
1961	11	0.755	−0.122
1962	12	0.770	−0.114
$f_{\text{II}} = 0.7408$		$(\log f)_{\text{II}} = -0.1305$	
1963	13	0.783	−0.106
1964	14	0.806	−0.094
1965	15	0.811	−0.091
1966	16	0.816	−0.088
1967	17	0.825	−0.0835
1968	18	0.830	−0.081
$f_{\text{III}} = 0.8118$		$(\log f)_{\text{III}} = -0.0906$	

The coefficients can now be calculated.

Modified exponential

$$c^{18/3} = \frac{f_{III} - f_{II}}{f_{II} - f_{I}} = \frac{0.8118 - 0.7408}{0.7408 - 0.5398} = 0.3532$$

$$c = 0.8408$$

$$a = \frac{f_I f_{III} - f_{II}^2}{f_I + f_{III} - 2f_{II}} = \frac{(0.5398)(0.8118) - 0.7408^2}{(0.5398) + (0.8118) - (2)(0.7408)} = 0.8506$$

$$b = \frac{n}{3} \frac{1}{c + c^2 + \cdots + c^{n/3}} \frac{(f_I - f_{II})^2}{2f_{II} - f_I - f_{III}}$$

$$= \frac{6}{0.8408 + 0.8408^2 + \cdots + 0.8408^6} \frac{(0.5398 - 0.7408)^2}{(2)(0.7408) - 0.5398 - 0.8118}$$

$$= 0.546$$

Therefore, according to the modified-exponential expression, the fraction of market captured by detergent will be

$$f_e = 0.8506 - (0.546)(0.8408)^t$$

Gompertz equation

$$c^{18/3} = \frac{(\log f)_{III} - (\log f)_{II}}{(\log f)_{II} - (\log f)_{I}} = \frac{-0.0906 - (-0.1305)}{-0.1305 - (-0.275)} = 0.276$$

$$c = 0.807$$

$$a = \frac{(\log f)_I (\log f)_{III} - (\log f)_{II}^2}{(\log f)_I + (\log f)_{III} - 2(\log f)_{II}}$$

$$= \frac{(-0.275)(-0.0906) - (-0.1305)^2}{-0.275 + (-0.0906) - (2)(-0.1305)}$$

$$a = -0.0754$$

$$b = \frac{n}{3} \frac{1}{c + c^2 + \cdots + c^{n/3}} \frac{[(\log f)_I - (\log f)_{II}]^2}{2(\log f)_{II} - (\log f)_{III} - (\log f)_I}$$

$$= \frac{6}{0.807 + 0.807^2 + \cdots + 0.807^6}$$

$$\times \frac{[-0.275 - (-0.1305)]^2}{(2)(-0.1305) - (-0.0906) - (-0.275)}$$

$$b = 0.396$$

Therefore, the Gompertz representation would be

$$\log f_G = -0.0754 - 0.396(0.807)^t$$

These trend curves can now be extrapolated in order to predict the future behavior of the soap-detergent market competition.

First, the fraction of market due to detergent will be calculated for 1969 and 1970 based on the historical pattern between 1951 and 1968. The predictions of the modified-exponential trend curve will be for 1969, the nineteenth demand year

$$f_e = 0.8506 - (0.546)(0.8408^{19}) = 0.830$$

and for 1970

$$f_e = 0.8506 - (0.546)(0.8408^{20}) = 0.834$$

compared with the actual fractional market captured of 0.842 and 0.848, respectively.

The Gompertz would predict for 1969

$$\log f_G = -0.0754 - (0.396)(0.807^{19}) \quad \text{and} \quad f_G = 0.828$$

and, for 1970

$$\log f_G = -0.0754 - (0.396)(0.807^{20}) \quad \text{and} \quad f_G = 0.830$$

Both trend curves predict performance slightly poorer than actually occurred. An examination of Fig. 8-2 shows that both the trend curves predict performance that would have been expected had the trend established from 1964 to 1968 continued.

Next, the maximum fractional market penetration to be expected for detergents will be examined by calculating the asymptotic limit of both of the trend curves

$$f_e = 0.8506 - (0.546)(0.8408^{\infty}) = 0.8506$$

$$\log f_G = -0.0754 - (0.396)(0.807^{\infty}) = -0.0754$$

$$f_G = 0.841$$

Both trend curves indicate that the present market penetration of detergents is very close to the maximum possible. Substantial increases in detergent sales can be expected, therefore, only if the total market for soaps and detergents increases or if detergents are developed that permit their penetration into new markets.

8-5 GROWTH OF THE GENERAL ECONOMY

The use of a modified-exponential curve, e.g., the Gompertz equation, implies that a demand ceiling exists. Even for long-established chemicals, however, the demand normally continues to increase because of the growth in the general economy. Thus, the long-term growth in the general economy is reflected as a continuing increase in the demand ceiling. The growth should be equal to the growth rate of the economy as a whole. When a trend curve indicates that a market is near its ceiling or saturation point and qualitative and judgment considerations indicate an expansion because of the general growth of the economy, it is simple to correct the demand figures to reflect the expansion in the ceiling market. If the demand figures are $D_1, D_2, \ldots, D_n$ and the general economy is growing at g percent annually, the corrected demand figures that reflect this growth in the general economy are

$$D_{1c} = \frac{D_1}{1 + g/100}$$

$$D_{2c} = \frac{D_2}{(1 + g/100)^2}$$

$$\ldots \ldots \ldots \ldots \ldots$$

$$D_{nc} = \frac{D_n}{(1 + g/100)^n}$$

These data should be fitted to the trend curve. If, for example, the Gompertz equation applies to the curve

$$\log D_{tc} = a - bc^t$$

the actual demand would be

$$D_t = D_{tc}\left(1 + \frac{g}{100}\right)^t$$

8-6 CORRELATION ANALYSIS

In this forecasting procedure an associative prediction is attempted by linking the historical consumption pattern to one or more of the usual business indicators, e.g., the gross national product or the industrial production index. If such a link is possible, and it is most likely in the case of a long-established chemical product, forecasting can be reasonably reliable. This is because attempts are made by government not only to forecast these business indicators quite accurately but also to influence them by fiscal and economic policy. Van Arnum (1964) presented a correlation for phenol production in which the production data were linked to an adjusted Federal Reserve Board Index of Industrial Production. His 4-year forecast turned out to be remarkably accurate, and the predicted phenol market was within 1 percent of the actual sales. The 6-year forecast, however, was 10 percent higher than the actual market.

Thus, although this method could be a reliable forecaster when an appropriate correlation can be found, it would prove to be woefully inadequate if applied to a new chemical product.

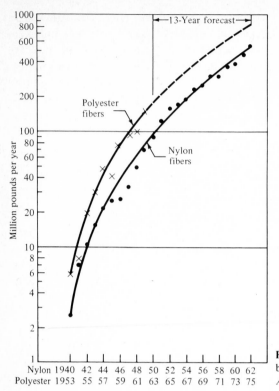

Figure 8-3 Forecasting polyester market by analogy with nylon. [*From Van Arnum (1964), by permission.*]

8-7 HISTORICAL ANALOGY

One of the tools that can be used in the case of a relatively new product is to relate the growth of the new product to that of a similar product when it was at a comparable stage in its market development. The market growth for polyester fibers was forecast (Fig. 8-3) by relating it to the market growth for nylon fibers at a similar period. Polypropylene development was predicted by Van Arnum with commendable accuracy by comparison with the growth of the linear polyethylene market. This technique is a risky one and should only be used along with other qualitative as well as quantitative procedures, especially in view of the fact that it would usually be resorted to in the case of new products which have only a short market history.

8-8 CONSUMPTION PATTERNS

Any chemical product would have a finite number of consumers, usually other chemical companies, as well as a finite, although possibly large, number of end uses. Forecasting by the technique of studying consumption patterns involves forecasting the future consumption for each consumer as well as the growth in the various end-use requirements. These individual forecasts are then summed up to give the forecast demand for the product. Errors will occur in each individual forecast, of course, but they have a tendency to cancel each other out. This method can give gross errors, however, if any major use or demand appeared or disappeared during the historical period under study or if a major substitute or competitor product appeared or disappeared. If such gross changes did occur, they should be examined separately and in great detail before they are included in the final overall forecast.

8-9 MARKET SIMULATION

Market simulation, or analog prediction, involves the utilization of all the quantitative and qualitative techniques discussed here. A simulation model is built that attempts to approximate the entire market condition, not only the consumption patterns described above. The market is viewed as a system in which all the inputs, outputs, interrelationships, and feedback mechanisms are considered and analyzed. Market simulation uses the qualitative techniques that rely on judgment, experience, and insight as well as the quantitative techniques that in essence are primarily extrapolation techniques.

A market-simulation model would include several additional important factors that have an important bearing on demand and production. One such factor is the relationship between price and demand as well as the relationship between price and supply or production. A complete model would also include forecasts of industrial capacity for at least several years in advance and the effects of increased production capacity on price as well as supply.

A market-simulation model permits the effect of various factors on future market development to be studied. Such a model can be used to examine the effect of a whole series of assumptions to be investigated for their effect on the future. Questions like the effect of a possible price change, of an increase in capacity by a competitor, or of a new competitive or alternative product can be answered easily and quickly. It must be borne in mind, however, that the answers to these questions and the accuracy of forecasts by any simulation model are no better than the accuracy and reliability of the data and assumptions fed into the model.

8-10 A WORD OF CAUTION

Although market forecasting is an essential element in any commercial or manufacturing operation, it involves making assumptions about future events. Long-term forecasting is, at best, a hazardous operation, and the uncertainties are such that particular caution should be observed, especially if attempts are made to predict beyond 5 years. As many different forecasting approaches as possible should be adopted and the results compared.

The quantitative methods described can serve as a guide to the forecaster. In some cases, however, several different trend equations could smooth the data equally well, but they do not necessarily represent the true rate of growth of the market. Extrapolation of a trend should not result in absurd or unlikely results that cannot be supported by a general qualitative assessment of the market.

If the market forecaster is fortunate, the demand predicted will be close to the true demand. An error on the high side may lead to excessive plant capacity and unnecessarily high capital costs and overhead expenses. An underestimate of demand may result in difficulties in meeting the market requirements and excessively high unit manufacturing costs as operating personnel attempt to push production beyond design capacity.

8-11 FORECASTING PLANT CAPACITY

Having succeeded in forecasting demand, we now are faced with an additional problem: What should be the initial capacity and the size and timing of later increments of capacity in the face of an increasing demand? The problem arises because of the balance that can be struck between the economy of scale and a delay in capital expenditure.

The economy of scale results from the fact that the installed cost of equipment can usually be expressed by $C = C'D^m$. The scale-up factor m is specific to the type of equipment or to the process. It ranges between 0.2 and 1.0 for items of equipment and between 0.5 and 0.8, in general, for process plants, the usual value being 0.6 to 0.7. If the scale-up factor were zero (cost independent of capacity), it would always pay to install capacity now that would be sufficient to supply the ultimate demand. If the value were unity (capacity obtained, for example, by multiple units), there is no incentive to install any overcapacity now. An additional unit would be installed each time installed capacity was insufficient to meet demand. Were capital available free of charge and in unlimited quantity, it is clear that one would always install capacity now to meet the ultimate demand irrespective of the value of the scale-up factor.

Coleman and York (1964) considered the case where the demand curve can be represented by an exponential growth curve with a limiting, constant demand. We shall deal with the slightly simpler case of a linear demand curve that reaches an ultimate constant demand.

Look at the following situation. Our company has decided to enter the market to supply an existing chemical. The plan is to begin marketing D_0 tons/yr and to increase our sales linearly with time until we are selling D_u tons/yr. The demand schedule to supply is

$$D = \begin{cases} D_0 + at & \text{for } t \le D_u - D_0/a \\ D_u & \text{for } t > D_u - D_0/a \end{cases}$$

The demand must be met at all times. What should the capacity of the initial and subsequent plants be? In devising our model for this situation we shall neglect taxes. We shall assume that a production unit has a useful life of L years and that the salvage value is equal to the cost of dismantling and removal. In addition, we shall assume that the service will be supplied in perpetuity. This will permit units to be added at a time close to the end of the project, thus permitting a reasonable model for an optimal solution.

This is not an unrealistic situation insofar as present-worth considerations are concerned because at usual rates of return the present value of an expenditure 15 or 20 years in the future is relatively low. This can be easily demonstrated. If, as we saw in Example 7-1, you wish to receive a 12 percent return on your money, you should be willing to pay $713.18 now to receive $100 per year for 20 years. You should be willing to pay $784.33, that is, only an additional $71.15, for you and your heirs to receive $100 per year in perpetuity.

The installed cost of the plant is $C_I = Ck_i^m$, where k_i is the capacity of the ith addition. The capitalized cost to supply this capacity in perpetuity will be

$$K = C_I + \frac{C_I}{e^{iL} - 1} = C_I \left[\frac{e^{iL}}{e^{iL} - 1} \right]$$

Therefore, the capitalized installed cost can be expressed as $C'k_i^m$, where C' is the product of C and the bracketed term in the above equation.

The present value of the complete capacity-supply strategy will be

$$\text{PV} = C'(k_0^m + k_1^m e^{-it_1} + \cdots + k_j e^{-it_j} + \cdots + k_n^m e^{-it_n}) \qquad (8\text{-}9)$$

where n is the number of capacity additions, and where k_1 is needed at time $t = k_0 - D_0/a$ and k_j is needed at $t_j = \bar{k}_j/a$. Where

$$\bar{k}_j = \sum_{j=0}^{j-1} (k_j) - D_0$$

the final capacity addition will be

$$k_n = D_u - \sum_{j=0}^{n-1} k_j$$

Substituting these terms, we get

$$\text{PV} = C(k_0^m + k_1^m e^{-i\bar{k}_1/a} + \cdots + k_j^m e^{-i\bar{k}_j/a} + \cdots + k_n^m e^{-i\bar{k}_n/a}) \qquad (8\text{-}10)$$

To minimize the present value we set the partial derivatives of this equation with respect to each of the k_j's equal to zero and solve the resulting equations simultaneously. The solutions are:

For $n > 1$:
$$\left(\frac{k_0}{k_1} \right)^{m-1} = e^{-(i/a)(k_0 - D_0)} \left(\frac{ik_1}{am} + 1 \right) \qquad (8\text{-}11a)$$

$$\left(\frac{k_j}{k_{j+1}} \right)^{m-1} = e^{-(i/a)k_j} \left(\frac{ik_{j+1}}{am} + 1 \right) \qquad (8\text{-}11b)$$

$$\left(\frac{k_{n-1}}{k_n} \right)^{m-1} = e^{-(i/a)k_{n-1}} \left(\frac{ik_n}{am} + 1 \right) \qquad (8\text{-}11c)$$

For $n = 1$:
$$\left(\frac{k_0}{D_u - k_0} \right)^{m-1} = e^{-(i/a)(k_0 - D_0)} \left[\frac{i}{am} (D_u - k_0) + 1 \right] \qquad (8\text{-}11d)$$

By solving the equations for the optimum k_j's we have obtained the lowest present value for a given number of additions n of production capacity. The optimum number n must be determined by calculating the present worth for $n = 0, 1, 2, \ldots$ with the optimum values for k_j for each n. The optimum value of n, and hence the optimum strategy for capacity addition, will yield the lowest present worth.

Example 8-2 Your company is building a new rural plant and intends to transfer a number of its processes to this site. At the beginning of operations it is expected that the boiler plant will need to supply 30 tons/h of steam. Steam demand will increase at the rate of 6 tons/(h · yr), and the ultimate demand is not expected to exceed 100 tons/h. What would you recommend as the appropriate strategy for provision of this steam demand? The installed cost of steam capacity varies with the 0.7 power of the capacity. Use 15 percent as the interest rate.

SOLUTION The present value will be calculated for several strategies.

Strategy 1 Provide the ultimate demand now

$$\frac{PV}{C'} = 100^{0.7} = 25.12$$

Strategy 2 Provide part of the demand now and the remainder with one addition of capacity. From Eq. (8-11d) the optimum strategy for $n = 1$ is

$$\left(\frac{k_0}{100 - k_0}\right)^{0.7-1} = \exp\left[-\frac{0.15}{6}(k_0 - 30)\right]\left[\frac{0.15}{(6)(0.7)}(100 - k_0) + 1\right]$$

Solving this for k_0 yields 69.4 tons/h for the initial installation. A second installation of 30.6 tons/h would be added after 6.57 years

$$\frac{PV}{C'} = (69.4^{0.7} + 30.6^{0.7}e^{-(0.15)(6.57)}) = 23.54$$

At this point in the analysis strategy 2 is preferable to strategy 1.

Strategy 3 Provide two additions subsequent to the initial provision of steam capacity, $n = 2$ from Eqs. (8-11a) and (8-11b)

$$\left(\frac{k_0}{k_1}\right)^{0.7-1} = \exp\left[-\frac{0.15}{6}(k_0 - 30)\right]\left[\frac{0.15k_1}{(0.7)(6)} + 1\right]$$

$$\left(\frac{k_1}{100 - k_1 - k_0}\right)^{0.7-1} = \exp\left(-\frac{0.15}{6}k_1\right)\left[\frac{0.15}{(0.7)(6)}(100 - k_0 - k_1) + 1\right]$$

These are to be solved simultaneously for k_0 and k_1. The results are $k_0 = 65.9$ and $k_1 = 20.0$ tons/h, and, to meet the ultimate demand, $k_2 = 14.1$ tons/h.

$$\frac{PV}{C'} = (65.9^{0.7} + 20^{0.7}e^{(-0.15)(5.98)} + 14.1^{0.7}e^{(-0.15)(9.32)}) = 23.65$$

We observe that strategy 3 is inferior to strategy 2 although the difference in present value is extremely small. The preferred strategy, therefore, would be to install a boiler with a capacity of 69.4 tons/h now to be followed by an additional boiler of 31.6 tons/h capacity after 6.57 years. The recommendation would, of course, be rounded off to an initial installation of 70 tons/h.

The primary question, of course, is what capacity to install now. In view of uncertainties in forecasts the question of later capacities and timing is almost a trivial one. When an addition of capacity is being considered, management would want to base it on new and updated forecasts.

8-12 CHEMICAL-PRICE FORECASTING

On a long-term basis substantial increases in production rate will generally be accompanied by lower prices. This is especially true in a dynamic, growing industry such as the chemical industry.

Stobaugh (1964) studied the price–production-volume relationships for 16 organic and petrochemicals over a 10-year period. For these particular chemicals he concluded that a doubling in production rate was accompanied by a price drop of 30 to 40 percent. The average of all petrochemical prices had dropped by almost 25 percent over a 6-year period while the production rate had almost doubled.

Three major factors are involved and result in a decline in the selling price, namely, increases in the scale of production, improvements in technology, and changes in raw-material costs. As the market grows, the size of each new plant is increased. As a result of the scale factor, these larger plants will have lower overhead and capital costs per unit of production. Improvements in technology take place and result in more efficient plants; hence they reduce unit production and capital costs even further. Production increases in turn will result in increased demand for raw materials, and they, too, will drop in price as larger plants with improved technology are built to supply the increased raw-material demand.

Twaddle and Malloy (1966) quantified price projections by modeling production, investment, and cost elements on a time basis. They base their projections on the reasonable assumption that prices will continue to fall as long as some producer can earn a return above his cost of capital by building a modern, efficient plant. This would set a floor under the price, and this floor price would be that which is required for a new, efficient producer to earn a return that is just sufficient to cover his cost of capital. In their model the selling price is projected as the sum of a price floor and a margin over the price floor. The margin would tend to decay at a rate proportional to the margin and would thus follow an exponential decay law. The price floor would also be expected to decrease according to an exponential decay law as a result of technological improvements. Reduction in the value of the price floor would become increasingly difficult, however, as further technological improvements become progressively harder to achieve. In accordance with these assumptions, a reasonable model for projecting the change in selling price with time would be

$$\text{Selling price} = P_M e^{-k_M t} + P_F e^{-k_F t} \tag{8-12}$$

The first term represents the decaying margin and the second represents the decaying floor.

The price-floor curve is based on cost estimates for a modern plant both at the beginning and at the end of the projected project life with a rate of return just sufficient to cover the cost of capital. These base plants must, of course, assume the largest and most efficient plants likely to be built. Price-floor decay rates in the range of 1 to 4 percent per year appear to be typical.

The margin over the price floor can decay at a much larger rate, and decay rates of 15 to 18 percent per year would not be unusual for a chemical enjoying a rapid growth rate. Sufficient price and cost data would normally be available on either the chemical in question or on related chemicals to permit a reasonable estimate of margin decay rate.

8-13 PROJECTING PRODUCTION RATES

The Twaddle-Malloy model also takes into account the fact that production rates can be expected to increase in a process unit, rapidly at first and then at a declining rate. Hirschman (1964) showed that the production rate can be expected to follow a learning curve

$$\frac{Q}{Q_d} = 1 + \left(\frac{Q_\infty}{Q_d} - 1\right)(1 - e^{-k_L t}) \tag{8-13}$$

The learning curve shows that the production rate will increase from the design capacity Q_d to some ultimate capacity Q_∞. Values of $Q_\infty/Q_d = 2$ and $k_L = 0.1$ characterize learning in catalytic cracking units as well as several other process units.

This increase in production rate results primarily from the progressive removal of bottlenecks and from the increasing skill and competence of the operating personnel in operating the equipment at the most efficient conditions. Safety margins are normally included in the design of critical equipment to ensure that the process unit will operate according to the design capacity. This suggests that the expected performance of these critical items should be higher than the design capacity. Equipment that was not considered to be critical in design or for which design criteria are well established would be designed without safety margins and hence would represent the limiting factors in exceeding the design target. Removal of these bottlenecks can result in marked improvement in production capacity.

Increases in production rate from learning requires additional investment for debottlenecking, and this additional investment must be allowed for. Information about the cost of debottlenecking can be determined from information on past debottlenecking experience.

8-14 PROJECTING COSTS

Variable costs tend to decline over the life of a process unit as a result of several factors. Raw-material costs will usually decline for reasons mentioned earlier. Yields will tend to increase and requirements for utilities, catalysts, and related items will tend to decrease as a result of learning. The reduction in variable costs can be expected to follow an exponential decay law, and the rate of decay can be obtained by estimating the variable costs at the beginning and end of the project.

Labor, maintenance, plant overheads, selling, and research and administrative expenses can be expected to grow at a constant percentage rate. These rates can be estimated from various statistical sources, e.g., Labor Department figures for wage rate increases.

Working-capital requirements will also vary during the life of the plant. Realistic allowances can be made for the variation in working-capital requirements by breaking it down into its basic components and allowing for the variation in values of these components with time.

8-15 PROJECTING DEMAND

An important element of the Twaddle-Malloy model is the projected market demand. They assume, in parallel with the other elements of the model, a growth rate that decays exponentially from the current level to some long-run equilibrium rate. The equilibrium rate can be the expected growth rate for the national economy. The growth rate

$$g = g_\infty + (g_0 - g_\infty)e^{-k_g t} \tag{8-14}$$

will result in a projected demand of

$$D = D_0 \exp\left[g_\infty t + \frac{(g_0 - g_\infty)(1 - e^{-k_g t})}{k_g}\right] \tag{8-15}$$

The demand for the individual firm is based on the assumption that its market share begins at f_0 and increases gradually to a long-run equilibrium fraction f, in accordance with

$$f = f_0 + (f_\infty - f_0)(1 - e^{-k_f t}) \tag{8-16}$$

8-16 TWADDLE-MALLOY ILLUSTRATIVE EXAMPLE

Twaddle and Malloy exercised their model on a hypothetical though typical situation. They assumed a monomer to be produced for a rapidly growing plastic. The market was assumed to be growing at 30 percent per year initially with the growth rate expected to drop to 15 percent per year in 4 years and to line out at 4 percent per year. The company expected to enter the market with a 4 percent share and to expand to 10 percent ultimately. They present in their paper all the other relevant information to estimate production costs, price floor, decay rates, etc.

The results of this study with respect to the expected net present worth as a function of initial plant size are shown in Fig. 8-4. The upper curve is the net present worth that would be expected in a static economic situation (no changes in production costs, selling prices, production rates, etc.). The optimum size plant to build under the assumption of an unchanging economic environment would have a capacity of 220×10^6 lb/yr and would return a profit of $60 million after repaying the cost of capital at 6 percent. The DCF rate of return at this optimum size would be 17.2 percent. A reasonably optimistic picture is obtained.

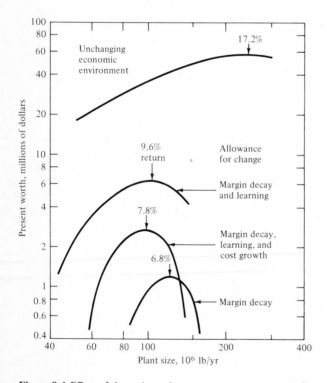

Figure 8-4 Effect of dynamic environment on present worth. [*From Twaddle and Malloy (1966), by permission.*]

The picture changes as dynamic effects are included in the evaluation. If only margin decay is permitted, i.e., selling-price decay, the optimum plant size would be 120×10^6 lb/yr, the NPW would be $1.3 million, and the DCF rate of return drops to 6.8 percent. The most realistic evaluation allows all the relevant dynamic factors to change—price, cost, and production capacity. The results now show that an NPW of $2.5 million would be expected or a DCF rate of return of 7.8 percent for a plant with the optimum capacity of 97×10^6 lb/yr.

The importance of considering the dynamic nature of the economic environment is emphasized if we consider what the result would have been if the optimum plant based on static considerations had been built whereas prices and costs changed as projected. If the 220×10^6 lb/yr plant had been built and the economic environment had developed as projected, the NPW would have been $-$3.8 million and the DCF rate of return would have been 4.4 percent, results that are not at all optimistic. Had this situation actually occurred, it is most likely that management would have attempted to force changes in some of the parameters used in the economic evaluation, e.g., to increase the company's share of the market by aggressive selling, etc.

It is interesting to examine the predictions of the Twaddle-Malloy model for selling price in the face of increasing production with the results observed by Stobaugh (1964) in his study of 16 organic and petrochemicals. In the hypothetical situation postulated for the Twaddle-Malloy model study, there was a 5.8-fold increase in production capacity for the monomer accompanied by a price drop from 23 to 8.1 cents per pound. According to the results gathered by Stobaugh, such an increase in production would have been accompanied by a decrease in price of about 50 percent, i.e., from 23 to 12 cents per pound. These results do not contradict the Twaddle-Malloy results, which were for a hypothetical, fast-growing monomer market, whereas the Stobaugh study was for more mature organic and petrochemicals. This comparison would tend to support the statement that although the situation postulated for the model study is hypothetical, it is also realistic.

8-17 ECONOMIC-PRICE FORECASTING

The market price of a chemical is the price one can obtain for the chemical in the marketplace. The economic price of a chemical is the sum of all of the input costs and the total value added. The input costs include such elements as raw materials, catalysts, and utilities. Value-added elements are labor, direct as well as indirect, distribution costs, insurance, depreciation, taxes, and profits. Liebeskind (1973) developed the input costs and value-added items from historical data, which were then extrapolated by use of existing forecasts for the several elements to yield the forecast value. Based on data to 1970, his forecast price for ethylene glycol in 1975 in terms of 1958 dollars was 6.02 cents per pound and 6.18 cents per pound in 1980.

Nathanson (1972) used an approach which in principle is similar to that described above but defines the economic price simply as the value added to only the raw-material cost. To extrapolate into the future he accepted the hypothesis that the value added over raw-material cost in constant dollars decreases at a constant percentage rate in relation to the cumulative volume of the chemical produced. In practical terms this means that a plot of the logarithm of value added vs. the logarithm of cumulative production should yield a straight line. Such a plot is presented in Fig. 8-5, where the value added over feedstock cost in the production of ethylene is presented as a function of the cumulative production of ethylene. The break in the curve in 1958 represented the start-up time of several new large ethylene plants.

Price forecasting now requires a projection of raw-material costs, production rates, and the inflation index. Nathanson projected prices for paraxylene based on data to 1968. The projected prices for 1970 and for 1975 were 6.33 and 5.12 cents per pound, respectively. The actual posted price in 1970 was 7.2 cents per pound.

No attempt will be made here to compare the forecast 1975 prices for ethylene glycol and for paraxylene with the actual prices. The year 1975 was an extremely dynamic one with respect to chemical prices, and the fluctuations in prices are such that they make nonsense out of both long-range and short-range forecasts. This dynamic situation is shown in Table 8-3. It should also be noted that the mid-1978 price for paraxylene was 12.5 cents per pound and the price for ethylene glycol was 24.5 cents per pound.

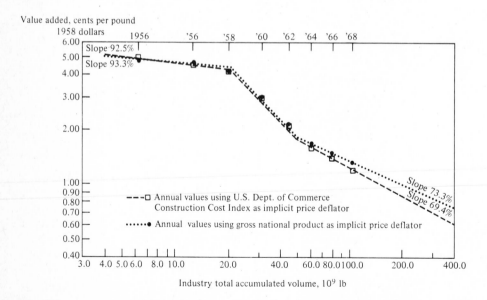

Figure 8-5 Cumulative ethylene production vs. value added over feedstock. [*From Nathanson (1972), by permission.*]

Table 8-3 Price indexes for paraxylene and ethylene glycol

Time	Price index†	
	Paraxylene	Ethylene glycol
June 1974	160	175
July 1974	375	330
Jan. 1975	354	300
July 1975	260	200

† Based on 1972 = 100.

Source: Eur. Chem. News, **27** (705): 34 (1975).

PROBLEMS

Some sources of information, additional to those cited in the problem section of Chap. 1, that may be useful are:

HPI Construction Box Score, appears semiannually in *Hydrocarbon Processing*
CE Construction Alert, appears semiannually in *Chemical Engineering*
Chemical Marketing Reporter, formerly *Oil, Paint, and Drug Reporter*

8-1 Fit a linear demand equation to the production data shown below. Base your equation on the first 10 years of data and predict the production for the following 2 years. Compare the actual production with the predicted production for the entire 12-year period.

Year	Production, 10^3 tons	Year	Production, 10^3 tons
1	40	7	68
2	47	8	82
3	54	9	97
4	56	10	101
5	56	11	101
6	70	12	102

8-2 Fit an exponential trend equation of the form

$$\log D = a + bt$$

to the production data shown in Prob. 8-1. Base your equation on the first 10 years of data and predict the production for the next 2 years. Compare the actual production with the predicted production for the entire 12-year period.

8-3 Calculate the constants a, b, and c in the Gompertz equation with data for sulfuric acid production in the United States using data for the years 1959 to 1973 inclusive. Compare the predicted with the actual production rates for several years subsequent to 1973. Would the predictions have been more accurate if the annual production had been linked to the Industrial Production Index by dividing the annual production by that year's index?

8-4 Sodium tripolyphosphate is used principally in detergents as a sequestrant and synergistic detergent agent. Obtain United States annual production figures from 1960 or earlier until at least 1975. Develop a demand-trend equation from the data until 1969. (*Hint:* Try log $D = a + bt$.) Using the trend equation, predict the production rates for the remaining years for which data were available and compare with the actual production. Explain the discrepancies, if any.

8-5 A market forecast for polyester fiber was made by analogy with the growth of the nylon-fiber market. The forecast was based on data until 1962, and the analogy and the forecast are shown in Fig. 8-3. Obtain production data for polyester fibers subsequent to 1962 and compare with the forecast of Fig. 8-3.

8-6 In Example 8-1 data for annual sales of soap and detergent were presented, and the trend of the soap-detergent market competition was examined. Soaps and detergents are consumer items, and it would appear reasonable to assume that a direct linkage exists between the number of consumers and soap and detergent sales. Examine the trend in per capita consumption of soaps and detergents and make sales predictions based on per capita consumption and population growth.

8-7 One of the major uses of ethylene glycol is as a permanent antifreeze in motor vehicles. Another important outlet is the manufacture of polyester resins. Can production figures for polyester resins along with data for motor vehicle registrations and/or new vehicle manufacture be used for reasonable forecasting of ethylene glycol production?

8-8 Obtain annual United States production figures for ammonia for the years 1940, 1950, 1960, and 1970. Compare with the capacity of the larger ammonia plants constructed during those years. Do the figures support the statement that for a number of basic chemicals the ratio between the capacity of the largest single plant built to the total annual capacity is approximately constant?

8-9 Repeat Prob. 8-8 for methanol.

8-10 The *turnover ratio* is the ratio of gross annual sales to the fixed-capital investment and is approximately equal to 1 for the chemical industry. This yields one of the rules of thumb for the rough approximation of plant cost, namely, that plant cost is approximately equal to the annual income from product sales. Assuming that plant cost is a function of capacity to the 0.6 power, that the market is expanding at the rate of 10 percent per year, and that the ratio of plant size built to total annual production is constant, show how the unit product selling price changes as a function of time over a 20-year period when the turnover ratio equals 1.

8-11 In his table 1 Nathanson (1972) presented historical data on paraxylene production and value added over gasoline cost during the period 1957–1968. Continue his table with data subsequent to 1968 and compare the results with the price projections made by Nathanson in his table 2.

8-12 Make a brief study of the effect of i and m on the optimum capacity-supply strategy by solving Example 8-2 for the following additional cases:

(*a*) For $i = 15\%$; $m = 0.5$ and 0.9

(*b*) For $m = 0.7$; $i = 7\%$ and 25%

NINE

DEALING WITH UNCERTAINTY

> Uncertainty about what is perceived by the senses can arise from one of two causes: Either the observer's competence in his field of endeavor is limited or else he has made things too easy for himself and let up on his investigation and observation.
>
> *Rabbi Sa'adia Gaon (882–942),*
> *"The Book of Beliefs and Opinions"*

One of the exciting and sometimes frustrating characteristics of the chemical engineering profession is that its practitioner is always confronted with uncertainty and never has adequate information. Although a process plant will be designed with the best design equations, correlations, and properties available, the engineer knows that the equipment will perform only approximately as designed. To compensate for this some kind of *efficiency factor* or *factor of safety* is used.

If uncertainty is great, the cost of factors of safety can be high, and it becomes desirable to develop quantitative measures for dealing with uncertainty. When we are dealing with the future in the form of projections and predictions, the problem can become even more complicated because the future cannot be known with certainty. In this chapter we shall see how analytical methods can be applied to assist in dealing with uncertainty.

9-1 DEALING WITH UNCERTAINTY IN ECONOMIC ANALYSIS

Economic analyses on which investment decisions are made are based on predictions and projections. They therefore have an element of uncertainty about them. If the forecasts are not realized, the decision made on the basis of these predictions and projections may turn out to be the wrong decision. As a result resources and capital are placed at risk. It is impossible to avoid uncertainty and risk when dealing with the future, and in this section some procedures for quantifying them will be presented.

One simple way of dealing with uncertainty is to alter the acceptable boundary values for the various indexes so that to be selected a more risky project would require a more attractive value than a less risky project. We have already seen that the attractive payback time for a high-risk project is considered to be 2 years or less whereas the payback time for a medium-risk project could be 5 years. Similarly, an acceptable DCF rate of return for a low-risk project might be as low as 15 percent whereas a value of 50 percent might be required for a high-risk, speculative project into a new market or for a new product.

9-2 SENSITIVITY ANALYSIS

The project evaluation carried out in Example 7-3 was based on single-valued estimates for the various cash-flow elements. The value of the information available from these cash-flow estimates can be increased substantially by carrying out a sensitivity analysis, in which the effects on the project of changes in these elements are explored. Such an analysis could explore, for example, the effects of changes in variable and fixed production costs, of delays in plant start-up, of different market growth patterns, and of scale of operation compared with a base-case estimate. Bauman (1964) lists the probable variations of a number of variables from forecast values over the life of a 10-year plant. Sales volume would probably vary from −50 to +15 percent, sales price from −50 to +20 percent, cost of investment from −10 to +25 percent, and construction time from −5 to +50 percent of the forecast values.

The results of a partial sensitivity analysis of Example 7-3 are presented in Table 9-1. The sensitivity analysis was made by assuming a 10 percent adverse change in the individual estimates of the cash-flow elements. From the results it can be seen that the net positive worth, or profitability, is most sensitive to the selling price and to the sales volume. A 10 percent reduction in the selling price would reduce the net positive worth by 44.3 percent, compared with the original NPW, and a 10 percent drop in sales volume, the next most sensitive item, would result in a 31 percent drop. The tax rate also has a significant effect on the profitability. A 10 percent rise in the depreciable investment, i.e., a 10 percent overrun on equipment, installation, and construction costs, would cause a 14 percent change in NPW, where part of the increased investment results in an additional tax credit via the increased depreciation.

Table 9-1 Effect of adverse changes on net postive worth in Example 7-3

10 percent adverse change in:	New NPW	Percentage change in NPW
Sales	$ 923,700	−31
Selling price	745,600	−44.3
Depreciable investment	1,149,800	−14.1
Fixed cost	1,281,800	−4.3
Variable cost	1,160,900	−13.3
Tax rate	1,072,700	−19.9

If the actual sales volume were 10 percent less than the volume originally projected and the product price obtained in the marketplace were 10 percent lower than the original estimate, the NPW would be reduced by more than 70 percent. Should an additional problem strike the project, e.g., a delay in construction or an increase in raw-material costs, the NPW would become negative. A project whose original evaluation showed an acceptable return on investment would earn far less than projected originally and might even become unprofitable if the economic elements upon which the evaluation was based were not realized.

A sensitivity analysis can be extended to include other percentage variations in estimates and the effect of scale of operations to obtain the optimum plant capacity. Such an analysis quantifies the effect of uncertainty on the project profitability but does not attempt to quantify the relative uncertainty of the various cash-flow elements. This must be left to subjective assessment, which need not be arbitrary, however.

9-3 QUANTIFYING UNCERTAINTY

When uncertainty arises, we can usually have recourse to an appropriate statistical technique in an attempt to quantify the uncertainty. Uncertainty in economic data and, as we shall see later, in other engineering data, can usually be expressed in terms of a probability distribution function. Such a function $p(V)$ is defined so that $p(V)\,dV$ is the probability that V, the variable element, will take on a value in the range of V to $V + dV$. Although a number of distribution functions can be used for describing uncertainty, for the present we shall concern ourselves only with the normal distribution function

$$p(V) = \frac{1}{\sigma\sqrt{2\pi}}\, e^{-(V-\bar{V})^2/2\sigma^2} \tag{9-1}$$

$\bar{V}$ is the most probable, or mean, value of V; σ^2 is the variance and is a measure of the spread about the mean value. For the normal distribution the probability is 0.68 that V will fall within $\bar{V} \pm \sigma$ and the probability is 0.954 that V will fall within $\bar{V} \pm 2\sigma$. If a variable is known with a high degree of certainty, we assign a small value of σ, the standard deviation, to it, whereas a parameter whose value is in doubt would be assigned a large value of σ, thereby producing a distribution function with a large spread. These concepts are illustrated in Fig. 9-1.

The functions we consider in engineering analysis usually depend upon a number of factors. In an economic analysis the cash flow, for example, is a function of sales volume, product price, depreciable capital, etc. In general,

$$\tilde{F} = f(V_1, V_2, \ldots, V_n)$$

In the sensitivity study we established a base case

$$\bar{F} = f(\bar{V}_1, \bar{V}_2, \ldots, \bar{V}_n)$$

and looked at the effects of small changes in the variable elements ΔV_j. We can define sensitivity coefficients as the change produced in the function $\tilde{F}$ by each change in a parameter

$$s_1 = \frac{f(\bar{V}_1 + \Delta V_1, \bar{V}_2, \ldots, \bar{V}_n) - \bar{F}}{\Delta V_1}$$

$$s_2 = \frac{f(\bar{V}_1, \bar{V}_2 + \Delta V_2, \ldots, \bar{V}_n) - \bar{F}}{\Delta V_2}$$

$$s_n = \frac{f(\bar{V}_1, \bar{V}_2, \ldots, \bar{V}_n + \Delta V_n) - \bar{F}}{\Delta V_n}$$

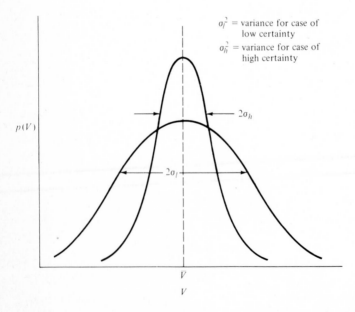

σ_l^2 = variance for case of low certainty

σ_h^2 = variance for case of high certainty

$2\sigma_h$

$p(V)$

$2\sigma_l$

$\bar{V}$

V

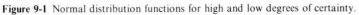

Figure 9-1 Normal distribution functions for high and low degrees of certainty.

If we proceed to the limit of vanishingly small changes in ΔV_j, these sensitivity coefficients become the partial derivatives of the function with respect to the variables

$$s_1 = \frac{\partial \tilde{F}}{\partial V_1}, \qquad s_2 = \frac{\partial \tilde{F}}{\partial V_2}, \qquad \cdots \qquad (9\text{-}1a)$$

It is usually convenient, however, to use the linear approximation for values near the base case, especially when one wishes to estimate the effects of changes and uncertainty

$$\tilde{F} - \bar{F} = \sum_{j=1}^{n} s_j (V_j - \bar{V}_j) \qquad (9\text{-}2)$$

Rudd and Watson (1968) show that if such a linear relationship holds and if the uncertainty of the variables V_j about the mean $\bar{V}_j$ can be described by a normal distribution with a variance σ_j^2, the uncertainty in $\tilde{F}$ can also be described by a normal distribution with a mean value of

$$\bar{F} = f(\bar{V}_1, \bar{V}_2, \ldots, \bar{V}_n) \qquad (9\text{-}3a)$$

and variance

$$\sigma_F^2 = \sum_{j=1}^{n} s_j^2 \sigma_{V_j}^2 \qquad (9\text{-}3b)$$

Example 9-1 We return to Example 7-3 and reexamine years 3 to 10 in more detail. The earlier sensitivity study showed that the project profitability was sensitive to changes in sales volume, sales price, and variable costs. The sales manager and the purchasing manager have now provided the following revised estimates for the range of sales volume and selling price for the years 3 to 10 and for raw material costs:

Estimate	Sales volume, 10^6 units/yr	Selling price per unit	Variable costs per unit
Lowest	4	$0.40	$0.10
Most likely	5	0.50	0.15
Highest	6	0.60	0.20

The NPW was also sensitive to depreciable investment, but we shall assume for the purposes of this example that the plant will be built on a fixed-price contract and that the depreciable investment and construction-time forecasts will be realized as projected. Calculate the cash flows and the possible effects of these estimates on the NPW.

SOLUTION The annual cash flow is expressed by

$$CF = \text{operating profit} - (\text{operating profit} - \text{depreciation}) \times \text{tax rate}$$

and

$$CF = [Vp - (F + Vv)] - [Vp - (F + Vv) - D]T$$

$$= [Vp - (F + Vv)][1 - T] + DT \tag{1}$$

where V = annual sales volume
p = unit selling price
v = unit variable cost
F = annual fixed cost
D = annual depreciation
T = tax rate

We shall assume that the mean value for the annual cash flow occurs when all the variable elements are at their most likely values

$$\overline{CF} = \{(5 \times 10^6)(0.5) - [200,000 + (5 \times 10^6)(0.15)]\}(1 - 0.45)$$

$$+ (230,000)(0.45) = \$956,000$$

The sensitivity of the cash flow to the three variables considered can be obtained from Eq. (9-1a)

$$s_V = \frac{\partial(CF)}{\partial V} = (\bar{v} - \bar{p})(1 - T)$$

$$= (0.5 - 0.15)(1 - 0.45) = 0.1925$$

$$s_v = \frac{\partial(CF)}{\partial v} = \bar{V}(1 - T) = 2.75 \times 10^6$$

$$s_p = \frac{\partial(CF)}{\partial p} = -\bar{V}(1 - T) = -2.75 \times 10^6$$

We now estimate the standard deviation of each of these elements on the assumption that there is a 95 percent probability that the actual values for the elements lie between the lowest and highest estimates, i.e., ± 2 standard deviations. On this basis

$$\sigma_v = \frac{6,000,000 - 4,000,000}{4} = 500,000$$

$$\sigma_v = \frac{0.6 - 0.4}{4} = 0.05$$

$$\sigma_p = \frac{0.20 - 0.10}{4} = 0.025$$

The variance of the annual cash flow can now be calculated using Eq. (9-3b)

$$\sigma_{CF}^2 = (0.1925^2)(500,000^2) + (2.75 \times 10^6)^2(0.05^2)$$
$$+ (-2.75 \times 10^6)^2(0.025^2)$$
$$= 32.9 \times 10^9$$
$$\sigma_{CF} = \$181,375$$

The annual cash flow will, with 95 percent confidence, be in the range of $956,000 \pm (2)(181,375)$ or from \$593,250 to \$1,318,750.

The complete results are presented graphically in Fig. 9.2, in which the ordinate represents the percentage probability that the actual value obtained will be less than the indicated value for the annual cash flow. There is a 96 percent probability, for example, that the annual cash flow during the years 3 to 10 will be less than \$1,300,000, whereas there is only a 10 percent probability that it will be less than \$670,000. Alternatively, we could say that there is a 90 percent probability that it will be greater than \$670,000.

The net present worth of the entire project has been calculated on the basis of the indicated annual cash flows, which are also presented in Fig. 9-2. The mean value for the NPW is at the 50 percent probability and is, of course, the value \$1,338,800, previously obtained in Example 7-3. From Fig. 9-2 we read that there is an 80 percent probability that the NPW will exceed \$900,000 (20 percent probability that it will be less than \$900,000) and an 80 percent probability that it will be less than \$1,770,000.

Although we have succeeded in quantifying the uncertainty, we are still faced with the problem of making a decision in the face of this, now quantified, uncertainty. We now address ourselves to this problem.

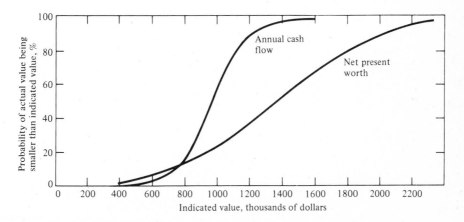

Figure 9-2 Probability of attaining annual cash flows and net present worth.

9-4 THE DECISION PROCESS

At every phase and stage of their professional careers engineers, are required to make decisions. They must decide on the pipe diameter for a certain service, the material of construction, corrosion allowance, whether or not a standby pump should be installed, daily operating instructions, selection from bids, investment, etc. The list is endless. Many decisions are intuitive. They are based on logical thought processes and on other factors such as experience, judgment, and the laws of science. The essential points we must remember about decisions are that they must be made when more than one possible course of action is available and that they are always made in the face of uncertainty. In some cases the uncertainty may be very small, so small, in fact, that for all practical purposes we are dealing with a certainty.

The decision process can be broken down into a series of steps (Bross, 1953):

1. We are faced with more than one possible course of action or strategy. It should be remembered that one course of action can also be "no action." These courses of action are identified by the decision maker. We shall symbolize them by $A_1, \ldots, A_i, \ldots, A_n$.
2. We predict the outcome of each course of action. Suppose, for example, we are to decide which proposal to select from among several vendors' bids for supplying a compressor. We would have to predict elements for each bid such as operating cost, maintenance cost, reliability, as well as other important variables. If we are faced with the decision of what the production capacity for a new plant is to be, it is obvious that we cannot predict outcomes with certainty. Costs, markets, prices, and competitor decisions are among the uncertain elements that will affect the outcome of any decision. These elements are referred to as states of nature and will be symbolized by $S_1, \ldots, S_j, \ldots, S_m$. As we have seen, we can predict or assign a probability that the variable or element will have a certain value. The important factor here is that each possible course of action may not have a unique outcome.
3. We attempt to quantify the desirability of each outcome by attaching a value to it. In the literature of decision theory the value is usually referred to as a *utility*, and we shall symbolize this by $U_{11}, \ldots, U_{ij}, \ldots, U_{mn}$.
4. We now make our decision by selecting the outcome with the highest utility. The decision criterion may appear to be obvious, i.e., select the action which yields highest utility. We shall soon see that it may not be as simple as it appears.

9-5 DECISION MAKING

From the discussion in the second step above it is apparent that three conditions are possible when we are called upon to predict the outcome of each course of action. As a result decision making is classified into three categories, distinguished as follows:

Decision making under certainty is the situation if each action leads without doubt or risk to a specific outcome. The outcome is the result of a causal series of events (if the bough breaks, the cradle will drop).

Decision making under risk is the classification if each action has several possible outcomes for which the probabilities are known.

Decision making under uncertainty is the situation if each action has several possible outcomes for which we do not even know the probabilities.

The framework for decision making can usually be represented in the form of a *decision matrix* (Luce and Raiffa, 1957). In the case of financial decisions the matrix is sometimes referred to as the *pay-off table* (Table 9-2).

Each possible outcome is represented by its appropriate location in the matrix. The outcome of decision or action A_i when the state of nature is S_j would be the utility U_{ij}. The steps involved in constructing the decision matrix, or payoff table, are parallel to the decision process as elaborated above:

1. Identify possible alternative courses of action
2. Identify the possible states of nature
3. Evaluate each outcome in terms of some measure of desirability

Decision Making under Certainty

The simplest case is that in which there is only one state of nature, and for this case the matrix collapses into a single column. The decision maker then simply selects from this column the alternative which has the greatest utility. Suppose that the decision to be made is whether or not to invest in a new production facility. The two alternatives in this simple case are "action" (invest) or "no action" (don't invest). It is certain that the alternative of "action" would result in additional income with a net present worth of $1,000,000. The other alternative is "no action," and then the capital that would have been invested in the

Table 9-2 Payoff table

Alternative action	State of nature				
	S_1	$\cdots$	S_j	$\cdots$	S_n
A_1	U_{11}	$\cdots$	U_{1j}	$\cdots$	U_{in}
$\cdots$	$\cdots$	$\cdots$	$\cdots$	$\cdots$	
A_i	U_{i1}	$\cdots$	U_{ij}	$\cdots$	U_{in}
$\cdots$	$\cdots$	$\cdots$	$\cdots$	$\cdots$	
A_m	U_{m1}	$\cdots$	U_{mj}	$\cdots$	U_{mn}

Table 9-3 Payoff table, no uncertainty

Alternative	Outcome NPW
Invest	$1,000,000
Don't invest	300,000

new plant could be invested elsewhere with a resulting NPW of $300,000. The payoff table for this situation is given in Table 9-3.

Our decision criterion is to maximize NPW, and it is obvious that the preferred decision is "invest."

Decision making under certainty is a trivial problem if we have identified the possible alternative courses of action correctly and evaluated each outcome correctly. We must be wary when we decide whether we are dealing with a situation of certainty, risk, or uncertainty. It is easy to fool ourselves into thinking we are dealing with a certainty when it really is a risk situation.

Decision Making under Risk

Decision making under risk is the situation that arises when each alternative course of action may result in more than one outcome depending upon the state of nature. Each state of nature will have a known or estimated probability of occurrence p_j. Since all the states of nature are mutually exclusive, their probabilities must sum to 1. A decision matrix with risk is shown in Table 9-4.

The decision criterion is not so obvious. Intuitively, for example, we might select the column with the maximum p_j and then select from that column the maximum utility U_{ij} so that the decision would be A_i. But the state of nature

Table 9-4 Decision matrix with risk

	Probability		
	p_1 $\cdots$	p_j $\cdots$	p_n
	State of nature		
Alternative	S_1 $\cdots$	S_j $\cdots$	S_n
A_n ...	U_{11} $\cdots$ ...	U_{1j} $\cdots$ ...	U_{1n} ...
A_i ...	U_{i1} $\cdots$ ...	U_{ij} $\cdots$ ...	U_{in} ...
A_m	U_{m1} $\cdots$	U_{mj} $\cdots$	U_{mn}

with the highest probability may not prevail. If, for decision A_i, the utilities for states of nature other than S_j are low, we made the wrong decision. The way out of this is to calculate the expectation or the expected utility for each alternative by summing the products of each utility and the probability of its occurrence

$$E_i = \sum_{j=1}^{n} p_j U_{ij}$$

The decision criterion is the action or alternative with the highest expected utility.

We shall illustrate this by amplifying our earlier trivial case. Suppose that in our original estimate and calculations for the NPW for the proposed project of building a new production facility, it was assumed that 30 percent of our production was going to be purchased by one of our "loyal" customers. This loyal customer uses our product as one of the raw materials in his production process. Rumor has it that our customer may be shutting down that particular production facility, in which event he would no longer purchase from us and would become "disloyal." Should that happen, the NPW of our project would drop to $200,000. In addition, our engineers tell us that they have not yet solved all the problems with respect to effluent and pollution control but are confident they will solve them within the 3-year deadline they were granted. If they do not solve these problems, our new production facility would have to shut down. After cash flows resulting from depreciation and tax considerations upon abandonment, the NPW for the project would be $-\$300,000$. After extended discussions and evaluations, our management estimates that the probability of plant shutdown because of inability to solve our effluent problems is 20 percent, that the probability of "loyal's" remaining loyal is 65 percent, and that the probability of "loyal's" becoming disloyal is 15 percent. The decision matrix, or payoff table under risk, appears in Table 9-5.

The decision criterion is the highest expectation, and the alternative action that produces the largest value for the expectation is "invest," which would then be the correct decision for this case.

Table 9-5 Payoff table under risk

Alternative	Probability			
	0.65	0.15	0.20	
	State			
	Loyal	Disloyal	Abandon	Expectation
Invest	1,000,000	200,000	−300,000	620,000
Don't invest	300,000	300,000	300,000	300,000

Suppose that the evaluation by management was that the probability of "loyal's" shutting down his plant was quite high and that the value assigned to the probability of his remaining loyal was only 0.15. The new situation would then be as follows:

Action	Expectation
Invest	$(0.15)(1,000,000) + (0.65)$ $(200,000) + (0.20)(-300,000)$ $= 220,000$
Don't invest	$(0.15)(300,000) + (0.65)$ $(300,000) + (0.20)(300,000)$ $= 300,000$

In this case the decision would be "don't invest."

It should be emphasized that the value obtained as the expectation will not be the utility resulting from the particular course of action chosen. The utility actually received will be that expected for the particular action chosen and the particular state of nature that prevails in the end. The expectation value, as a general measure of risk and its effect on a project, is a useful concept even though the expectation cannot actually be achieved for any particular project.

Decision Making under Uncertainty

Under conditions of uncertainty we have several alternative courses of action and several possible states of nature, but this differs from conditions of risk because we now know nothing about the likelihood or probability of each state of nature. Let us return to our preceding example, but now management is unable to assign any probability to loyal, disloyal, or abandon. The payoff table is shown in Table 9-6. In a decision problem under uncertainty, as under risk, a decision cannot be made until the elements of each row of utilities have been replaced by a single utility, thereby formally reducing the problem to one of certainty. Several criteria are commonly used.

Table 9-6 Payoff table, uncertainty

Alternative	State of nature		
	Loyal	Disloyal	Abandon
Invest	$1,000,000	$200,000	$-300,000
Don't invest	300,000	300,000	300,000

Equal-likelihood criterion This criterion, the *Bayes-Laplace criterion*, says that if we do not know anything about the probabilities for the various states of nature, the best thing we can do is to assume that they are all equal. This reduces our problem to one of decision making under risk. For the example, if we assume that each state of nature has equal probability, the expectations would be

Action	Expectation
Invest	$(1,000,000)(\frac{1}{3}) + (200,000)(\frac{1}{3}) + (-300,000)(\frac{1}{3})$ $= \$300,000$
Don't invest	$(300,000)(\frac{1}{3}) + (300,000)(\frac{1}{3}) + (300,000)(\frac{1}{3})$ $= \$300,000$

We note that the expectations are equal for the two alternative actions. The procedure now is to flip a coin and decide on that basis. Alternatively, we can try other criteria.

Maximin-utility criterion This criterion is completely pessimistic in orientation and attempts to avoid the worst outcome. The method requires that each row of utilities be replaced by a single value which is the minimum utility in that row. The matrix then collapses into a single column of row minima. The decision rule then is to select that alternative which results in the maximum value in this column. Hence, the term *maximin*, the maximum of the minima over the rows. The payoff table for the criterion is shown in Table 9-7. A_1 and A_2 represent "invest" and "don't invest," and S_1, S_2, and S_3 represent loyal, disloyal and abandon, respectively. Application of the maximin criterion leads to the preferred action A_2, "don't invest." It avoids the possibility of a $300,000 loss in present value by playing safe and accepting a $300,000 present value no matter what state of nature occurs. This policy was at the expense of a possible high return of $1,000,000. If S_1 were the state of nature prevailing in the end, we would regret that we had decided A_2 because if we had decided A_1, the NPW would have been $1,000,000 instead of $300,000, that is, "regrets" amount to $700,000.

Minimax regret criterion To apply this criterion we construct a table of regrets. To construct this table we consider each state of nature and determine the best outcome and its corresponding alternative. All other alternatives will yield lower

Table 9-7 Maximin criterion

	S_1	S_2	S_3	Row minima
A_1	$1,000,000	$200,000	$ -300,000	$ -300,000
A_2	300,000	300,000	300,000	300,000

Table 9-8 Regrets table

	S_1	S_2	S_3	Row maxima
A_1	$ 0	$100,000	$600,000	$600,000
A_2	700,000	0	0	700,000

utilities. The regret for each alternative for a given state of nature is the difference between the utility of the best outcome and the utility of that alternative, in
other words, what was lost by making the wrong decision for a given state of
nature. The decision criterion is to minimize the maxima of the regrets. The
regrets table for the preceding example is given in Table 9-8. The minimax of the
regrets row is A_1 so the preferred action using the minimax regrets is "invest,"
contrary to the decision reached by the maximin utility criterion.

Maximax criterion The two previous criteria were pessimistic. They looked at
the worst possible outcomes for each state of nature. The maximax criterion is
optimistic, and the decision criterion is to select the maximum utility from the
highest utilities in each row. From Table 9-9 the decision on this basis would be
A_1, "invest."

Hurwicz criterion The three previous criteria for decision-making under uncertainty were either pessimistic, and concentrated on the worst possible outcome
for any state of nature, or optimistic, looking only at the best outcomes. The
Hurwicz criterion attempts to give some weight to the best as well as to the worst
outcomes that can arise from each alternative.

Consider the row of utilities corresponding to the alternative A_i. The minimum utility for this row will be designated u_i and

$$u_i = \min_j u_{ij}$$

The maximum utility U_i will be

$$U_i = \max_j u_{ij}$$

The matrix is now collapsed into a single column by assigning to each alternative A_i the value

$$\alpha u_i + (1 - \alpha)U_i \qquad 0 \le \alpha \le 1 \tag{9-4}$$

Table 9-9 Maximax criterion

	S_1	S_2	S_3	Row maxima
A_1	$1,000,000	$200,000	$-300,000	$1,000,000
A_2	300,000	300,000	300,000	300,000

which is called the α index of A_i. The coefficient α is the pessimism-optimism index.

Let us consider now the two limiting cases of $\alpha = 0$ and $\alpha = 1$. If $\alpha = 1$, the α index becomes the row minimum and the case reduces to the maximin-utility criterion. If $\alpha = 0$, the α index becomes the row maximum and therefore focuses on the best outcome. This is the maximax criterion. The two extreme cases of $\alpha = 0$ and $\alpha = 1$ therefore, represent the optimistic and pessimistic limits, respectively. A value of α between 0 and 1 would represent some sort of a compromise between the optimistic and pessimistic approaches. Why α is referred to as the pessimism-optimism index should now be clear.

The problem of selecting α, or in other words, how much weight should be assigned to the best compared with the worst, still remains. It can be regarded as a characteristic of the decision maker.

The decision maker is still going to have to use good sense, sound business and engineering judgment, and intuition.

9-6 DESIGN UNDER RISK

In our discussion of decision making under risk we considered the case where randomness, expressed as probabilities, occurred only in the state of nature. We made the tacit assumption that we could evaluate the utility for each alternative action with certainty for each state of nature. Another type of decision problem under risk can arise when randomness appears in the utilities as well as in the states of nature. The decision matrix cannot be used in this case, as discussed by Siddall (1972).

If we know the function relating the utility to the state of nature and to the decision and also have available the appropriate probability density function, it is possible to obtain the expected value for the utility. The decision criterion is to choose the alternative action that maximizes the expected utility. The situation in which we have this type of randomness is typical of design problems in which there may be uncertainty in the values for the design parameters. We shall see how this situation can be dealt with by means of an example.

Example 9-2 You are involved in a "very hot project," and plant design and construction are to get under way before all the required process data are available. You are to design the reactor, which is to produce Q kg of product per hour. The reaction is known to be zero order, but there is some uncertainty about the reaction rate, which is somewhere in the range of R_H to R_L kg/(m^3 · h). If the reactor you design is able to produce at a rate in excess of Q, there will be no premium, but there will be a penalty of P_d dollars per kilogram-hour of shortfall between desired and actual production rate if the reactor cannot produce at the rate Q. What will be the optimal recommended design volume for the reactor?

SOLUTION Our problem is to estimate the expected cost of the reactor when we know the upper and lower limits of reactor volume Q/R but have no information concerning the expected likelihood of any particular value of R other than that $R_L < R < R_H$. We would therefore be justified, in using the equal-likelihood criterion and assuming equal probabilities for all values of R in the permitted range. Siddall (1972) shows that

$$\text{Expected value of utility} = \int_{-\infty}^{\infty} U(y)p(y)\,dy \qquad (1)$$

where $U(y)$ = utility function
$\quad\quad p(y)$ = probability density function
$\quad\quad y$ = dependent variable

The utility function for our problem is the cost of reactor as a function of reaction rate. We can assume that reactor cost will vary with volume in a power-law form

$$C(V) = C_R V^n = C_R\left(\frac{Q}{R}\right)^n$$

The appropriate probability density function for the equal-likelihood criterion is the uniform distribution function, which for our case would be

$$p(R) = \begin{cases} \dfrac{1}{R_H - R_L} & \text{for } R_L < R < R_H \\[2mm] 0 & \text{otherwise} \end{cases} \qquad (2)$$

We now select a volume V and calculate the costs if the reaction rate attained is R

$$C(V) = \begin{cases} C_R V^n & \text{if } RV > Q \\ C_R V^n + P_d(Q - VR) & \text{if } RV < Q \end{cases}$$

The second equation includes the penalty charge for a shortfall in production. The expected cost will be

$$\bar{C}(V) = \int_{R_L}^{Q/V} [C_R V^n + P_d(Q - VR)]\frac{dR}{R_H - R_L} + \int_{Q/V}^{R_H} C_R V^n \frac{dR}{R_H - R_L}$$

For the purposes of this example we shall assume that the reactor cost varies linearly with volume; that is $n = 1$. Integration now yields the expected cost

$$\bar{C}(V) = \frac{1}{R_H - R_L}\left[C_R V(R_H - R_L) + \frac{P_d Q^2}{2V} + P_d R_L\left(\frac{R_L V}{2} - Q\right)\right]$$

and

$$\frac{Q}{R_H} \leq V \leq \frac{Q}{R_L}$$

What we wish to extract from this expression is the reactor volume that will result in the minimum cost. This can be obtained by differentiating $\bar{C}(V)$ with respect to V and setting the derivative equal to zero. When this is done, we obtain

$$V^* = \sqrt{\frac{P_d Q^2}{2C_R(R_H - R_L) + P_d R_L^2}} \qquad \text{where} \qquad \frac{Q}{R_H} \le V \le \frac{Q}{R_L}$$

We now attach some numbers to get a quantitative feel for the situation. We are to produce 300 kg/h of product for our reaction, in which $R_H = 30$ kg/(m^3 · h) and $R_L = 20$ kg/(m^3 · h). The penalty due to lost sales revenue for not meeting the desired production rate is \$1500 per kilogram per hour. Reactor cost is estimated at \$4000 per cubic meter. Substituting these values, we get

$$V^* = \sqrt{\frac{(1500)(300^2)}{2(4000)(30 - 20) + (1500)(20^2)}} = 14.1 \text{ m}^3$$

This is compared with required reactor volume of 10 and 15 m^3 at the high and low reaction rates, respectively. If the penalty were reduced to \$500, for example, the optimum reactor volume would be 12.7 m^3. The sensitivity of the optimum volume to the several parameters can be obtained easily by differentiating the expression for optimum volume with respect to the desired parameters.

9-7 RELIABILITY AND REDUNDANCY

In the real chemical-plant world one cannot expect that the plant will behave precisely as it was designed to behave. Nor can we expect that the various elements and components of the plant will perform, even with the best of maintenance, without an occasional failure; a pump bearing will fail, a control valve will cease to operate, a batch of product will be off specification. The theory of probability can be used to quantify the problem of failure or, in more positive terms, to help us deal with the reliability of equipment or process components.

If an operation involves repetitive or cyclic acts, e.g., opening and closing a control valve or the operation of batch reactors, we can associate a reliability R with the operation. The reliability can be quantified by the number of past successes in operation divided by the number of trials T

$$R = \frac{\text{number of successful operations}}{\text{number of trials}} \tag{9-5}$$

Series Structure

Let us now consider a system comprising N components connected in series in which a failure in any one component shuts down the system

$$\bigcirc \rightarrow R_1 \rightarrow \bigcirc \rightarrow R_2 \rightarrow \bigcirc \rightarrow \cdots R_{n-1} \rightarrow \bigcirc \rightarrow R_n \rightarrow \bigcirc$$

For the first component

$$\text{Number of expected successes} = R_1 T$$

For the second component, the number of trials it will undergo will equal the number of successes of component 1. Therefore

$$\text{Number of expected successes of } 2 = R_2(R_1 T)$$

Similarly, for the nth component, the number of trials it will undergo will equal the number of successes of component $n - 1$, that is, $R_1 R_2 \cdots R_{n-1} T$, and

$$\text{Number of expected successes of component } n = T(R_1 R_2 \cdots R_{n-1} R_n)$$

$$= T \prod_{j=1}^{n} R_j$$

In accordance with the definition of R, the reliability of the entire system will be

$$R_{\text{sys}} = \prod_{j=1}^{n} R_j \tag{9-6}$$

For a series system it is seen that the reliability of the system will be less than or equal to the reliability of the most unreliable component.

As an example, let us assume that our system consists of three operations: a tricky batch reactor, whose product is discharged into a temperamental centrifuge, followed by an uncooperative reaction-crystallization step. From past performance records, we know that reactor performed well, i.e., succeeded, during three-fourths of the past trials, the centrifuge operated during seven-eighths of the trials, and the final step succeeded for only 40 percent of the trials. The chances of simultaneous success, i.e., that the operators will get a batch through successfully will be

$$R_{\text{sys}} = R_1 R_2 R_3 = (\tfrac{3}{4})(\tfrac{7}{8})(0.40) = 0.2625$$

Parallel Structure

For a parallel structure, as shown in Fig. 9-3, the system does not shut down if a component fails, but instead a redundant component is activated in place of the

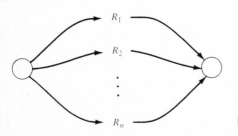

Figure 9-3 Parallel structure.

failed one. For a total of T trials, for component 1

$$\text{No. of successes} = R_1 T$$

Component 2 will undergo trials in accordance with those remaining after the failure of component 1

$$\text{No. of trials} = T - R_1 T$$

and
$$\text{No. of successes} = (T - R_1 T)R_2 = (1 - R_1)TR_2$$

For components 1 and 2

$$\text{No. of successes} = R_1 T + (1 - R_1)TR_2 = T[1 - (1 - R_1)(1 - R_2)]$$

For component 3

$$\text{No. of trials} = T\{1 - [1 - (1 - R_1)(1 - R_2)]\}$$

and
$$\text{No. of successes} = R_3 T\{1 - [1 - (1 - R_1)(1 - R_2)]\}$$

Finally, for components 1, 2, and 3

$$\text{No. of successes} = T[1 - (1 - R_1)(1 - R_2)(1 - R_3)]$$

In general, for an n-component system

$$\text{No. of successes} = T\left[1 - \prod_{j=1}^{n} (1 - R_j)\right]$$

and the system reliability therefore becomes

$$R_{\text{sys}} = 1 - \prod_{j=1}^{n} (1 - R_j) \tag{9-7}$$

To demonstrate we shall use the same values for reliability of elements as we used in the series example but for this example will assume that the three elements are in a parallel configuration

$$R_{\text{sys}} = 1 - (1 - \tfrac{3}{4})(1 - \tfrac{7}{8})(1 - 0.4) = 0.981$$

For a parallel configuration the reliability of the system will be greater than or, in the limit, equal to the reliability of the most reliable element. It is for this reason that standby units are used for elements that are prone to failure when their failure may result in hazardous consequences from the safety viewpoint or in undesirable economic consequences as a result of plant shutdown for repair or replacement of the failed component.

We saw earlier that economic evaluations and analyses are based on the time value of money. The preceding reliability analysis was based on number of successes and number of trials. To use the results of such a study in an economic analysis, we must bring the number of trials to an appropriate time basis. This can be done easily in the case of batchwise operations and has been done by Rudd (1962). It is much more difficult to use our definition of reliability for the purpose of deciding on spare or redundant equipment for continuous plant operations. For such an analysis, time will have to appear explicitly in the characterization of reliability or failure.

9-8 PROCESS AVAILABILITY AND REDUNDANCY

Cox (1976) deals with the problem of providing standby or redundant equipment from the reliability as well as the economic points of view. He characterizes
reliability in terms of time so that the time value of money can be considered in
economic analysis. This section will be based on his treatment. A more detailed
presentation of reliability theory is given by Siddall (1972).

The time that a component operates between a start-up and a failure is
considered to be a random variable. The distribution of this time-to-failure
random variable is described by a probability density function $f_f(t)$, where the
probability of failure between time t and time $t + dt$ is $f_f(t)\, dt$. The probability of
failure in the interval between start-up and time t is

$$p(\text{component failure}) = \int_0^t f_f(t)\, dt \tag{9-8}$$

and the complement will give the probability of successful operation during that
interval

$$p(\text{component success}) = 1 - p(\text{component failure}) \tag{9-9}$$

After the component has failed, it can be repaired and put back into service. We
assume that the same probability density applies to the repaired element as to
the component before failure. The mean time between failure (MTBF) is the
mean value of the probability density function

$$\text{MTBF} = \int_0^\infty t f_f(t)\, dt \tag{9-10}$$

Similarly, the time to repair or replace a failed component is also assumed to
follow a random distribution with a probability density function $f_r(t)$. The mean
time to repair (MTTR) can therefore be expressed as

$$\text{MTTR} = \int_0^\infty t f_r(t)\, dt \tag{9-11}$$

The mean availability of the component has been shown by Cox (1976) to be

$$A = \frac{\text{MTBF}}{\text{MTBF} + \text{MTTR}} \tag{9-12a}$$

and the unavailability

$$U = 1 - A \tag{9-12b}$$

If we now assume that the probability density function for time to failure,
and time to repair can be expressed, respectively, by the exponential functions

$$f_f(t) = \lambda e^{-\lambda t} \quad \text{and} \quad f_r(t) = \mu e^{-\mu t} \tag{9-13}$$

Then
$$\text{MTBF} = \frac{1}{\lambda} \quad \text{and} \quad \text{MTTR} = \frac{1}{\mu}$$

Thus, λ is a mean failure rate, and μ is a mean repair rate.

We shall deal with the case where we have two components in parallel, only one of them being in operation while the other is in standby condition. This is termed the *two-unit cold standby configuration*. It is assumed that when a unit fails, the failure is immediately sensed, the other unit goes into operation instantaneously, and repair operations begin. Only if both units are simultaneously in a failed condition would the process shut down.

The mean, or long-term, availability of one unit will be

$$A_1 = \frac{\mu}{\mu + \lambda} = \frac{1}{1 + \rho} = 1 - U_1 \quad \text{where} \quad \rho = \frac{\lambda}{\mu} \tag{9-14}$$

If it is assumed that only one repair crew is available to repair a failed component, Cox shows that the availability of the two-unit configuration is

$$A_2 = \frac{1 + \rho}{1 + \rho + \rho^2} = 1 - U_2 \tag{9-15}$$

and this equals the probability that at least one of the two units will be operable. The probability of both units being available is

$$p(\text{no failed unit}) = \frac{1}{1 + \rho + \rho^2} = 1 - U_r \tag{9-16}$$

where U_r represents the probability that a repair action is underway.

9-9 EVALUATING THE PROFITABILITY OF REDUNDANCY

Redundant equipment will be installed, leaving aside irreducible factors, if it can be justified by profitability. Let us consider a piece of process equipment whose failure would shut down tne process and assume that ρ for the component is 1/20. In accordance with the equation developed above, the component and therefore the process would have an unavailability of 0.0476. In other words, for almost 5 percent of the time the plant would be down and revenue would be lost due to loss of production. If a second unit were installed in parallel and put into operation when the first fails, the fractional unavailability of the plant would now be reduced to 0.0024; that is, would be down only 0.24 percent of the time. In the profitability analysis that now remains to be done, we shall address ourselves to the question: Is the additional capital investment for redundant equipment justified by the additional revenue to be expected? The procedure to be followed in principle was used in the solution of Example 7-2.

We first define several terms. Let I equal the operating income per year (sales revenue − operating expenses), where all operating expenses other than maintenance cost are included. M, the maintenance expense per year of down time, can be expressed as $\bar{F}C$, where $\bar{F}$ is the cost of maintenance per year per unit of capital investment and C is the investment for the standby unit. The incremental annual cash flow due to one redundant unit when the tax rate is T will be

$$CF_1 = [I(A_2 - A_1) - M(U_r - U_1)][1 - T] + \frac{CT}{L} \qquad (9\text{-}17)$$

The subscript I indicates one standby unit. The final term in Eq. (9-17) is the tax credit due to depreciation and assumes straight-line depreciation over a depreciable life of L years.

The NPW due to the redundant or standby unit will equal the present worth of this annual incremental cash flow less the investment in the standby unit, which we assume is made at time zero

$$NPW_1 = \left|[I(A_2 - A_1) - M(U_r - U_1)][1 - T] + \frac{CT}{L}\right| \frac{1 - e^{-it}}{i} - C \qquad (9\text{-}18)$$

The ratio of this incremental NPW to the incremental investment will give us a measure of the profitability that can be credited to the addition of a standby unit. Cox (1976) calls this ratio the *dimensionless incremental present worth*

$$DIPW_1 = \frac{\{[I(A_2 - A_1) - M(U_r - U_1)](1 - T) + CT/L\}[(1 - e^{-it})/i]}{C} - 1 \qquad (9\text{-}19)$$

which with Eq. (9-17) yields

$$DIPW_1 = \frac{CF_1(1 - e^{-it})}{Ci} - 1 \qquad (9\text{-}20)$$

If we define R_I as CF_1/C, the ratio of incremental annual cash flow to incremental investment, then

$$DIPW_1 = R_I \frac{1 - e^{-it}}{i} - 1 \qquad (9\text{-}21)$$

$DIPW_1$ is plotted vs. time in Fig. 9-4 for several values of R_I and for i equal to 0.15 and 0.25. When $DIPW_1$ is positive, the significance is that by time t the installation of a standby unit will have produced a profit above and beyond the cost of the capital employed as represented by the interest rate i.

Let us examine the breakeven requirements on the reasonable assumption that the project life will be equal to the depreciable life. We can conclude from Eq. (9-21) that at breakeven

$$R_I \frac{1 - e^{-iL}}{i} = 1 \qquad (9\text{-}22)$$

For the installation of the redundant unit to be profitable, i.e., to return more than the cost of capital,

$$\frac{CF_1}{C} = R_I \geq \frac{i}{1 - e^{-iL}} \qquad (9\text{-}23)$$

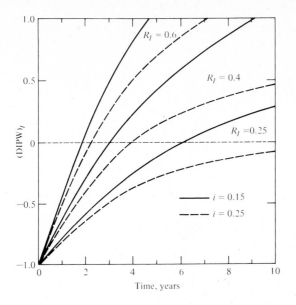

Figure 9-4 Dimensionless incremental present-worth behavior with time. [*From Cox (1976), by permission.*]

Figure labels:
- $R_I = 0.6$
- $R_I = 0.4$
- $R_I = 0.25$
- $i = 0.15$
- $i = 0.25$
- (DIPW)$_I$
- Time, years

This equation also verifies what should be intuitively obvious: under no circumstances will a standby configuration be profitable if the ratio of the incremental cash flow to the incremental investment is less than i, the cost of capital. This is also seen in Fig. 9-4 for $R_I = 0.25$ at $i = 0.25$. The curve is always below DIPW$_I$ equal to zero and only approaches this value asymptotically.

There is a paucity of information in the literature on reliability and failure rates of chemical plant equipment. Even fewer data on time to repair are available although they can be easily generated from internal data and experience records of any company. In the following example we show that although the intuitive judgment of the engineer can be used to reach a well-founded decision about the profitability of a standby configuration, the decision can be completely quantified if data on mean time between failure and mean time to repair are available.

Example 9-3 A process you are designing will have an operating income, without maintenance expenses included, of \$2.5 million per year. A solvent feed pump is crucial to the operation, and its failure would cause a plant shutdown. A spare pump, installed, and its associated piping cost \$3500. Annual maintenance costs are 5 percent of investment. The plant is to operate continuously with an on-stream factor of 0.95. An interest rate of 15 percent is required, and the tax rate is 50 percent. Project life is 10 years, and the plant will be completely depreciated during that time by straight-line depreciation. Would you recommend the installation of a standby pump?

SOLUTION At the breakeven point

$$R_I = \frac{0.15}{1 - e^{-(0.15)(10)}} = 0.1931$$

and from Eqs. (9-19) and (9-20)

$$R_I = \frac{[I(A_2 - A_1) - M(U_r - U_1)](1 - T) + CT/L}{C}$$

Substituting for A_1, A_2, M_1, U_r, and U_1, we have

$$R_I$$

$$= \frac{\left[I\left(\frac{1 + \rho}{1 + \rho + \rho^2} - \frac{1}{1 + \rho}\right) - FC\frac{1 + \rho}{\rho}\left(1 - \frac{1}{1 + \rho + \rho^2} - 1 + \frac{1}{1 + \rho}\right)\right](1 - T)}{C}$$

$$+ \frac{T}{L}$$

This can be simplified to

$$R_I = \frac{\{I\rho/[(1 + \rho + \rho^2)(1 + \rho)] - FC\rho/(1 + \rho + \rho^2)\}(1 - T)}{C} + \frac{T}{L}$$

We now have numerical values for all the variables except ρ. Substituting the known values, we obtain

$$0.1931 = \frac{\{(2.5 \times 10^6\rho)/[(1 + \rho + \rho^2)(1 + \rho)] - (0.05)(3500\rho)/(1 + \rho + \rho^2)\}(0.5)}{3500}$$

$$+ \frac{0.5}{10}$$

ρ can now be calculated by trial and error or successive-approximation procedures. (The value of ρ is in general much smaller than 1, so that the first approximation can be made by neglecting ρ^2 and ρ with respect to 1.) The trial-and-error calculation yields 0.000401 as the value of ρ. An installed spare will be profitable, therefore, if ρ is equal to or larger than 0.000401.

We recall that ρ is the ratio of the MTTR to the MTBF. Unfortunately, however, we have no statistical information on the MTTR or the MTBF. With respect to the MTTR we discuss this with our maintenance engineer, who tells us that repairs to that particular type of pump would require—depending, of course, on the type of failure and the availability of spare parts—about 3 to 4 h. He would want to add at least 1 h to that because he might not have a maintenance crew available at the time of breakdown. He is willing to "guarantee" that he could put the pump back in service in, on the average, 5 h. We accept this as the MTTR. For the value we obtained for ρ of 0.000401, this would mean that, at breakeven, the pump would have to operate for 12,469 h between failures. Since 1 year of plant operation is $(0.95)(24)(365) = 8322$ h, the pump would have to operate for 1.5 years between failures. Intuitively, one would expect that a pump failure would be quite likely to occur more frequently than that. Consequently, the recommendation would be to install the spare.

9-10 PLANNING PRODUCTION CAPACITY WITH UNCERTAIN DEMAND

Planning for future investment in plant capacity in view of a changing demand requires that the additional capacity due to new plant or to plant expansions be considered as well as their timing. Generoso and Hitchcock (1968) consider the optimal route for plant expansion in view of the fact that capacity of existing plant can usually be increased by a number of ways. When strategies for increasing production capacity are being generated, alternatives such as construction of new plant and scrapping an existing plant with replacement by a larger, more efficient plant must be considered as well as the possibility of product purchase from competitor producers in order to meet the demand and maintain the loyalty of existing customers. Economic evaluation and selection from among the competing strategies can be done by calculating the net present worth for each.

The magnitude of the problem is considerably increased when forecasts are uncertain and presented, in all likelihood, as a number of separate sets, each set with its probability. Rose (1977) and Rose et al. (1974) developed a dynamic-programming search algorithm that optimizes the set of decisions from a list of possible capacity expansions and selects the appropriate timing to maximize profitability. Before proceeding to an example based on their work, we shall undertake a brief and superficial introduction to dynamic programming, a complete presentation being beyond the scope of this book.

Dynamic Programming

Aris (1964) points out that dynamic programming applies to the situation in which many decisions have to be made to maximize or optimize the overall performance of a system. The system must be one in which distinct stages can be recognized and decisions at later stages do not affect the performance of earlier ones. This is precisely the situation which will exist when the strategy of provision of production capacity is being considered. In this case the distinct stages would be time, and each stage could be, for example, 1 year.

Dynamic programming is an application of the principle of optimality as stated by Bellman (1957): "*An optimal policy has the property that whatever the initial state and initial decision are, the remaining decisions must constitute an optimal policy with regard to the state resulting from the first decision.*"

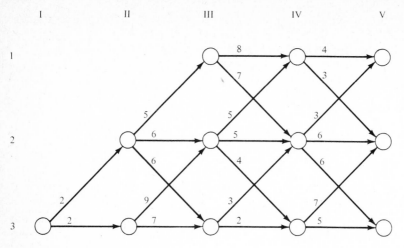

Figure 9-5 Network routing problem.

This principle, which Bellman calls intuitive, was stated in a more understandable but perhaps slightly irreverent manner by Aris (1964): "*If you don't do the best you can with what you happen to have got, you'll never do the best you might have done with what you should have had.*"

We shall solve a problem in network routing by using the principle of optimality and then show how this can be related to the problem of planning capacity. Our example is a slight variation of one presented by Wilde and Beightler (1967), and the data for it are presented in Fig. 9-5. The problem facing us is to move a shipment from node I-3 to any node in column V by the optimum, i.e., cheapest, route. The numbers above each arrow are the costs of shipping between the two nodes. The constraints are that we can only proceed in the I toV direction and that no backtracking is permitted. The decisions possible at the nodes or decision points are to take the left (L), straight (S), or right (R) path, in accordance with the route possibilities available at the node.

The problem can be solved quite easily by applying the principle of optimality and looking at the problem backward. Let us assume, for example, that our shipment arrived at IV-1. What route would you now choose to get to V (do the best you can with what you happen to have got)? By inspection, the best route is "right," and that would add 3 cost units to those already accumulated up to IV-1. This procedure is repeated for nodes IV-2 and IV-3. We now back up to node III-1. Again, do the best you can. If we decide "straight," this would cost 8 units and bring us to IV-1, where, as we already know, the optimum additional cost to V is 3 units. Therefore, the decision "straight" costs us $8 + 3 = 11$ units to get to V. The decision "right" costs us 7 units to get to IV-2, where the optimal route to E adds an additional 3 units. So the best we can do from III-1 is "right," and the cheapest cost to V from that node is 10 units. The complete solution to the problem is shown in Fig. 9-6, where the optimal cost from each node to V is shown in the node circle and the optimum path by a double arrow.

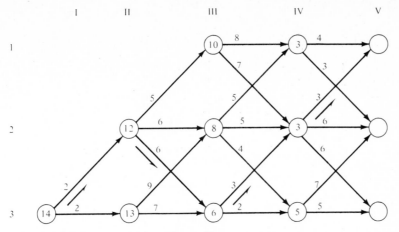

I   II   III   IV   V

Figure 9-6 Solution to network routing problem.

The optimum path is I-3, L to II-2, R to III-3, L to IV-2, and L to V-1. The shipping cost will be 14 units.

The advantage of dynamic programming is a reduction in computation requirements. In this simple network routing problem, had we decided to enumerate and evaluate all the route possibilities, we would have found 29 different routes for the shipment to move from I-3 to V with 4 node-to-node calculations per route, i.e., a total of 116 calculations. By using the principle of optimality, we reduced our problem to 21 calculations. The reduction was achieved by removing each nonoptimal route from consideration as soon as it was detected.

Let us now see how we can reduce a simple problem of introducing capacity in the face of an increasing demand into a network problem. We begin at time zero with an existing plant (A) and are to arrive at an end state after 3 years of (existing plant + plant B + plant C). The alternative decisions possible at each node (year) are: "Build plant B" (left), "do nothing" (straight), or "build plant C" (right). This is shown as a network diagram in Fig. 9-7. At node III-1, for example, the situation would be that at year 1 we now have plants A and C. The decision could be either "do nothing," which would bring us to III-2 or else "build," which would bring us to II-2. There is no freedom of decision at column 2 because we must get to II-3 in accordance with the constraints. To complete the analogy with the network routing example, it only remains for us to insert the NPWs for each node-to-node move. The computation requirements for this particular case would be 12 node-to-node calculations by the dynamic-programming approach compared with 18 node-to-node calculations if each strategy were to be enumerated and evaluated. The interested reader is referred to Rose (1976) for a full description and presentation of the dynamic-programming research algorithm for planning production capacity for an uncertain future demand. Example 9-4 is taken in a slightly different form from Rose (1977).

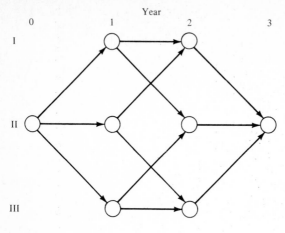

Figure 9-7 Provision of capacity as a network routing problem; constraints are as follows: year 0 = plant A, year 3 = plants A + B + C; left arrow = build plant B, straight arrow = do nothing, right arrow = build plant C. To avoid confusion between year 2 and column 2 the relative positions of arabic and roman numerals have been transposed from those in Figs. 9-5 and 9-6.

Example 9-4 A new product has emerged from the research department and its sales will depend on how well the product is accepted by the market and whether any disadvantage becomes apparent after it has been introduced. The product should take hold in certain market areas, but there are varying degrees of likelihood of its acceptance in other areas. The widely divergent sales forecasts are presented in Fig. 9-8. Each forecast has a 0.25 probability. The alternatives for supplying the product, which is expected to sell for $400 per ton,† are presented in Table 9-10. The policy which is to be followed is to be a two-stage expansion, i.e., an initial plant now, which will be followed

† The original problem was stated in pounds sterling, which have been changed to dollars for simplicity.

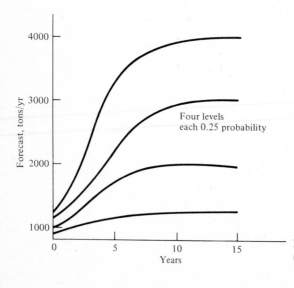

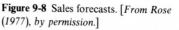

Figure 9-8 Sales forecasts. [*From Rose (1977), by permission.*]

Table 9-10 Alternatives for supplying product, data summary

Size code	Plant size	Capacity, tons/yr	Capital, thousands	Annual fixed cost, thousands	Variable cost per ton
1	Large	3750	$2000	$350	$35
2	Medium	2750	1650	325	35
3	Small	2000	1430	275	35
4	Mini	1200	1100	200	35

Source: From Rose (1977), by permission.

at some later date by additional capacity. Alternatively, a new plant could be built to replace the initial plant, which would be scrapped. The first examination to be made will be on the basis of a policy of satisfying demand for product, after the initial plant, only if it is economic to do so. A subsequent examination will be made for the case where the policy will be to satisfy demand at all times.

SOLUTION The dynamic-programming search algorithm (Rose, 1976) was used to calculate the optimum strategy for each sales forecast level and for each initial decision alternative. The optimum strategy for sales-forecast level 2 when the initial decision is plant 4 is shown in Table 9-11, along with

Table 9-11 Typical optimum investment plan

Year	Decision Build	Decision Scrap	Capacity, tons/yr	Demand, tons/yr
1	4	0	1200	1200
2	0	0	1200	1400
3	0	0	1200	1600
4	0	0	1200	1810
5	0	0	1200	2080
6	0	0	1200	2340
7	2	4	2750	2580
8	0	0	2750	2760
9	0	0	2750	2880
10	0	0	2750	2940
11	0	0	2750	2974
12	0	0	2750	2980
13	0	0	2750	2990
14	0	0	2750	3000
15	0	0	2750	3000

† Sales forecast, level 2, first decision plant 4.
Source: From Rose (1977), by permission.

Table 9-12 Decision matrix, shortfall allowed

Initial decision (plant)	Forecast level (equal probability)				Expected value
	1	2	3	4	
1 Large	$131,780	$ 7,540	$ −91,100	$ −177,710	$ −32,370
2 Medium	93,250	44,050	−43,130	−129,730	−8,890
3 Small	78,330	32,440	3,650	−82,950	7,870
4 Mini	84,050	16,440	−1,940	−16,280	28,710

Source: From Rose (1977), by permission.

the production capacity and forecast demand for each year. The decision matrix for this case, where a shortfall in production capacity compared with demand is permitted, is given in Table 9-12. The utilities in the matrix are NPWs in dollars.

By inspection of this table it is apparent that the appropriate initial decision based both on expected value and on minimizing the maximum loss is to build 4, the miniplant.

If it were possible to ascertain by a reliable forecast how the market would develop, we would be able to make a more profitable decision. If, for example, level 1 were the correct one, the decision would be one large plant with an NPW of 131,780. We can calculate the value of having such "perfect" information. The value of this information will be the difference between the expected value of the perfect-information decisions and the expected value of the best single decisions. For this case

$$\text{Value of perfect forecast} = \frac{131,780 + 44,050 + 3650 + 16,280}{4} - 28,710$$

$$= \$20,230$$

This sum, then, is the "regret" for not having an accurate forecast and can provide a quantitative basis for a decision whether a more extensive sales-forecasting program should be undertaken and how much can be allocated to such a program.

The policy discussed thus far was based on providing capacity, only when it was economic to do so. In Table 9-11 we saw that one optimal strategy had a shortfall during 10 years of the 15-year project life. Such a marketing policy is clearly impossible for a new product and would be highly undesirable for any product. In Table 9-13 we see the decision matrix resulting, again for a two-level expansion for each initial decision, the policy now being that no shortfall is allowed.

Table 9-13 Decision matrix, no shortfall allowed

Initial decision (plant)	Forecast level (equal probability)				Expected value
	1	2	3	4	
1 Large	$13,780	$ 7,540	$−91,100	$−177,710	$−32,370
2 Medium	80,670	44,050	−43,130	−129,730	−12,035
3 Small	66,490	8,270	3,650	−82,950	−1,135
4 Mini	50,830	−16,320	−36,270	−16,280	−4,510

Source: From Rose (1977), by permission.

The optimal initial decision for the new policy would be to build 3 (small plant) for the maximum expected value, but the maximin decision would still be to build 4 (miniplant) as the initial decision. It should be noted that the NPWs are, in general, lower for this case than for the preceding one. Also, as one would expect, the expected value of perfect forecast information is higher for this case compared with the preceding one and is $41,930.

In concluding this example it is well to point out that although a complete strategy for providing production capacity can be produced, the only decision of current importance is the initial decision: How big should the next plant be? The strategy approach is necessary to arrive at a well-founded preferred initial decision. Forecasts, technologies, and economic considerations will change, and the second decision will be taken in the light of the situation prevailing at that time. In other words, another strategy will have to be selected at that time and rather than adhering slavishly to our original "second" decision, we would then select the appropriate new "initial" decision. Management will, of course, want the most likely strategy for long-range planning considerations, but the only operative part of the strategy will be the initial decision.

PROBLEMS

9-1 The cash flow for the project described in Prob. 7-9 is uncertain, primarily because of uncertainties in the sales projections. Assume that the uniform cash flows presented in Prob. 7-9 are the mean values and that the actual cash flows will, with a 0.95 probability, lie between ± 20 percent of the mean values. Make an analysis of the net present worth on the basis of this additional information.

9-2 Two streams, A and B, are to be blended continuously. The desired flow rates are 400 m^3/h of A and 600 m^3/h of B. The estimated tolerance of the flow control loop is ± 5 percent of the set-point value. Assume that the distributions are normal and that the tolerances represent ± 2 standard deviations. What is the probability that the actual total flow rate will be (a) 1000, (b) 950, and (c) 980 m^3/h?

9-3 You are designing a heat exchanger in which hot water is to be cooled with air. The following design equations, nominal conditions, and tolerances are available:

$$U_0 = CV^{0.8} \text{ J/(m}^2 \cdot \text{s} \cdot {}^\circ\text{C)}, \text{ overall heat-transfer coefficient}$$

$$V = \text{hot water velocity, m/s}$$

$$d_o = \text{pipe OD, cm}$$

$$\text{Design air temperature} = 35^\circ\text{C}$$

$$\text{Hot-water temperature} = 80 \pm 10^\circ\text{C}$$

$$C = 200 \pm 40 \qquad V = 3 \pm 0.5 \qquad d_o = 2.50 \pm 0.03$$

Assume that the distributions are normal and that the tolerances represent ± 2 standard deviations.

(a) What is the probability that the actual heat-transfer rate will exceed the nominal rate?

(b) What is the probability that the actual heat-transfer rate will be at least 80 percent of the nominal rate?

9-4 MPA and PPA, which have identical selling prices, can both be made in the same plant by appropriate changes in operating conditions. Two processing schemes are available: scheme A, in which MPA production is favored and PPA is produced at a penalty cost, and scheme B in which PPA production is favored. Our marketing department is confident that it can sell 100 tons/day of MPA + PPA. The market estimates are (1) 75% MPA, 25% PPA with a 0.6 probability; (2) 50% MPA, 50% PPA with a 0.3 probability; and (3) 25% MPA, 75% PPA with a 0.1 probability. Production costs are $300 and $400 per ton for MPA and PPA, respectively, for proposal 1 and progress linearly to $280 and $500 per ton for proposal 3 when produced via scheme A. The comparable figures for scheme B are $420 and $320 per ton for proposal 1 progressing linearly to $460 and $300 per ton for proposal 3.

(a) Which processing scheme would you recommend, and what would be the expected cost per ton of production?

(b) Repeat part (a) if the sales forecast were proposal 1 with a 0.8 probability and proposal 2 with a 0.2 probability.

(c) Repeat part (a) if the sales were proposal 2 with a 0.2 probability and proposal 3 with a 0.8 probability.

(d) What suggestions could you make to the marketing department with respect to market development goals?

9-5 Our marketing department has had second thoughts about the forecasts concerning the sales of MPA and PPA (Prob. 9-4). Although the market studies still show that the sales mix to be expected is proposal 1, 2, or 3, as before, the marketing people are now reluctant to attach probability figures to these mixes. They still are supremely confident that they will be able to sell 100 tons/day of MPA + PPA. What processing scheme would you now recommend?

9-6 Your company is to build a plant to produce a peroxide free-radical catalyst. Marketing has lined up several clients who are willing to sign long-term purchase contracts at a price for the catalyst of $20 per kilogram. At that price the contracts would guarantee sales of 40,000 kg/yr. Marketing estimates that additional spot sales of up to 20,000 kg/yr can be made at $25 per kilogram. Investment in the catalyst plant is $10 for each kilogram per year of production capacity. Variable production costs are $7 per kilogram and annual fixed charges are 40 percent of plant investment. Since the catalyst has a relatively short shelf life, it must be assumed that any unsold production cannot be stored. The handling and disposal of unsold production would cost $3 per kilogram. What catalyst production capacity would you recommend, and what would be the expected maximum profit?

9-7 A plant is to be built to process an agricultural product. The quantity of product the farming community will make available to the processing plant is uncertain, but it is estimated that it will range from 10,000 to 25,000 tons per season if the farmers are paid $50 per ton. Plant investment would be $20 per design ton per year, and processing costs would be $10 per ton. Annual fixed costs including all capital charges will amount to 50 percent of the investment.

(*a*) If the processed product sells for $100 per ton (1 ton of product is obtained per ton of processed feed), what plant capacity would you recommend and what would be the expected annual profit?

(*b*) If the price paid to the farmer were $60 per ton, it is expected that the supply to the plant would be in the range of 20,000 to 25,000 tons/yr. What plant capacity would you recommend, and what would be the expected annual profit?

9-8 Your company now has a plant under construction to produce product A. The plant is expected to come on stream in 1 year. Marketing would like to begin to develop our market by buying A in bulk and marketing it under our label. They are uncertain how much market penetration they can expect during the year before our production comes on stream but estimate that they can sell between 100,000 and 250,000 tons of A during this period at $150 per ton. They also have found a supplier who is willing to sell us A at $110 per ton provided that we sign a purchase contract within 10 days. Marketing wants to be able to satisfy any demand that they are able to develop. If the demand they generate exceeds the original purchase-contract quantity, they will be able to make additional spot purchases of A at $140 per ton. In either case we estimate that our additional costs for repackaging, overheads, etc., will amount to $20 per ton of purchased A. If we should have any unsold A left when our plant comes on stream, it will represent an additional loss of $20 per ton because of the reduction in production requirements in our new, efficient plant. How much A should you recommend be purchased at $110 per ton? What will the expected profit be?

9-9 An organic chemical intermediate is needed as part of a processing sequence. Since the intermediate decomposes rapidly, it will be necessary to produce it on site in a batch reactor. The cost of running a batch is C_R, where C_R includes chemicals, labor, utilities, fixed costs, etc. The probability that a batch of intermediate will be of specification grade is R. If off-specification material is obtained, i.e., failure occurs, the downstream operation cannot proceed and this involves an additional cost of fC_R. To overcome this problem it has been suggested that a number of batch reactors be operated to ensure minimum interruption of the downstream processing operations.

(*a*) Show that the number of batch reactors that would minimize the total cost is

$$n = \frac{\ln\left(1/\{f\ln\left[1/(1-R)\right]\}\right)}{\ln\left(1-R\right)}$$

(*b*) How many reactors would be required if $C_R = \$1000$, $f = 25$, and $R = 0.6$?

9-10 In a certain plant an annual shutdown after 8000 operating hours is scheduled to permit equipment repairs and overhauling, heat-exchanger cleaning, corrosion measurements, minor process changes, etc. Repaired and overhauled equipment is assumed to behave like new. From records on a certain type of pump it is known that the failure probability density function can be expressed as an exponential probability density function and that the mean time between failures is 12,000 h. What is the probability that a pump will fail between the scheduled shutdowns?

9-11 A compressor crucial to a certain process costs $50,000 installed and ready to operate. From statistical information on the performance of this type of compressor the MTBF is 18,000 h, and the MTTR is 60 h. The annual operating income for the process (without annual maintenance expenses of 5 percent of investment) is expected to be $20 million. The plant is to operate for 8 kh/yr and then will be shut down for annual maintenance, repairs, and overhaul. The project life assumed in the economic analysis was 15 years with straight-line depreciation over this period and taxes at 48 percent. The interest rate was 15 percent. Would you recommend installation of a standby compressor?

9-12 It is obviously expecting too much of a market survey to be able to provide us with perfect information. Even the most extensive market survey could provide us only with imperfect information, i.e., forecasts to which we could assign probabilities. From our past experience with extensive forecasts let us assume for Example 9-4 that if either level 1 or 4 were forecast, the probability is 0.7 that the forecast level is correct and 0.3 that the adjacent level is correct. If levels 2 or 3 were forecast, the probability is 0.7 that the forecast level is correct and 0.15 that each of the adjacent levels is the correct forecast. Using the data in Table 9-12, prepare a new decision matrix in which the elements of the matrix are the new expected values based on the imperfect information. What will the expected value of this information be?

CHAPTER

TEN

TACTICS AND STRATEGY OF PROCESS SYNTHESIS

> "And God Saw that it was good" (*Genesis*, 1:10) Does not his saying this after the act [of Creation] imply that he did not foresee its nature, perfection and beauty? ... No! The words are meant as a precept for mankind. It is not proper for a man to praise his own work before it is complete, while it may yet be found wanting and belie him.
>
> *"New Zohar" on Genesis, 19*

In Chap. 2 we touched upon some of the elements of synthesis, and in subsequent chapters we discussed a number of tools and techniques that are available to the chemical engineer for the analysis of chemical processing schemes. We are now in a position to discuss some of the more formal and systematic procedures for the synthesis of chemical processes. We shall first discuss several approaches to process synthesis whose aims are to lead to a better understanding of the process and from that to synthesis. This will be followed by programmed strategies, a brief introduction to optimization methods and to process simulation. We then present one of the computer-aided process synthesizers, in which the computer and the design engineer interact to reach the goal of a process synthesis.

10-1 QUALITATIVE APPROACHES TO PROCESS SYNTHESIS

Heuristics

Heuristics is a branch of study whose aim is to study the methods and rules of invention and discovery. Heuristic, as an adjective, means "serving to discover" (Polya, 1945). It proceeds along empirical lines using rules of thumb and pertains to exploratory methods of problem solving. Heuristic rules, or heuristics, are rules of thumb that result from generalizations based upon many observations. We have already encountered some heuristics in, for example, the rules of thumb for distillation-column sequences for the separation of multicomponent mixtures.

Heuristic reasoning is not to be regarded as final and strict but only as provisional and plausible. The heuristic method seeks the solution to a problem by means of plausible but fallible guesses. Following Pappus, whom we quoted in Chap. 2, "we assume what is required to be done as already done" by some plausible route, but we must determine whether this plausible solution is, indeed, a solution in the set of plausible solutions.

Heuristic rules can be invoked to assist in the establishment of processing conditions and configurations. Some of these rules can be cited as examples: the optimum reflux ratio is in the range of 1.2 to 1.4 times the minimum reflux ratio; the temperature of cooling water being returned to cooling-water towers should not exceed 120°F (50°C); minimum approach temperature in a heat exchanger should be limited to 20°F (10°C); etc. These rules are useful to the designer and can help in the reduction of the number of conditions to be examined.

Let us assume that we are faced with the task of performing a separation. We saw earlier that King (1971) listed 42 separation processes. An exhaustive study of each of these processes with the aim of reaching the optimum process would require a formidable investment of time, effort, and money. The field of candidate processes can be narrowed considerably if we invoke heuristics, some of which were discussed by King.

The first and obvious rule of thumb is that the separation process must be capable of effecting the desired separation, i.e., the principle upon which the separation is based must be relevant to the system to be separated. Ion exchange cannot be used, for example, for the separation of nonionic compounds. An additional heuristic is that processes that do not require extreme processing conditions are to be preferred over those which do. In addition, systems and components that cannot tolerate extremes of temperature or pressure, for example, cannot be separated by processes in which such extremes would be required to effect the separation.

The list of heuristics includes additional items: avoid solid phases, which are difficult to handle; favor the process with large separation factors; favor energy as a separating agent over a mass separating agent since the mass separating agent (solvent, stripping gas, etc.) must be removed in an additional separation operation; favor equilibrium separation processes over rate separation processes; favor processes for which experience and know-how are available. An expensive separation process can be considered for a high-value product but not for a low-value product. The process must also be compatible with the scale of operations contemplated. Some separation processes can handle only kilogram, gram, or even milligram quantities, and they are obviously unsuitable for processing tonnage quantities. When a number of components are to be separated into only a few products, the process that will separate the components directly into the desired product groups will be the preferred one. Finally, unless there is a good reason why distillation cannot be used or is not well suited as a separation process, distillation will usually prove to be the leading candidate process.

Although several of these heuristics may conflict with each other in any real situation, they at least are capable of narrowing the field and of reducing the number of processes to be considered. We shall return to heuristics when we examine some programmed synthesis techniques.

Morphological Analysis

Morphology is the study of the structure or form. In our case morphological analysis involves the rigorous examination and evaluation of all possible alternatives for each part of a problem. It aims to produce the optimum solution by virtue of having considered and evaluated every possibility. The analysis provides a logical structure that can replace random thinking, is comprehensive, helps to ensure that all alternative solutions are being considered, and presents the possibility of new process syntheses resulting from combinations of alternatives that had not been previously considered.

A simple analysis is shown in Fig. 10-1. In Fig. 10-1*a* solutions to a one-stage process problem are indicated, and elimination of alternatives is to be accomplished by the successive application of criteria A, B and C. The first part of the operation, where the alternatives are generated, is termed *divergence*, and the second part, where evaluation and elimination takes place, is termed *convergence*. In Fig. 10-1*a* alternatives 2 and 6 were eliminated by criterion A, alternatives 5 and 7 by criterion B, and alternatives 1, 4 and 8 by criterion C. Thus, the surviving solution is alternative 3.

In Fig. 10-1*b* a three-stage process is shown, and one alternative from each group of five alternatives provides the solution. Criterion A eliminated alternatives I-4, II-2, and III-2; B eliminated I-1, II-3, II-5, and III-3; and C eliminated I-3, I-5, II-1, III-4, and III-5; so the surviving process solution is I-2, II-4, and III-1.

A number of morphological analyses reported in the literature were cited and briefly described by King (1974). We shall examine in some detail the morphological analysis applied by Bridgwater (1968) to the production of iron from iron ore or scrap iron. A comprehensive list of all known methods of separation was first compiled, but processes involving vaporization were eliminated from consideration. Standard mechanical methods of separation such as filtration, electrostatic separation, and centrifugation were not considered but, where appropriate or required, were assumed to be part of, and included in, the

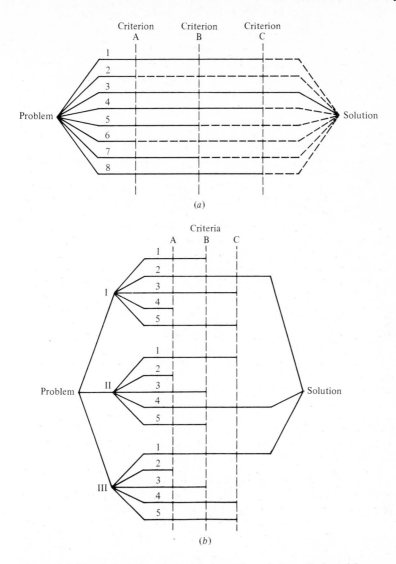

Figure 10-1 Morphological analysis: (*a*) alternative solutions eliminated by successive application of criteria; (*b*) analysis of the three-stage process.

individual mass-transfer separation methods. Twenty-seven possible separation processes remained as candidates for consideration as stages of a chemical process for the separation of iron. It was assumed that the maximum number of acceptable stages for the process would be limited to four. If all permutations of these 27 methods in a four-stage process were to be considered, a total of 531,441 complete routes would have to be evaluated, a formidable if not impossible task.

The next step of the morphological analysis was the elimination of candidate processes on the basis of two criteria: available knowledge of the method and feasibility for the system involved. The application of the "information" criterion eliminated 7 methods, and feasibility considerations eliminated 9 additional methods. The remaining 11 surviving candidate methods would still require about 15,000 route evaluations for convergence to be obtained.

Additional criteria were then applied. The feedstock, iron ore or scrap iron, is solid and, as a result, the only feasible first-stage method from the surviving candidates would be leaching. Purity requirements for the product were such that the only feasible final-stage separation methods would be electrolysis or reduction. At this stage of the analysis, electrolysis could have been in either a fused or aqueous phase, and reduction could have been in a liquid or a solid phase. Compatibility rules were then established. These rules were based on processing requirements, phase relations, and logic. Typical compatibility (really incompatibility) rules were, for example, that leaching cannot be followed directly by fused electrolysis, crystallization by evaporation, etc., and, in addition, repetition of a method was not permitted. After the application of these criteria, only 56 routes remained to be examined: 2 possible routes for a two-stage process; 14 possible routes for a three-stage process; and 40 possible routes for a four-stage process.

The penultimate stage of the analysis was an economic evaluation of the surviving separation methods. This was done by assuming a maximum allowable processing cost per stage, this cost being based on the value of the iron product. The costs of four methods exceeded this maximum and were eliminated by the cost criterion. Now only six candidate processing routes remained for final evaluation:

Leaching and aqueous electrolysis
Leaching, crystallization, and aqueous electrolysis
Leaching, crystallization, and solid reduction
Leaching, crystallization, thermal decomposition, and solid reduction
Leaching, precipitation, thermal decomposition, and solid reduction
Leaching, evaporation, thermal decomposition, and solid reduction

An exhaustive evaluation of all of the original candidate-routes for two-, three-, and four-stage processes would have required that 551,880 processing routes be evaluated. The number of complete routes to be evaluated was reduced to 6 by the application of three sets of criteria to the candidate methods during the convergence stage of the morphological analysis: information and feasibility criteria, processing and compatibility rules, and economic criteria.

It was gratifying for Bridgwater (1968) to note that the Research Council of Alberta (RCA) process, as developed for ore or scrap-iron leaching to produce pure iron powder, was indeed included as one of the final six routes reached in the morphological analysis that he performed. The steps in the RCA process are leaching, crystallization, and solid reduction, but the complete RCA process includes, of course, the mechanical separations that were eliminated from separate consideration in the early stages of the morphological analysis.

Although in many cases the results of a morphological analysis are obvious solutions that are very likely those to be found intuitively by the experienced design engineer, the technique requires that all possibilities and alternatives be covered and considered by the designer and ensures that they will be. The detailed examination of the fundamentals of a process as demanded by the morphological analysis also creates favorable circumstances and background for radical innovation "not necessarily as a direct result of rigorous analysis, but possibly sparked off by well-disciplined thought," as Bridgwater says.

10-2 FUNCTIONAL ANALYSIS AND EVOLUTIONARY PROCESS SYNTHESIS

We have already seen some elements of a functional analysis in our earlier discussion of the attribute table (Woods and Davies, 1972). Functional analysis, which leads to a thorough understanding of a process, its configuration, and equipment items, is a necessary part of evolutionary process synthesis or design. In functional analysis the overall goals of a process are stated, and each step of the process is then examined and analyzed to see how it relates to the overall goals. In evolutionary process synthesis a design that meets the overall goals is first established. A functional analysis of each part of the design is then made, and opportunities are sought for design improvement on an evolutionary basis.

Functional Analysis

We shall perform a functional analysis of a urea process. This analysis will restrict itself to the major process steps, but a more detailed analysis can be left as a problem for the reader. The largest consumer of urea, a major petrochemical, is the fertilizer industry, in which urea represents an important supplier of nutrient nitrogen. Urea, NH_2CONH_2, is produced commercially by the direct dehydration of ammonium carbamate, NH_2COONH_4, at elevated temperature and pressure. Ammonium carbamate is obtained by reaction between ammonia and carbon dioxide:

$$2NH_3 + CO_2 \rightleftharpoons NH_2COONH_4$$
$$NH_2COONH_4 \rightleftharpoons NH_2CONH_2 + H_2O$$

The product stream from the urea-synthesis reactor always contains unreacted carbamate and excess ammonia.

In the once-through process, which is no longer extensively used, the unconverted carbamate is decomposed into NH_3 and CO_2 by heating the reactant effluent at low pressure. The NH_3 and CO_2 are separated from the urea solution, the NH_3 is usually converted into ammonium sulfate or nitrate, and the CO_2 is vented. An important group of commercial processes uses internal recycle of carbamate. In this processing scheme, the unreacted carbamate and the excess ammonia are stripped from the urea-synthesis-reactor effluent by means of gaseous CO_2 or NH_3 at reactor pressure, and the recovered NH_3 and CO_2 gas is condensed and returned to the reactor. One of the commercial realizations of this scheme is the Stamicarbon process, shown in Fig. 10-2 as a flow diagram.

A functional analysis of the major steps in the Stamicarbon CO_2 process is presented in Table 10-1. The overall goal of the process is the production of prilled urea from NH_3 and CO_2. This type of analysis can be extended to include not only the major process steps but also each individual item of equipment. As a result of the analysis questions can be raised about process conditions and alternative equipment configurations. When that point is reached, we are entering into evolutionary process synthesis.

Evolutionary Process Synthesis

Most design engineers approach a process synthesis or design problem through an evolutionary strategy. A process configuration is established that can reach the desired goal, the configuration and conditions are examined and analyzed, and changes are made that improve the design. The revised process is again analyzed, additional improvements are made, etc. Each revised version is an improvement on its predecessor.

To demonstrate the procedure, we rely primarily on the example by King et al. (1972), in which they used evolutionary design in the optimization of a demethanizer tower in an ethylene plant. The function of the tower is to remove hydrogen and methane from ethylene and heavier components as one of the steps in product separation and purification. Operation is generally at a pressure of 400 to 500 lb/in^2 abs with ethylene as the refrigerant. The specifications for the demethanizer, as presented by King et al. (1972), are shown in Table 10-2. The temperature of the vapor distillate product, or tail gas, cannot exceed $-150°F$ because this gas is to be used for additional heat-exchange purposes in other parts of the overall process. The factor x in Table 10-2 represents the amount of ethylene lost in the tail gas and is a dependent variable.

The evolutionary synthesis was begun by establishing the simplest process that could be conceived. This scheme I is shown in Fig. 10-3. An economic evaluation of this scheme was then made. As can be seen in Table 10-3, the total cost was $1,894,000 per year, the major contributors to this cost being ethylene losses at $1,025,000 and $-150°F$ refrigeration at $555,000. The ethylene loss in the tail gas could be decreased by reducing the overhead temperature. This led to the first improvement shown as scheme II (Fig. 10-4), in which the tail gas is expanded to produce an autorefrigeration effect. As a result, the separator-drum temperature was reduced to $-143°F$, and the $-150°F$ refrigeration duty was slightly decreased. The result is a slight decrease in ethylene loss and in refrigeration costs; the total annual cost is reduced to $1,754,000.

The $-150°F$ refrigeration and ethylene losses are still the predominant expense items, and both can be reduced if the amount of overhead reflux can be reduced. This can be achieved if the feed temperature is reduced. This is shown as

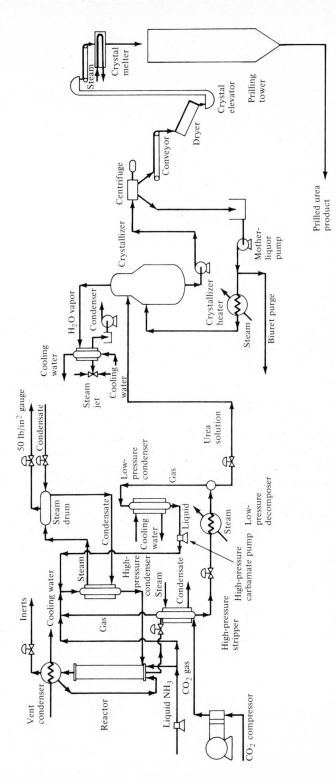

Figure 10-2 Stamicarbon CO$_2$ stripping process for urea manufacture. [*From Kirk and Othmer (1970), by permission.*]

Table 10-1 Functional analysis of Stamicarbon CO_2 stripping process for urea manufacture

Item	Function	Relation to overall goal
Reactor	Converts feed material into urea and ammonium carbonate	Creates product for recovery and purification
Vent condenser	Condenses NH_3 and CO_2 in reactor gas	Recovers and recycles reactants from inert purge stream
High-pressure stripper	Strips excess NH_3 and decomposes unreacted carbamate from reactor effluent	Reduces cost and increases efficiency by recovering and recycling unreacted materials in reactor effluent
High-pressure condenser	Condenses NH_3 and CO_2 from high-pressure stripper	Permits recycle of unreacted NH_3 and CO_2 to reactor
Low-pressure decomposer	Decomposes carbamate remaining in high-pressure-stripper effluent	Produces urea solution
Low-pressure condenser	Condenses NH_3 and CO_2 from low-pressure decomposer	Permits recovery and recycle of NH_3 and CO_2
High-pressure carbamate pump	Pumps low-pressure condensate to high-pressure condenser	Permits recycle of unreacted NH_3 and CO_2 to reactor via high-pressure condenser
Crystallizer	Crystallizes urea	Produces urea crystals from 70–75% urea solution
Condenser-steam jet	Condenses H_2O vapor from crystallizer	Produces vacuum to permit crystallizer operation
Biuret purge	Purges biuret from the system	Removes biuret, an undesirable material that forms at conditions prevailing in crystallizer, to ensure product quality
Centrifuge	Removes mother liquor from crystallizer effluent	Reduces cost by reducing amount of water to be removed in dryer
Dryer	Removes water from centrifuged product	Produces dry, crystalline product
Melter	Melts crystalline product	Produces molten feedstream for prilling tower
Prilling tower	Produces urea prills	Reaches final goal

268

Table 10-2 Demethanizer specifications

	Feed			Vapor distillate product	Bottoms product
	Total	Liquid	Vapor		
Pressure, lb/in^2 abs	460			65	460
Temperature, °F	-20			-150	~ 50
Flow, lb mol/h:					
Hydrogen	1,770	45	1725	1770	
Methane	2,980	375	2605	2965	15
Ethylene	3,080	1246	1834	x	$3080 - x$
Ethane	740	377	363	—	740
Propylene	1,290	1007	283	—	1290
Propane	140	113	27	—	140
Total	10,000	3163	6837	$4735 + x$	$5265 - x$

Murphree vapor efficiency in tower = 50%

	Ethylene refrigeration	
	Level, °F	Cost
	-90	\$2.10 per 10^6 Btu
	-150	\$3.80 per 10^6 Btu

Cost of steam in reboiler = 50 cents/10^6 Btu
Cost of ethylene lost in tail gas = 2.0 cents/lb
Cost of recycled propane = 0.265 cent/lb (available at -80°F, 460 lb/in^2 abs)

Source: From C. J. King, D. W. Gantz, and F. J. Barnes, *Ind. Eng. Chem. Process Des. Develop.*, **11**: 271 (1972), by permission. Copyright by the American Chemical Society.

scheme III (Fig. 10-5), in which the feed has been cooled to -80°F with -90°F refrigerant. The annual expense due to ethylene loss has now been reduced to \$800,000 and total refrigeration expenses to \$489,000.

In scheme IV (Fig. 10-6) the -90°F refrigeration load was reduced by first performing a vapor-liquid separation on the feed stream. The vapor stream was cooled to -80°F, and the vapor and liquid feed streams were introduced into the tower separately, each at its appropriate location. Although the total annual cost decreased as a result, we note from Table 10-3 that it was necessary to increase the overhead reflux flow to effect the separation and an increase in expenses due to ethylene loss was a direct result. To decrease the overhead reflux flow scheme V (Fig. 10-7) was generated, in which the separated vapor feed was further chilled with -150°F refrigerant. The expense due to ethylene loss is still the predominant factor. The final scheme in this sequence is scheme VI (Fig. 10-8), in which the overhead reflux flow was reduced by supplying intermediate reflux near the top of the tower. This was accomplished by withdrawing

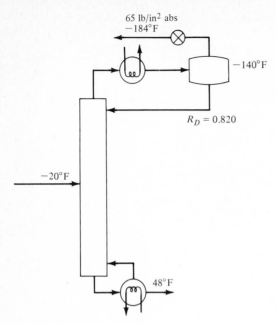

65 lb/in^2 abs
$-184°$F

$-140°$F

$R_D = 0.820$

$-20°$F

48°F

Figure 10-3 Demethanizer scheme I. [*From King (1974), p. 15, by permission.*]

Table 10-3 Comparison of demethanizer designs

	Scheme					
	I	II	III	IV	V	VI
Overhead reflux flow, lb mol/h	4,059	4,309	1,667	1,962	491	341
Temperature of final separator, °F	-140	-143	-146	-145	-150	-160
Ethylene in tail gas, lb mol/h	209	185	163	166	136	71
Annual costs:						
Heat exchangers	$96,000	$179,100	$227,000	$201,000	$212,000	$211,000
Distillation column	70,000	70,000	70,000	70,000	70,000	70,000
C_2H_4 refrigeration, $-90°$F	—	—	295,000	208,000	208,000	208,000
$-150°$F	555,000	548,000	194,000	234,000	238,000	280,000
Steam to reboiler	48,000	52,000	76,000	61,000	61,000	65,000
Propane recycle	—	—	—	—	—	—
Ethylene loss	1,025,000	905,000	800,000	812,000	665,000	348,000
	$1,894,000	$1,754,000	$1,662,000	$1,586,000	$1,454,000	$1,182,000

Source: C. J. King, D. W. Gantz, and F. J. Barnes, *Ind. Eng. Chem. Process Des. Develop.*, **11:** 271 (1972), by permission. Copyright by the American Chemical Society.

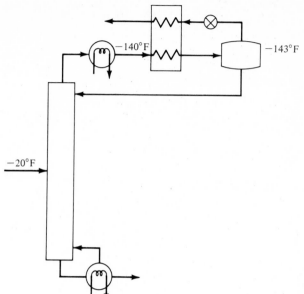

Figure 10-4 Scheme II: tail-gas expansion provides additional cooling. [*From King (1974), p. 13, by permission.*]

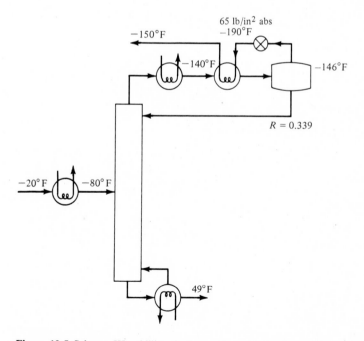

Figure 10-5 Scheme III: chilling feed provides additional overhead reflux. [*From King (1974), p. 14, by permission.*]

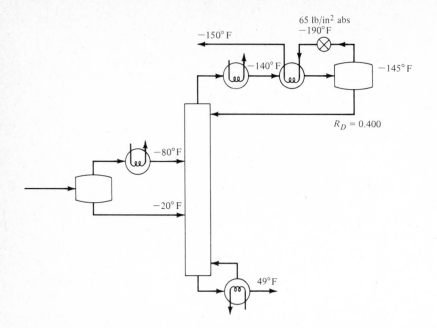

Figure 10-6 Scheme IV: feed separation combined with feed prechilling. [*From King (1974), p. 14, by permission.*]

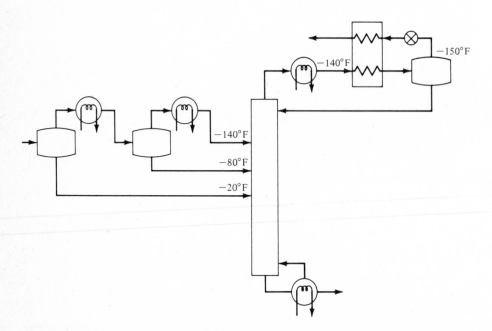

Figure 10-7 Scheme V: additional feed chilling and separation. [*From King (1974), p. 14, by permission.*]

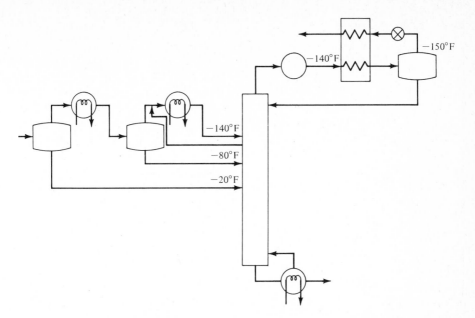

Figure 10-8 Scheme VI: elimination of need for external refrigeration in column overhead system. [*From King (1974), p. 15, by permission.*]

vapor from a plate with a temperature above $-140°F$, passing this vapor through a $-150°F$ chiller, and returning it to the tower partially liquefied. By providing enough condensation in this intermediate condenser this scheme also completely eliminated the need for the refrigerated overhead condenser.

This evolutionary process synthesis resulted in a design in which the total annual cost for the demethanizer was reduced to \$1,182,000 from the annual cost of \$1,894,000 for the original process scheme. The final process shown in this evolutionary-process-synthesis example was based on the original plan for effecting the demethanization, and no new separation concepts were introduced. The changes and improvements made from scheme to scheme were all based on sound knowledge and understanding of the distillation operation.

Evolutionary process synthesis represents a systematic, logical approach to process synthesis. It is also the procedure adopted by the experienced human design engineer, who has (among others) one important advantage over the evolutionary strategy. The experienced engineer also does a morphological analysis, at least in thinking about the problem, if not in the formal manner outlined earlier. Thus, the engineer is more likely to arrive at additional conceptual approaches than a synthesis strategy based only on evolutionary improvement of a primitive design that "solves the problem."

10-3 PROGRAMMED STRATEGIES FOR PROCESS SYNTHESIS

The qualitative procedures discussed above were qualitative only in the sense that the design engineer generated alternatives without recourse to quantitative procedures. The evaluation stages of the synthesis, whether they were reached by heuristics, morphological analysis, or evolution, are generally carried out with the help of a computer. The volume of computation required in the evolution is such that an optimal or near optimal solution could not otherwise be reached.

Efforts have been made to produce structured methods for process synthesis that could be programmed for, and implemented by, the digital computer. Several of these techniques will be discussed. Generally, each of the techniques is used in combination with one or more of the other techniques in order to perform an optimal process synthesis.

Process Decomposition

Even relatively simple processing systems are characterized by a large number of equipment configurations and operating conditions. A search over all possible sets of configurations and operating conditions presents an almost impossible combinatorial problem. Rudd (1968) proposed a strategy for process synthesis in which a design problem for which no previous technology exists is broken down into a sequence of smaller subdesign problems. These smaller tasks either are within existing technology or present simpler synthesis problems.

In accordance with Rudd, the original synthesis problem is defined by a set of task constraints X, which represent raw-material availability, product specifications, temperature, etc. The region of available technology is represented by R. If X is contained within available technology, i.e., if $X \subset R$, no synthesis problem exists and the economics are denoted by $E(X)$. If, however, X does not lie within R, the desired performance cannot be achieved with one piece of technology and a process must be synthesized. The desire is to optimize system performance, and subtasks X_j are sought such that

$$O^*(X) = \underset{X_j}{\text{Opt}} \left[\sum E_j(X_f) \right] \tag{10-1}$$

and

$$X = \bigcup_j (X_j) \tag{10-2}$$

$$X_i \cap X_j = 0 \qquad \text{for all } i \neq j \tag{10-3}$$

$$X_j \subset R \qquad \text{for all } j \tag{10-4}$$

Let us assume that our original problem is decomposed or torn into two smaller problems S_1 and S_2 such that

$$S_1 = X_1 \cup T \qquad S_2 = X_2 \cup T$$

$$X = X_1 \cup X_2 \qquad X_1 \cap X_2 = 0 \tag{10-5}$$

where T is a set of arbitrarily imposed tear constraints. They unite the two subtasks S_1 and S_2 to accomplish the original task specified by X. The basic problem is to select between the alternate structures that arise from the decomposition into two subtasks. If this can be done, the entire synthesis can be performed by sequentially decomposing the subtasks until existing technology is reached, i.e., when

$$S_j \subset R \qquad (10\text{-}6)$$

The basis for this selection is

$$O^*(X) = \underset{\substack{\text{tear location} \\ \text{tear values}}}{\text{Opt}} \; [O(S_1) + O(S_2)] \qquad (10\text{-}7)$$

$$= \underset{\substack{S_1 \text{ and } S_2 \\ \text{selection}}}{\text{Opt}} \; \{\text{Opt} \; [O^*(S_1) + O^*(S_2)]\} \qquad (10\text{-}8)$$

The original task X is divided into tasks $X_1 \cup T$ and $X_2 \cup T$, and the terms $O^*(X_1 \cup T)$ and $O^*(X_2 \cup T)$ are the optimal objective functions that can be obtained by the solution to the subtasks for given values of $X_1 \cup T$ and $X_2 \cup T$. The tear constraints are free to be adjusted. The interior optimization adjusts T to optimize the sum of the optimum objective function of the two parts, and the exterior optimization is over the distribution of X between X_1 and X_2 and constitutes an optimization over the structure of the system.

The implementation of Eq. (10-8) requires knowledge of the optimal objective function that can be obtained for any task, but this information is available only for tasks within the region of existing technology; then

$$O^*(S_j) = E(S_j) \qquad (10\text{-}9)$$

where $S_j \subset R$. If the objective function is known, Rudd's synthesis procedure is in accordance with the algorithm shown in Fig. 10-9.

When we are dealing with a new process-design problem, it is obvious that we cannot know the optimal objective function $O^*(X)$ because this is precisely the economic objective function for the optimal solution to a problem whose solution we are seeking. Thus, the algorithm in Fig. 10-9 cannot be implemented. We can assume, however, that in any given area of technology there will be sufficient experience available to permit an estimated optimal objective function $O^{(1)}(X)$.

Such an estimated optimal objective function can be used in the synthesis algorithm, in a manner reminiscent of evolutionary design, to compose an estimated optimal system. This then forms the basis for the improvement of initial estimate $O^{(1)}$ to form $O^{(2)}$, which is an improved estimate of the optimal objective function O^*. The iteration on the optimal objective function with the synthesis algorithm used may converge to the synthesis of the optimal system as the sequence $O^{(1)}, O^{(2)}, O^{(3)}, \ldots$ converges to O^*. The iteration plan is outlined in Fig. 10-10 as the accumulation of experience.

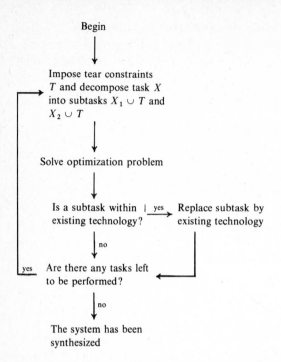

Begin

Impose tear constraints T and decompose task X into subtasks $X_1 \cup T$ and $X_2 \cup T$

Solve optimization problem

Is a subtask within existing technology? — yes → Replace subtask by existing technology

no

yes — Are there any tasks left to be performed?

no

The system has been synthesized

Figure 10-9 Rudd's process-synthesis algorithm. [*From Rudd (1968), by permission.*]

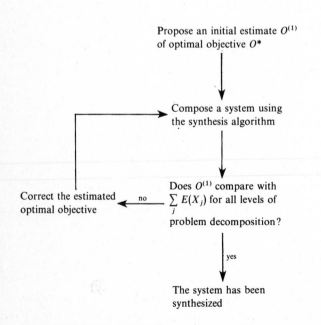

Propose an initial estimate $O^{(1)}$ of optimal objective O^*

Compose a system using the synthesis algorithm

Correct the estimated optimal objective ← no — Does $O^{(1)}$ compare with $\sum_{j} E(X_j)$ for all levels of problem decomposition?

yes

The system has been synthesized

Figure 10-10 Rudd's accumulation of experience. [*From Rudd (1968), by permission.*]

Hendry et al. (1973) point out that even though the estimated optimal objection function is in agreement with the actual objective function of the composed system, this does not guarantee that the synthesized system will be the optimal structure. If the estimated optimal objective function accurately represented the objective function of one particular feasible structure and unfairly penalized all other feasible structures, it would be possible for the structure whose objective function is accurately represented to be selected as being optimal. Therefore, agreement between estimated and observed objective functions cannot guarantee that some other structure with superior system economics does not exist. Thus, the proposed convergence criterion of Rudd's synthesis algorithm may not always lead to an optimal system structure.

Although certain difficulties can arise in establishing system optimality, the strategy of process decomposition provides a means by which reasonable solutions can be obtained for large problems which would otherwise appear unmanageable.

Heuristic Synthesis

We have already encountered heuristics in our discussion of qualitative procedures. We now discuss several applications of heuristic methods in programmed process synthesis.

Masso and Rudd (1969) attempted to overcome the difficulties in the selection of optimal tear locations and tear values in the process-decomposition technique by the application of heuristic techniques. They introduced the requirement that decomposition of the original problem or any unsolved part of it must produce at least one subproblem X_j which is immediately soluble with available technology, $(X_j \cup T_j) \subset R$, such that an exact economic evaluation $E(X_j)$ is available. Structuring then proceeds according to

$$O^*(X) = \underset{(X_j \cup T_j) \subset R}{\text{Opt}} \left(\underset{T_j}{\text{Opt}} \{O^*[(X_j \cup T_j) \subset R] + O^*(\bar{X}_j \cup T_j)\} \right) \quad (10\text{-}10)$$

where $\bar{X}_j$ is the complement of X_j; that is, $X = X_j \cup \bar{X}_j$, and $X_j \cap \bar{X}_j = 0$. The result of this modification is that optimum structuring must now proceed according to an optimal sequence of subtask selection decisions. Obviously, Eq. (10-10) requires knowledge of $O^*(\bar{X}_j \cup T_j)$, the optimal objective function for the unsolved subproblem. However, rather than estimating this with some sequence $O^{(1)}, O^{(2)}, O^{(3)} \ldots$, Masso and Rudd proposed that structuring decisions should be made using heuristic selection of X_j and T_j at each stage. Thus, a heuristic selection of the subtask to be solved immediately with available technology and heuristic estimation of the variables linking this subtask to the remaining unsolved problem were made.

Masso and Rudd implemented this heuristic decomposition synthesis procedure as an iterative learning scheme designed to promote convergence of a sequence of trial structures to the optimum processing system. They found it was necessary to include a random selection rule in the set of decomposition heuristics not only to guard against possible systematic exclusion of certain types of structures but also to indicate insensitivity of the optimal structure to the particular heuristics used.

Rudd (1968) and Masso and Rudd (1969) illustrated their procedures by application to the synthesis of heat-exchanger networks, one of the categories to which programmed synthesis techniques are usually applied. The generation of process flowsheets, however, is more complex than the computer-aided synthesis of a subsystem such as a heat-exchanger network. The latter type of subsystem involves only the interconnection between specified kinds of processing equipment. The creation of a flowsheet involves, in addition, a determination of the tasks to be accomplished and the specification of the type of technology to execute these tasks in order to convert some particular raw materials into desired products. We shall return to this point in a later discussion of a computer-aided process synthesizer.

10-4 OPTIMIZATION METHODS FOR PROCESS SYNTHESIS

Various optimization and mathematical programming techniques are used to search for the optimal set among the alternative structural configurations and operating conditions. The number of combinations of equipment, temperature, pressure, concentrations, etc., is so large for even a small process plant that exhaustive enumeration of all cases and the calculation of the objective function for each case followed by direct selection of the optimum represents an impossible computational task. It is the purpose of the various optimization techniques not only to reach the optimal solution but also to reach it efficiently.

In optimization we seek to maximize or minimize a function of a number of variables with the variables subject to certain constraints. Until approximately 30 years ago the only mathematical methods available for handling optimization problems were classical differential and variational calculus. Since World War II there has been a rise in interest in optimization methods for dealing with problems not solvable by classical methods. Two classes of methods have been developed, optimum-seeking procedures and mathematical programming.

Although we now briefly discuss a number of optimization methods, a presentation of optimum-seeking procedures and mathematical programming in any depth is well beyond the scope of this book. A number of textbook treatments can be cited for study by the reader. Wilde (1964) presents single-variable and multivariable optimum-seeking procedures and also discusses the effects of experimental errors. In the work by Wilde and Beightler (1967) a unified theory of optimization is presented in a compact, readable form. Peters and Timmerhaus (1968) give a brief introduction to linear and dynamic programming to serve as a basis for further study and applications. Beightler and Phillips (1976) discuss the relatively new technique of geometric programming and present a number of applications. Avriel et al. (1973) is a more advanced treatise in which optimization methods and their applications to design are discussed.

10-5 OPTIMUM-SEEKING PROCEDURES

Optimum-seeking procedures are strategies to guide the search for the optimum of any function about which full knowledge is not available. Such functions obviously will arise when direct observations must be made on a physical system. In process synthesis optimum-seeking strategies can be used to guide us in the choice of values for the variables to permit an economic search of the response surface instead of performing an exhaustive evaluation over the entire response surface.

10-6 MATHEMATICAL PROGRAMMING

Mathematical programming developed as a branch of optimization theory to deal with maximization and minimization problems that arise in the decision sciences such as management science, operations research, and engineering design. It should not be confused with computer programming, although the solution of many algorithms arising in mathematical programming would not have been possible without the computer. We briefly describe several methods that are used in process synthesis, namely, linear programming, dynamic programming, geometric programming, and branch-and-bound methods.

Linear Programming

Linear programming is applicable to a large class of problems which involve linear objective functions subject to linear inequality and equality constraints. Linear-programming algorithms search the extreme points, which are the extremes of the region of feasible solutions. We shall see from a graphical solution to a simple linear optimization problem that the optimum solution occurs at an extreme point.

> **Example 10-1** A fertilizer-blending plant has a market for two grades of fertilizers, 10-8-5 and 7.5-10-15. (Fertilizers are specified by percentages of three major nutrient elements, N, P, K, where N is nitrogen, P is equivalent P_2O_5, and K is equivalent K_2O.) The expected profit on the first grade is $20 per ton and $30 per ton on the second grade. The plant has available 1500 tons of equivalent nitrogen, 1200 tons of equivalent P_2O_5, and 1500 tons of equivalent K_2O. How much of each grade of fertilizer should be made to realize maximum profit?
>
> SOLUTION The problem statement can be cast into a linear optimization form, where x_1 and x_2 are the tons of first-grade and second-grade fertilizers,

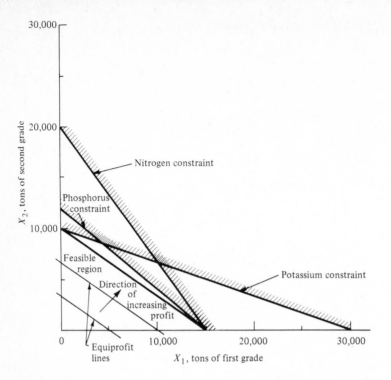

Figure 10-11 Graphical representation of fertilizer-optimization problem.

respectively,

$$\max P = 20x_1 + 30x_2$$

Subject to $0.1x_1 + 0.075x_2 \leq 1500$ nitrogen constraint

$$0.08x_1 + 0.10x_2 \leq 1200 \text{ phosphorus constraint}$$

$$0.05x_1 + 0.15x_2 \leq 1500 \text{ potassium constraint}$$

$$x_1, x_2 \geq 0$$

This is shown in graphical form in Fig. 10-11, which presents the linear constraints and equiprofit lines. It is obvious that the maximum profit will be reached at the extreme point where the equiprofit lines leave the feasible region at the intersection of the phosphorus and potassium constraints. The optimum production is 4286 tons of first-grade and 8571 tons of second-grade, yielding a maximum profit of $342,850. All the phosphorus and potassium would be used, but 429 tons of nitrogen would remain unused.

In general, the solution of a linear-programming problem requires a search only over the extreme points. Extremely efficient algorithms have been developed which take advantage of this property, and problems involving thousands of variables can be solved routinely and by rote. A generalization of linear programming is *convex programming*, in which the objective function and feasible-solution set are permitted to be convex.

Dynamic Programming

Dynamic programming can be used to transform an N-decision one-state initial-value optimization problem into a set of one-decision one-state problems. It transforms a large serial structure into a sequence of smaller problems and thus is reminiscent of process decomposition. The required computational effort is greatly reduced, but the technique requires that there be no recycle of information. A brief discussion of dynamic programming was presented in Chap. 9.

Geometric Programming

Geometric programming was developed for solving algebraic nonlinear programming problems subject to linear or nonlinear constraints. It can appear to be almost magical in its efficiency when, as can happen in certain cases, it can locate optima only by inspection of the exponents in the objective function. The optimal values of the independent variables are not sought directly by geometric programming. Instead, the optimal way to distribute the costs among the elements of the objective function is first sought. The optimal cost is then easily calculated. Only then does one determine the policy needed to reach the optimal cost.

As an example of geometric programming we shall use a hypothetical chemical plant postulated by Wilde and Beightler (1967). In this plant raw materials are mixed with a recycle stream from a recirculating compressor. The mixed stream is compressed and fed to a reactor followed by a separator. The product is one stream from the separator; the other stream is recycled to the mixer via the recirculating compressor. The annual cost figures are $1000x_1$ for the compressor; 10^4 for the mixer; $4 \times 10^9/x_1 x_2$ for the reactor; $10^5 x_2$ for the separator; and $1.5 \times 10^5 x_2$ for the recirculating compressor. x_1 is the operating pressure, and x_2 is the recycle ratio. The problem is then to choose x_1 and x_2 so as to minimize the annual cost y

$$\min y = 1000x_1 + \frac{4 \times 10^9}{x_1 x_2} + 2.5 \times 10^5 x_2$$

The classical method to determine the optimum conditions would be to differentiate the cost equation with respect to x_1 and x_2 and set the derivatives equal to zero. The asterisk in the following equations refers to the optimum.

$$\left(\frac{\partial y}{\partial x_1}\right)^* = 1000 - \frac{4 \times 10^9}{x_2^* x_1^{*2}} = 0$$

$$\left(\frac{\partial y}{\partial x_2}\right)^* = -\frac{4 \times 10^9}{x_1^* x_2^{*2}} + 2.5 \times 10^5 = 0$$

These nonlinear equations would then be solved to determine x_1^* and x_2^*.

In geometric programming the optimal distribution of costs would first be found. The optimal weights are first defined

$$w_1 = \frac{1000x_1^*}{y^*} \qquad w_2 = \frac{4 \times 10^9}{x_1^* x_2^* y^*} \qquad w_3 = \frac{2.5 \times 10^5 x_2^*}{y^*}$$

We now note that the first partial derivative $(\partial y / \partial x_1)^*$ is

$$w_1 \frac{y^*}{x_1^*} - \frac{w_2 y^*}{x_1^*} = 0$$

and the second partial derivative $(\partial y / \partial x_2)^*$ is

$$-w_2 \frac{y^*}{x_2^*} + w_3 \frac{y^*}{x_2^*} = 0$$

Because the weights must sum to unity, we therefore have,

$$w_1 + w_2 + w_3 = 1$$

$$w_1 - w_2 \qquad = 0$$

$$-w_2 + w_3 = 0$$

By inspection,

$$w_1 = w_2 = w_3 = \tfrac{1}{3}$$

at the optimum. Thus, the optimal solution will occur when the compressor cost, the reactor cost, and the combined separator and recirculator cost are equal. This optimal distribution is also totally unaffected by the cost coefficients but would be affected by the exponents on the independent variables x_1 and x_2.

The minimum cost can now be calculated even though we do not yet know the optimal pressure and recycle ratio. It can be calculated from

$$y^* = \prod \left(\frac{c_j}{w_j} \right)^{w_j}$$

where the c_j's are the coefficients in the cost equation. Thus,

$$y^* = \left(\frac{1000}{1/3} \right)^{1/3} \left(\frac{4 \times 10^9}{1/3} \right)^{1/3} \left(\frac{2.5 \times 10^5}{1/3} \right)^{1/3} = 3 \times 10^6$$

The optimal values for x_1 and x_2 can now be calculated. The compressor cost must be $(\tfrac{1}{3})(3 \times 10^6) = 1 \times 10^6$, and because the cost equals $1000x_1$, x_1 must equal 1000 atm. The combined separator-recirculator cost was also 1×10^6, which equals $2.5 \times 10^5 x_2$; thus x_2 equals 2.5. It was not necessary to solve any nonlinear equations in this case. This is generally true in geometric programming although nonlinearities may be encountered.

Branch-and-Bound Methods

Branch-and-bound methods involve a decomposition of the original problem to generate sets of subproblems. Branching consists of dividing sets of solutions into subsets, and bounding consists of establishing bounds on the value of the objective function over subsets of the solution. Each set of subproblems is checked to see whether it has solutions within the bounds. If the solutions to a given subset are outside the bounds, that subset need no longer be considered. Thus, by bounding the problem certain sets can be eliminated and need not be investigated. The power of the branch-and-bound methods lies in the reduction of the number of solutions that need be enumerated.

10-7 CHEMICAL-PROCESS SIMULATION

Process simulation is the representation of a chemical process by a mathematical model, generally as a computer program. The model can be solved to obtain information about the performance of the process. Process simulation is now an accepted tool for understanding chemical processes, and, in addition, some of the simulation programs are useful tools in design and process synthesis. Motard et al. (1975) review this field and the progress that has been made since the first process simulator was published in 1958.

A process simulator can, among other things, be used to predict the effects of changing conditions and capacity, to do mass and energy balances, and to optimize the operation quickly and safely. It can be used to provide depth of knowledge about complete system behavior, to improve control, and to facilitate cost calculations and planning of operations. It can also be a valuable training tool for operators and engineers.

The input data for a typical process simulator contain the process structure, the design and operating parameters of the units, the composition and state of the process feed streams, and an estimate of any recycle streams. The executive program determines the order of calculation, calls the unit module subroutines, and supplies them with the necessary input data. The unit module or unit computation is the separate mathematical model for each chemical processing step. The unit module subroutines calculate the output from the units. These calculations require physical and thermodynamic data that are preferably supplied by the engineer but more usually are calculated with the help of estimation parameters that are part of a thermodynamic- and physical-properties package. The output unit presents the information obtained in the form of an easily readable technical report. Process-simulation programs referenced by Motard et al. (1975) include PROVES, PRIMER, GEMCS, CONCEPT, CAPES, PACER, SLED, FLOWTRAN, FLOWPACK, CHEEP, and CHESS.

There are several problems common to all these programs, e.g., the analysis of information flow, detection and treatment of recycle, and ordering of the calculations. We will now discuss these topics.

Information Flow

The unit module subroutine (UMS) is a set of mathematical manipulations for calculating output information from specified input information. The term *module* is used to indicate that the UMS must be written so that the calculation is independent of the source of the input information and of the use of the output information. In much of our study of chemical engineering we performed such subroutines or developed mathematical models in our problem solving.

Symbols adopted for several UMS are shown in Fig. 10-12. It should be noted that the arrows now refer to information flow. These symbols can be assembled into an information flow diagram for a simple process, as shown in Fig. 10-13a. In the process a feed enters a distillation tower. The bottom product leaves the process, but the overhead is mixed with a second feed stream before entering the reactor along with a recycle stream. The reactor effluent enters a settler whose underflow leaves the process. The overflow is distilled in a two-tower sequence. The overhead from the first tower is recycled to the reactor. The bottom product enters the second tower. Both its product streams leave the process.

The flows of information between units are shown as directed lines between symbols, and the symbols and lines are separately numbered. Numbering is arbitrary, with the restriction that no two symbols and no two lines may have the same number. The information flow diagram generally will resemble the process flow diagram although some units and some streams may not appear in both diagrams. Units such as pumps and heat exchangers are not shown, for example, in Fig. 10-13a.

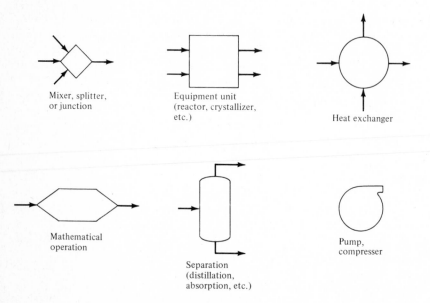

Mixer, splitter, or junction

Equipment unit (reactor, crystallizer, etc.)

Heat exchanger

Mathematical operation

Separation (distillation, absorption, etc.)

Pump, compresser

Figure 10-12 Some typical unit module symbols.

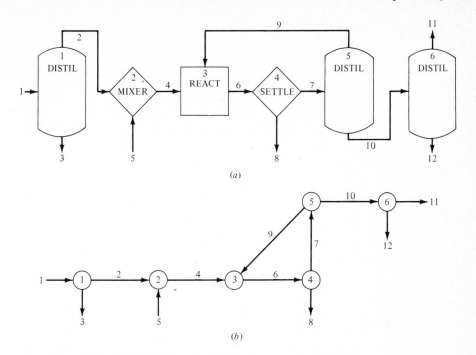

Figure 10-13 Information flow diagram: (*a*) equipment units and (*b*) digraph.

The information flow can also be presented as a directed graph, or digraph, as shown in Fig. 10-13*b*. The equipment units are represented by nodes and the streams by directed edges. The node represents, of course, not only the equipment but also the UMS. Thus, the parameters of output streams from the node can be calculated from information on the parameters of the input streams.

The information flow diagram must now be encoded in a numerical form appropriate for the computer. Two methods of encoding will be presented, the process matrix and the adjacency matrix.

Process Matrix

A process matrix for Fig. 10-13 is shown in Table 10-4. Each unit is given one row, which contains the unit or node number, the name of the UMS for the unit, the input streams to the unit as positive numbers, and the output streams as negative numbers. The order of the input and output stream numbers is important. The user's guides for those process simulators which encode via a process matrix will specify the order to be used. The process matrix encodes the entire information flow diagram by showing what streams link what units, the name of the UMS for each unit, and the order of input and output stream for each unit. Thus, the information flow diagram can be completely reconstructed from the process matrix.

Table 10-4 Process matrix for Fig. 10-13

Unit No.	UMS	Associated stream numbers		
1	DISTIL	1	-2	-3
2	MIXER	2	5	-4
3	REACT	4	9	-6
4	SETTLE	6	-7	-8
5	DISTIL	7	-9	-10
6	DISTIL	10	-11	-12

Adjacency Matrix

The adjacency matrix for Fig. 10-13 is shown in Table 10-5. It is a square matrix in which a given row and column number corresponds to a given unit or node. A 1 shows that there is a connection between the unit given by the column number and that given by the row number. A zero or no entry shows that no connection exists. Since no unit is connected to itself, the diagonal elements of the matrix are zero. From Table 10-5 we read that unit number 1 is connected to unit 2, unit 2 to unit 3, unit 3 to unit 4, unit 4 to 5, and unit 5 to units 3 and 6.

The matrix has little content; no feed or product streams appear, and stream numbers are suppressed. The adjacency matrix is useful, however, to find recycle, as we shall see.

Ordering the Calculations

By inspection of Fig. 10-13 we can see that unit 1 can be directly calculated because the input is known. After that unit 2 can be calculated because its inputs will be known. Units 3, 4, and 5 form a recycle loop and cannot be directly calculated, and unit 6 can be calculated only after 5 has been calculated. The

Table 10-5 Adjacency matrix for Fig. 10-13

From unit number	To unit number					
	1	2	3	4	5	6
1		1				
2			1			
3				1		
4					1	
5			1			1
6						

computer must be appropriately programmed to enable it to extract this same sort of information from the encoded information matrix so that it will be able to order the calculations. It must be able to separate the units into those which are (or will be) directly calculable and those which are parts of recycle loops. It must also be able to plan the calculations of the separate recycle loops.

If there is no recycle in the information flow diagram, the units can be calculated sequentially starting from a unit which has only known feed streams entering it. Such a process is called a *serial* or a *cyclic process*, and a part of a process which has no recycle can be called a *serial set*. A recycle process is one in which an output stream of a unit affects at least one of its input streams. A recycle set is a set of units which are connected so that there is some path following the information arrows from every unit or node to every other unit or node. Several recycle loops may be present in a recycle set. If all units cannot be calculated sequentially, there is at least one recycle loop present and the process is a recycle process.

We shall see first how the process matrix and then how the adjacency matrix can be used to distinguish between the serial sets and recycle sets. If we examine the rows of Table 10-4, we note first that unit 1 can be calculated because all its feed streams (actually only one stream) are known. This makes streams 2 and 3 known. In the second row, unit 2, we note that the unit is calculable after unit 1 has been calculated because stream 5 is a feed stream and stream 2 will be known. Unit 3, however, cannot be calculated although stream 4 will be known after unit 2 is calculated. One of its feed streams, stream 9, comes from unit 5. Since some of the units are uncalculable, we have at least one recycle set.

We can continue the examination of Table 10-4 to learn whether any serial sets follow all recycle sets. To find such sets we scan to see whether there are any units whose output streams leave the process (they are not connected to any other unit) or are connected to such a serial set. Unit 4 does not qualify although stream 8 is not connected to any other unit because stream 7 is connected to unit 5. Unit 5 does not qualify because output stream 9 is an input to unit 3. Unit 6 is in a serial set because neither of its output streams is connected to any other unit. It is the only unit in the serial set, however. When units 1 and 2 at the beginning of the process and unit 6 at the end of the process are removed, units 3, 4, and 5 remain. Any units remaining after the serial sets at the beginning and end of the process have been removed may be part of a recycle set or of a serial set joining recycle sets.

The adjacency matrix can be used not only to detect recycle sets but also to identify them. Each element of the adjacency matrix a_{ij} can have the value of unity or zero. If $a_{ij} = 1$, it means that a stream goes from the unit corresponding to row i to the unit corresponding to column j. If there is no such stream, then $a_{ij} = 0$.

In Table 10-5 we note that only zeros appear in column 1. This means that no unit is connected to unit 1 and only feed streams enter it. Unit 1 can be listed as calculable and eliminated from the matrix; i.e., row 1 and column 1 are struck out. The new matrix is shown in part (a) of Table 10-6. With unit 1 removed the

Table 10-6 Elimination of columns and rows

(a) Unit 1 removed

From unit number	To unit number				
	2	3	4	5	6
2		1			
3			1		
4				1	
5		1			1
6					

(b) Unit 2 removed

From unit number	To unit number			
	3	4	5	6
3		1		
4			1	
5	1			1
6				

(c) Unit 6 removed

From unit number	To unit number		
	3	4	5
3		1	
4			1
5	1		

column corresponding to unit 2 has only zeros and can be listed as calculable after unit 1. It also can be removed from the matrix, as shown in part (b) of Table 10-6. The row corresponding to unit 6 has only zeros. This means that unit 6 is not connected to any other unit and can be listed as calculable when the other units have been calculated. Thus unit 6 is eliminated, and the surviving matrix is part (c) of Table 10-6. It contains only units 3, 4, and 5 and no zero rows or columns. The remaining units are therefore either in recycle sets or in a serial set between recycle sets.

We shall call this surviving matrix A. Norman (1965) showed that A^n, where n takes on the values of 2, 3, ... up to a limit equal to the number of rows in A, gives the paths that exist from any unit to any other unit through n streams. This means that any unit i with a diagonal element equal to unity has a path via n streams from unit i and back to itself. Thus, all the recycle loops can be found.

The arithmetic used in calculating the powers of A is boolean, i.e.,

$$0 + 0 = 0 \qquad 0 \times 0 = 0$$
$$0 + 1 = 0 \qquad 0 \times 1 = 0$$
$$1 + 1 = 1 \qquad 1 \times 1 = 1$$

Reachability Matrix

The next step in the partitioning of the adjacency matrix is to calculate the reachability matrices. The reachability matrix is the sum of the adjacency matrix and its powers. Thus, R_1 is the adjacency matrix, $R_2 = A + A^2$, $R_3 = A + A^2 + A^3$, etc., until R_n. The reachability matrix lists the connections from unit i to unit j via a number of streams equal to the index on R. R_n therefore records whether any connection exists from unit i to unit j in the entire digraph. Such a connection would be indicated by $a_{ij} = 1$.

In the transpose of R_n the rows become columns and the columns become rows. Thus, R_n^T records connections between unit j and unit i. In a recycle set there must be a connection not only of $i \to j$ but also from $j \to i$. In other words, only if $i \rightleftarrows j$ will unit i be in a recycle set. Since this information is available from the intersection of R_n and R_n^T, the intersection presents an ordered picture of the recycle sets. The intersection is obtained by retaining unity only when it appears in both elements of R_n and R_n^T.

The calculation of the adjacency matrix and its powers, the reachability matrices, and their intersection will be demonstrated in the following example, which is slightly more complex than the simple process examined earlier.

Example 10-2 A digraph for part of a process is shown in Fig. 10-14. The input and output serial sets have already been eliminated by the deletion of empty columns and the respective rows and the empty rows and the respective columns from the original adjacency matrix. Identify the recycle and serial sets in the digraph.

SOLUTION The adjacency matrix for the digraph of Fig. 10-14 is shown in Table 10-7, and we note that no empty columns or rows exist. Thus, the input and output serial sets had, indeed, been eliminated. Six units are

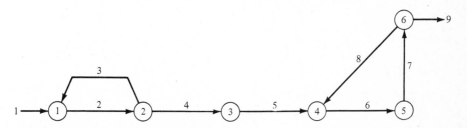

Figure 10-14 Information digraph for Example 10-2.

Table 10-7 Adjacency matrix and its powers

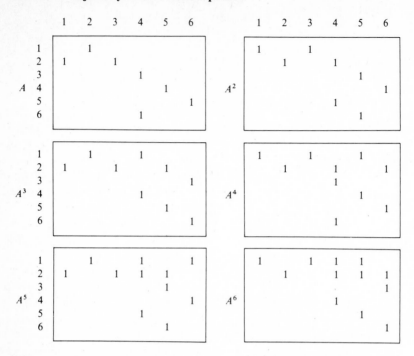

A

	1	2	3	4	5	6
1		1				
2	1		1			
3				1		
4					1	
5						1
6				1		

A^2

	1	2	3	4	5	6
1	1		1			
2		1		1		
3					1	
4						1
5				1		
6					1	

A^3

	1	2	3	4	5	6
1		1		1		
2	1		1		1	
3						1
4				1		
5					1	
6						1

A^4

	1	2	3	4	5	6
1	1		1		1	
2		1		1		1
3				1		
4					1	
5						1
6				1		

A^5

	1	2	3	4	5	6
1		1		1		1
2	1		1	1	1	
3					1	
4						1
5				1		
6					1	

A^6

	1	2	3	4	5	6
1		1		1	1	1
2			1	1	1	1
3						1
4			1			
5					1	
6						1

Table 10-8 Reachability matrix to R_3

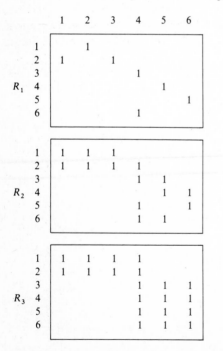

R_1

	1	2	3	4	5	6
1		1				
2	1		1			
3				1		
4					1	
5						1
6				1		

R_2

	1	2	3	4	5	6
1	1	1	1			
2	1	1	1	1		
3				1	1	
4					1	1
5				1		1
6				1	1	

R_3

	1	2	3	4	5	6
1	1	1	1	1		
2	1	1	1	1		
3				1	1	1
4				1	1	1
5				1	1	1
6				1	1	1

Table 10-9 Reachability matrices R_6 and R_6^T and their intersection

		1	2	3	4	5	6
	1	1	1	1	1	1	1
	2	1	1	1	1	1	1
	3				1	1	1
R_6	4				1	1	1
	5				1	1	1
	6				1	1	1

		1	2	3	4	5	6
	1	1	1				
	2	1	1				
	3	1	1				
R_6^T	4	1	1	1	1	1	1
	5	1	1	1	1	1	1
	6	1	1	1	1	1	1

		1	2	3	4	5	6
	1	1	1				
	2	1	1				
	3						
$R_6 \cap R_6^T$	4				1	1	1
	5				1	1	1
	6				1	1	1

present, and $n = 6$. The powers of the adjacency matrix up to A^6 are also presented in Table 10-7. By scanning for unity on the diagonals it is apparent that recycle loops are present that include units 1 and 2 and units 4, 5, and 6.

The reachability matrices are calculated next; R_1, R_2, and R_3 are shown in Table 10-8, and R_6, R_6^T, and their intersection are shown in Table 10-9. All members of one recycle set will appear in a nonzero row of $R_6 \cap R_6^T$, and any unit with a zero row will be a serial set connecting a pair of recycle sets. From Table 10-9 we observe that units 1 and 2 are in one recycle set, units 4, 5, and 6 are in another recycle set, and unit 3 is a serial set connecting the two recycle sets. Unit 3 will be calculable when the recycle set of units 1 and 2 has been calculated. Norman (1965) points out that cycles can occur that correspond to multiple passages around the cycle. To overcome this problem the individual members of a cycle are replaced by a pseudo node once the cycle is detected.

The steps involved in partitioning the adjacency matrix to isolate the serial sets and the recycle sets are summarized:

1. Units or nodes without inputs are eliminated by deleting empty columns and respective rows. The units eliminated are listed as being calculable.

2. Units or nodes without outputs are eliminated by deleting empty rows and the respective columns. They are listed as calculable after the recycle sets, if any, are calculated.
3. Steps 1 and 2 are repeated until no more empty columns or rows are found.
4. The powers of the resultant matrix are computed sequentially to a power equal to the number of units or nodes in the matrix n. Values of unity on the diagonals indicate recycle loops.
5. The reachability matrix R_n is calculated.
6. R_n^T, the transpose of R_n, is calculated.
7. The interection matrix $R_n \cap R_n^T$ is determined.
8. All members of a recycle set are listed by each nonzero row in step 7.
9. Units that are not in any of the recycle sets are scanned for. These units are then in serial sets between pairs of recycle sets.
10. Each serial set from step 9 is listed as calculable after its preceding recycle set is calculated.

Calculating the Recycle Set

The problem of calculating the recycle sets can be exemplified by referring to Fig. 10-14. Stream 3 is calculable if we know stream 2, the input to unit 2, but we do not know stream 2, the output from unit 1, because we cannot calculate stream 2 until after we know stream 3, which is one of the inputs to unit 1. The usual approach of the chemical engineer to solving a problem of this type is the trial-and-error, or iterative, method.

In the iterative calculation values must be given to what are termed the *cut streams*. These are the streams whose variables have not yet been calculated and must therefore be given starting values to begin the iteration procedure. The iteration is continued until the assumed and calculated values converge to some satisfactory error or accuracy criteria.

Although in many practical problems or small problems of the size examined earlier the appropriate locations of the tearing streams are either intuitively obvious or easily deduced, the problem of choice exists in large or complicated cyclic nets. Motard et al. (1975) reviewed some of the tearing algorithms that can be used to locate the optimal tearing, or cut, streams. A simple algorithm of the probability type was presented by Kehat and Shacham (1973). In all the cases they examined their algorithm produced a number of tears that was either equal to or less than the number of tears produced by other, more complex algorithms. The probability tearing algorithms are based on assumptions that give a large probability that each tear will cut a maximum number of internal cycles. The tearing algorithm searches for nodes of the digraph that have a minimum number of input streams and a maximum number of output streams. The steps of the algorithm are as follows:

1. List all the units or nodes in the maximal cyclic set with their input and output streams.

2. Delete nodes without input streams (initially none) and list them in the order of node calculations.
3. Scan for nodes with a minimum of input streams and rank in the order of number of output streams.
4. Mark as tear streams the input stream or streams to that unit with the minimum number of input and maximum number of output streams. Delete these streams and repeat step 2.
5. If more than one such node was found, repeat step 4 for all these nodes and list separately all the ordered nodes until each list contains all the other compared nodes. The list with the largest number of compared nodes is the optimal list and is added to the output list. If the lists are of the same size, it does not matter which is used first.
6. Repeat step 2 until no more nodes remain in the list established in step 1.

The algorithm is demonstrated with the same example used by Kehat and Schacham (1973). The digraph for a cyclic net is shown in Fig. 10-15, and the application of the tearing algorithm is presented in Table 10-10. Two tear streams are obtained, stream 7 to unit 9 and stream 13 to unit 6. All 10 nodes can be calculated when values of the parameters of streams 7 and 13 are given starting values.

Calculation of the Variables

The simulation problem is now reduced to the problem of determining the values of the variables in the tear streams. The values of the variables are given initial estimates to begin the iterative calculations. In chemical engineering systems the calculation of the streams including the calculated recycle values in each iteration usually involve a set of nonlinear equations.

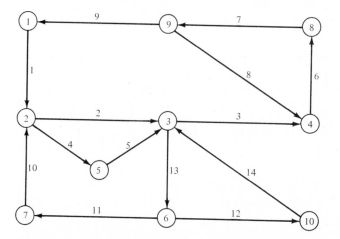

Figure 10-15 Ten-node cyclic set.

Table 10-10 Tearing the cyclic net of Fig. 10-15

Node	Input streams	Output streams	Cut stream	Order	Calculable output streams
1	9	1			
2	1, 10	2, 4			
3	2, 5, 14	3, 13			
4	3, 8	6			
5	4	5			
6	13	11, 12	13	2	11, 10, 12, 14, 2, 4, 5, 3, 13, 6
7	11	10			
8	6	7			
9	7	8, 9	7	1	7, 8, 9, 1
10	12	14			

The simplest iterative method is that of successive substitution, in which the new estimated values of the variables in the tear streams are the last values calculated from the previous iteration. This method must converge for every physically realizable recycle system from an initial estimate which is close enough to the solution (Motard et al. 1975). The convergence may be slow, however, and convergence-acceleration methods may be used.

Convergence-acceleration methods have been reviewed by Motard et al. (1975), who cite three methods used for process recycle systems. The *bounded Wegstein* and the *dominant-eigenvalue* methods accelerate the convergence by reducing the effective magnitude of the maximal eigenvalue of the iteration matrix. The third method, the *quasi-Newton*, or *secant*, *method*, provides improved estimates of the tear-stream variables by a correction vector which includes an approximation to the inverse of the jacobian matrix.

10-8 COMPUTER-AIDED PROCESS SYNTHESIS

Most simulation programs work in the simulation-performance-rating mode. In such a calculation mode all system inputs and unit design parameters are specified, but the engineer who is faced with the problem of synthesizing a process would prefer a simulation program that works in the design mode; i.e., the system inputs and design parameters are calculated from specified outputs. Unfortunately, the design mode of calculation is usually less stable numerically than the simulation mode. The usual solution is to perform the design calculation by iterated simulation.

It should be apparent that process synthesis requires interaction between synthesis and analysis. Chemical engineers are well equipped and command techniques and methods for analyzing specific and well-defined processing steps and processes. They also have at their disposal an impressive array of alternative routes for performing many operations. To blend these into a whole—process synthesis—requires the systematic aspects and the intuitive aspects to be combined. We now discuss AIDES (Adaptive Initial Design Synthesizer), a computer program (Siirola and Rudd, 1971) which interacts with the designer via a remote computer terminal. The program is designed to allow the computer and the engineer to do what each is best at, the computer handling the systematic and the engineer the intuitive aspects.

Siirola and Rudd (1971) list 12 alternating synthesis and analysis steps which are involved in a process synthesis. Each synthesis step is followed by an analysis step, which determines the implications of the synthesis step and then leads to the subsequent synthesis step. The synthesis begins with a sequence of chemical reactions and ends with process elements of the flowsheet, material and energy balances, equipment size, and the estimation of capital requirements and operating costs. AIDES implements most of these steps while interacting with the design engineer.

The alternating synthesis and analysis steps follow.

Synthesis step 1: chemical reaction path It is assumed that an attractive sequence of chemical reactions is at hand to convert readily available raw materials into more valuable products.

Analysis step 1: stoichiometric material balance An estimate of reaction conversions is necessary. The material balance indicates amounts of raw materials and products and excesses of each species in the system and also compares expected sales income with raw-material costs. Thus, preliminary economic feasibility can be evaluated.

Synthesis step 2: component matching Primary matches are made between places where species are known to exist (raw materials, reactor effluents) and destinations (reactor feeds and products). Secondary matches are also made to account for contaminants, excess requirements of some species, etc. These matches result in proposed interconnections in the process structure.

Analysis step 2: match evaluations The possibilities of multiple sources and destinations and alternative routes from species source to destination are examined. One of the factors considered is the estimated difficulty of separations that might be required between species.

Synthesis step 3: species allocation The decisions are made where each species in each source is to be sent. In this key synthesis step species allocation is usually made so as to minimize the difficulty of separation. The species routes from each source to each destination now become information channels, and information on composition, concentration, and flow rate is generated.

Analysis step 3: information-difference detection Differences between the values of information at source and at destination are detected.

Synthesis step 4: nonseparation-task identification If destinations with multiple sources exist, a mixing task is specified before each such destination. If there are multiple destinations from single sources, stream-splitting tasks are specified. When pressure, temperature, or similar differences are detected, the appropriate property-changing task is specified.

Analysis step 4: physical-property evaluation When concentration differences are detected in analysis step 3, it is necessary to search for physical-property differences that can be exploited to effect the required species separation. For each possible physical property that might be exploited the appropriate stream conditions must be evaluated. If source-stream properties are not appropriate for the separation, a property-changing task may have to be specified.

Synthesis step 5: separation-task identification Selection is made from the separator-task candidates. Multiple designs may be generated if in synthesis step 3 alternative species allocations are investigated.

Analysis step 5: separation-task feasibility If the proposed separation task is found to be infeasible for any reason, it is rejected and a return is made to synthesis step 5 to obtain another candidate task. When a task is found to be acceptable, it is included in the process structure and new source streams are assigned to the effluents from the separation task. Information differences involving these new channels must be detected by analysis step 3.

Synthesis step 6: task integration This step is concerned with integrating the task specifications into pieces of existing technology. Inverse tasks are examined such as combining in one heat exchanger a heat-removal task on one stream with a heat-addition task on another stream. Thermodynamic and other constraints are also considered. Finally, the tasks are integrated into actual processing equipment.

Analysis step 6: final evaluations Overall material and energy balances are made, the equipment is sized, and cost estimates are made.

The AIDES program described by Siirola and Rudd (1971) performs nine functions, from analysis step 1 through analysis step 5. Although it produces a system of tasks as an output, it does not translate them into the actual processing equipment.

As an example Siirola and Rudd synthesized a process for the manufacture of ethyl acetate from ethyl alcohol. The chemical reactions specified were

$$C_2H_5OH + O_2 \xrightarrow{\text{cat}} CH_3COOH + H_2O$$

$$CH_3COOH + C_2H_5OH \rightleftharpoons H_2O + CH_3COOC_2H_5$$

We present only part of the process synthesis here, dealing with the identification of two separation tasks, the separation of an acetic acid–water stream and of an ethanol-water stream. The printout of the beginning of the acetic acid–water separation task is shown in Fig. 10-16. The engineer first requested an explanation of the legal commands and then asked the computer to consider a separation task. The computer requested solubility data for an effluent stream (4 from reaction 100R). The computer then considered several separation phenomena and scored the separations on the basis of heuristic considerations. It then suggests a distillation operation to produce new stream 6 and 7. The engineer approved, and the computer moved on to the next task.

In Fig. 10-17 the computer first suggested distillation for the ethanol-water separation. This was rejected by the engineer. The computer then suggested crystallization, which the engineer also rejected. This AIDES program was not programmed for other types of separation operations, such as azeotropic distillation, that might have been suitable for this separation task, and so the computer stated that there was no means of separation acceptable to the engineer and asked the engineer to acknowledge this fact. The engineer asked the computer to return to the task and accepted the distillation method.

This computer-aided synthesizer, based on computer-engineer interaction, is designed to define the decisions to be made in the creation of a flowsheet, to organize the information that is available to evaluate the alternatives, and to determine the consequences that flow from the decisions made. The program itself does much of the evaluations of alternatives, preliminary screening, and determination of the consequences of the decisions. A number of the decision and scoring routines are practical applications of some of the heuristics we discussed earlier.

10-9 CONCLUDING REMARKS

In this chapter we have presented a number of approaches to the analysis, simulation, and synthesis of chemical process. It should be clear that any engineer confronted with real design situations other than the simplest ones will have to contend with myriad factors beyond those that can be represented by any programmed system. Many of these factors cannot even be represented quantitatively with any confidence. King (1974) points out that

> programmed synthesis methods cannot uncover a new alternative and hence do not lend themselves to truly open-ended problems. Because of the inherent complexity, the typical open-endedness, the defiance of totally quantitative description, and the frequent emphasis upon novelty in chemical process design situations, it is important to seek methods which will add structure and logical direction to process design and engineering but at the same time will fall short of yielding a programmed synthesizer.

AFTER THE SPECIFICATION OF EACH SEPARATOR OR TEE THE SYSTEM WILL
PAUSE FOR A COMMAND. DO YOU WISH AN EXPLANATION OF LEGAL COMMANDS
- YES -
ANY OF THE FOLLOWING WORDS MAY BE ENTERED

        QUIT - TERMINATES THE PROGRAM
        STOP - TERMINATES THE CURRENT PROBLEM
        PASS - CAUSES THE SYSTEM TO CONTINUE
        JUST A LEFT ARROW - OPERATION CHOSEN IS ACCEPTABLE
        SKIP - OPERATION CHOSEN IS NOT ACCEPTABLE. THE SYSTEM
               WILL SEEK ANOTHER OPERATION.

        RETN - CAUSES THE SYSTEM TO RECONSIDER ALL SEPARATIONS
               USED TO RECOVER OPERATION REJECTED BY SKIP.
        YSCR - CAUSES OPERATION SCORES TO BE PRINTED
        NSCR - SUSPENDS OPERATION SCORE PRINTING
        YPOL - CAUSES OPERATION SCORE POLYNOMIAL TO BE PRINTED
        NPOL - SUSPENDS OPERATION SCORE POLYNOMIAL PRINTING
        CELL - PRINTS THE AMOUNT OF SPACE REMAINING TO THE SYSTEM

ENTER SYSTEM COMMANDS
- NPOL -
OPERATION SCORE POLYNOMIAL PRINTING SUSPENDED
- PASS-

```
MATCHES FROM STREAM NUMBER   4 FROM OPERATION  4 - 100R
     1692.9      N2
      631.8      H2O
      450.0      HAC

WHAT IS THE SOLUBILITY OF    N2 IN  HAC
= 0.
WHAT IS THE SOLUBILITY OF    HAC IN ANY GAS
= 0.
WHAT IS THE SOLUBILITY OF    HAC IN  H2O
= 100.

FOR S/L SPLIT BETWEEN HAC  AND H2O AT 16.5 C SCORE IS -42.721
WHAT IS THE SOLUBILITY OF  H2O IN HAC
= 100.

FOR L/G SPLIT BETWEEN H2O  AND HAC  AT 100.1 C  SCORE IS  25.952
FOR S/G SPLIT BETWEEN  N2  AND HAC  AT  16.5 C  SCORE IS-219.617
OPERATION 7 O-L/G SEPARATION BETWEEN H2O AND  HAC AT 100.1 C
PRODUCES STREAM 6 AT 118.1 C  AND STREAM 7 AT 100.0 C.

OPERATION  8  HEATER  BETWEEN STREAM 4 AT 40.0 C
AND STREAM  8  AT 100.1 C
```

Figure 10-16 AIDES acetic acid–water separation task. [*From J. J. Siirola and D. F. Rudd, Ing. Eng. Chem. Fundam.,* **10**:353 (1971), by permission. Copyright by the American Chemical Society.]

```
MATCHES FROM STREAM NUMBER  1  FROM OPERATION  1 - 1001

  100.0    H2O
  900.0    ETOH

WHAT IS THE SOLUBILITY OF H2O IN ETOH
- 100.

FOR S/L SPLIT BETWEEN H2O AND ETOH AT -0.1 C SCORE IS -157.442
FOR L/G SPLIT BETWEEN ETOH AND H2O AT 78.6 C SCORE IS  47.401
OPERATION 22 D-L/G SEPARATION BETWEEN ETOH AND H2O AT 78.6 C
PRODUCES STREAM 26 AT 100.0 C AND STREAM 27 AT 78.5 C
- SKIP

REJECTED L/G SEPARATION BETWEEN ETOH AND H2O AT 78.6 C
OPERATION 22 S-S/L SEPARATION BETWEEN H2O AND ETOH AT -0.1 C
PRODUCES STREAM 26 AT 114.6 C AND STREAM 27 AT 0.0 C
- SKIP

REJECTED S/L SEPARATION BETWEEN H2O AND ETOH AT -0.1 C
NO ACCEPTABLE SEPARATIONS FOR SYSTEM FOR STREAM 1 ACKNOWLEDGE
- RETN

OPERATION 22 D-L/G SEPARATION BETWEEN ETOH AND H2O AT 78.6 C
PRODUCES STREAM 26 AT 100.0 C AND STREAM 27 AT 78.5 C
- ..

OPERATION 23 HEATER BETWEEN STREAM 1 AT 25.0 C
```

Figure 10-17 AIDES ethanol-water separation task. [*From J. J. Siirola and D. F. Rudd, Ind. Eng. Chem. Fundam., 10:353 (1971), by permission. Copyright by the American Chemical Society.*]

Real-life engineers cannot be replaced. To cope with all the elements unique to their problems they have at their disposal a wealth of personal experience, the opportunity to consult with colleagues, vendors, and contractors, and the ability to draw upon all the information available in books, journals, and reports.†

PROBLEMS

10-1 Perform a functional analysis of the major steps in the production of (a) beet sugar, (b) sulfuric acid from sulfur, and (c) nitric acid from ammonia. Flowcharts and brief descriptions of these processes can be found in R. N. Shreve and J. A. Brink, "Chemical Process Industries," 4th ed., McGraw-Hill, New York, 1977.

10-2 Desalination of seawater can be achieved by a freezing process. Potable water is produced by melting the ice crystals after they have been washed free of adhering brine. Perform a morphological analysis of the ice-crystal separation and washing operations. List a number of alternatives for performing these operations, briefly discuss their applicability to the ice-brine system, and consider the possibility of combining the separation and washing operations. After completing the analysis compare with A. F. Snyder, Freezing Methods, chap. 2 in K. S. Spiegler (ed.), "Principles of Desalination," Academic, New York, 1966.

10-3 An installation to supply 1.5 MJ/s of refrigeration is being designed. The design calls for an ammonia vapor-compression system in which the ammonia is to evaporate at $-20°C$ and condense at $40°C$. Cooling water is available at $25°C$.

(a) Calculate the minimum power requirement and condenser duty for a primitive design involving a condenser, expansion valve, evaporator, and single-stage compressor.

(b) Improve the design by an evolutionary procedure. Calculate the minimum power requirements and heat-exchanger duties and list any additional equipment requirements for each design. Make at least two improved designs.

10-4 Liquid methane at 25 lb/in² is to be produced by a Linde cycle from methane at 400 lb/in² abs and 80°F. The simple Linde cycle proposed is as follows. Feed methane is mixed with recycle methane at 400 lb/in² and compressed to 1000 lb/in². This gas is cooled to $-100°F$ and then expanded through a valve into a vapor-liquid separator at 25 lb/in². The liquid is the desired product. The gas is compressed to 400 lb/in² in a three-stage compressor with intercooling to 80°F and mixed with the fresh methane feed. You are to evolve an improved cycle by evolutionary procedure involving at least two improvements, where the improvement criterion is reduction in energy requirements. Present flowcharts for the simple cycle and for each improved cycle and calculate the total energy requirements per pound of liquid methane for each cycle. In your calculations you may assume that insulation is perfect; unlimited cooling water at 75°F is available; compression is reversible and adiabatic; approach temperatures are 5°F; and energy requirements

† A more complete presentation of chemical-plant simulation, simulation programs, and problems in simulation is beyond the scope of this book. The interested reader is referred to such sources as Crowe et al. (1971), which describes and illustrates the strategy of simulation and presents a comprehensive case process; J. D. Seader, W. D. Seider, and A. C. Pauls, FLOWTRAN Simulation: An Introduction, CACHE Corporation, Cambridge, Mass., 1977, in which the FLOWTRAN system is described in detail; and user's guides such as R. C. Motard and H. M. Lee, "CHESS User's Guide," latest edition, Department of Chemical Engineering, University of Houston, Texas. Problems ranging from short exercises to help learn the mechanics of simulation to comprehensive design problems are presented in J. P. Clark (ed.) "Exercises in Process Simulation using FLOWTRAN," CACHE Corporation, Cambridge, Mass., 1977. The exercises should be solvable on the commonly used simulation systems.

for refrigeration can be calculated on the basis of an ideal Carnot refrigeration cycle working at an efficiency of 70 percent. Do not consider an isentropic expander as a replacement for an isenthalpic Joule-Thomson expansion.

10-5 You have been assigned the task of recommending changes that would reduce the steam and cooling-water requirements for the distillation operation described in Prob. 5-5. Your first step was to determine whether any immediate results can be achieved with the existing system. At the unit you note that it is well insulated, condenser fouling (Prob. 5-6) has been overcome, maintenance is excellent, steam traps are functioning, and no steam leaks are evident. Your conclusion is that salvation must be sought elsewhere. Back at your desk you study the flowsheets and note that two streams, both at 35°C, enter the column feed tank, one stream containing 40 mol % benzene, the other containing 55 mol % benzene. This mixed stream is pumped directly to the column feed tray, the feed preheater in the original design having been deleted. The overhead and bottom products are cooled to 40°C and then transferred to storage tanks. All heat for the unit is supplied by steam and all cooling by cooling water.

(*a*) Prepare a flowchart for the existing installation.

(*b*) Make a series of revisions in the process and discuss how each revision results in an improved design.

(*c*) Prepare a flowsheet for your final design. Calculate the steam and cooling-water requirements for your final design and compare with the requirements for the present design.

10-6 One of our customers can be supplied with our product from either of our three plants. The customer does not have any preference for source but demands that all our sales originate from the same plant. Our production costs are identical at each plant, but shipping routes and costs can differ. Our traffic department has supplied the shipping costs and route data presented in Fig. P10-6. The numbers above the arrows represent node-to-node unit shipping costs. From which plant should we supply the customer, what is the optimum route, and what will the minimum shipping cost be?

10-7 Specialty and special-purpose solvents can be produced by blending petroleum fractions of appropriate aromaticity and boiling range. One solvent manufacturer provides a limited range of special-purpose solvents but intends to terminate this operation. The storage tanks were carefully gauged with the following results: blending solvent A, 70 m³; blending solvent B, 100 m³; blending solvents C and D, 60 m³ each. These four solvents are to be disposed of either as is or as specialty solvents. Two specialty solvents can be produced with these blending solvents:

1. Paint and varnish thinner: 15 vol % A, 15 vol % B, 30 vol % C, 40 vol % D
2. Printer's-ink solvent: 30 vol % A, 50 vol % B, 20 vol % C

The profit realizable by sale of specialty solvents is $50 per cubic meter of thinner and $30 per cubic meter of printer's-ink solvent. Blending solvent sold as is would realize no profit and would be sold

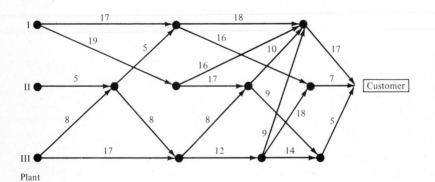

Figure P10-6 Shipping routes and costs.

at cost price. How much of each speciality solvent should be made to maximize profit, and what quantities of blending solvents will remain to be disposed of at cost price?

10-8 Much of the November issue of *Hydrocarbon Processing* in odd-numbered years is devoted to brief descriptions and abridged process diagrams for a number of active and new petrochemical processes. Select a process in which recycle is included and construct an information flow diagram, process matrix, and adjacency matrix for the selected process. Find the serial sets of units and the recycle sets in the process, find the number of tear streams, and specify the sequence in which mass-balance calculations, for example, can be performed.

ELEVEN

CASE STUDIES

> Specific cases require generalization and general rules require specification.
>
> *"Zohar" on Exodus, 3*

Many of the principles that have been discussed in the preceding chapters are illustrated in the five case studies on the following pages. The cases, with one exception, are real, and each one is presented in a slightly different form. In the first two cases we discuss the production of methanol and styrene and analyze some of the reasons why the processes are operated as they are. We next show how the commercial production of ethylene glycol has changed over the years as a result of process developments and changing raw-material costs. Next we look into the case of a runaway reaction, analyze what happened, and draw conclusions. The final case study is of a hypothetical but realistic proposal to expand production capacity for an existing product.

11-1 METHANOL PRODUCTION

Methanol is one of the most important petrochemicals. Its synthesis from hydrogen and carbon monoxide dates back to well before World War II. More than 5 million tons per day of methanol are now produced, and single-train plants with capacities of 1000 tons/day and larger are common. A 250 ton/day plant was considered to be a large plant in 1962. Formaldehyde production is the largest consumer of methanol, and more than 45 percent of the methanol produced is converted into formaldehyde. Other important uses of methanol are as a methylating agent and in solvents.

Methanol can be used as a fuel that can be conveniently stored, transported, and distributed. One of its large-scale future uses may therefore be as an energy carrier and as an alternative to natural gas. It would be produced where natural gas is in surplus and transported to areas with a deficit of energy.

The methanol synthesis reaction is simple

$$CO + 2H_2 \rightleftharpoons CH_3OH$$

but additional reactions can occur, e.g., carbon deposition from CO

$$2CO \rightleftharpoons CO_2 + C(s)$$

and a number of additional reactions between CO and H_2

$$CO + H_2 \rightleftharpoons HCHO$$

$$2CO + H_2 \rightleftharpoons CH_4 + CO_2$$

$$CO + 3H_2 \rightleftharpoons CH_4 + H_2O$$

or, in general,

$$nCO + (2n + 1)H_2 \rightleftharpoons C_nH_{2n+2} + nH_2O$$

If the H_2O-producing reaction occurs, an additional series of undesirable secondary reactions can result in the production of higher alcohols, acids, and esters; olefins; and dimethyl ether. Thus, a complex mixture of organics is possible. A selective catalyst and appropriate reaction conditions must be chosen so that methanol formation will be dominant and the undesirable reactions will be kept to a minimum.

Until the late 1960s the methanol synthesis was performed at high pressures, typically 300 to 375 atm, and at temperatures in the range of 600 to 650 K over a $ZnO-Cr_2O_3$ catalyst. In 1967 the ICI low-pressure process became available for licensing. In the low-pressure process typical operating conditions are 50 atm and 500 to 550 K, and the catalyst is copper-based with metal oxide promoters.

The flowsheets for the high and low-pressure processes are similar. Figure 11-1 is a basic flow diagram. The manufacture starts with the preparation of synthesis gas, which may be made from natural gas or from other hydrocarbons such as naphtha, fuel oil, or crude petroleum. The hydrocarbon is steam-reformed in accordance with

$$C_nH_{2n+2} + nH_2O \rightleftharpoons nCO + (2n + 1)H_2$$

and the quenched and cooled gases are then compressed and fed as makeup gas to the converter. Carbon dioxide, obtained by scrubbing the reformer furnace flue gas, is typically added to the reformer feed gas to provide control over the ratio of H_2 to $2CO + 3CO_2$ in the converter makeup gas. This is the reactant ratio that should be considered because methanol can be synthesized not only

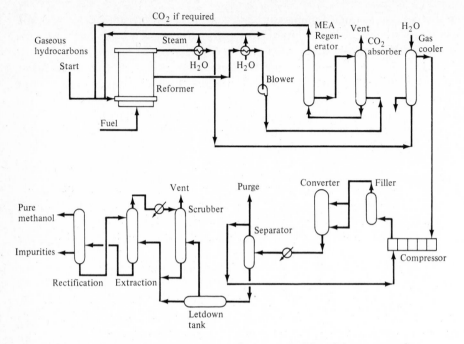

Figure 11-1 Flow diagram of methanol synthesis. [*From Strelzoff (1970), by permission.*]

from CO and H_2 but, at the conditions prevailing in the converter, also from CO_2 and H_2 in accordance with

$$CO_2 + 3H_2 \rightleftharpoons CH_3OH + H_2O$$

Typical compositions of steam-reformed natural gas with and without CO_2 additions to the reformer feed gas are shown in Table 11-1.

Table 11-1 Typical compositions of steam-reformed natural gas

	Without added CO_2 before reforming, mol %	With added CO_2 before reforming, mol %
H_2	75.50	68.55
CO	5.74	23.15
CO_2	14.95	5.00
CH_4	4.65	3.20
Ar	0.03	0.20
N_2	0.13	
$H_2/(2CO + 3CO_2)$	1.34	1.17

Source: From Strelzoff (1970), by permission.

The makeup gas is compressed and fed to the synthesis loop, which comprises a circulator, the converter with feed-effluent heat exchange, cooler, and separator. The crude methanol contains such impurities as acetone, aldehydes, dimethyl ether, and alcohols in small quantities. The impurities are removed by distillation. Part of the recycle gas from the separator is purged to maintain the inert-gas concentration in the synthesis loop at a pre-determined level. The purge gas would typically be burned in the reforming furnace as fuel.

Some further details of the methanol synthesis process will be described in the analysis and discussion which follow.

Thermodynamic and Kinetic Information

The equilibrium relationships can be expressed by

$$\Delta G° = -RT \ln K_f \qquad \text{and} \qquad K_f = K_\gamma K_p$$

The dependence of the standard free energy of reaction upon temperature for a number of typical reactions involving carbon monoxide and hydrogen is presented in Table 11-2. The values presented for the methanol synthesis reaction are quite close to those obtained from the equation presented by Dodge (1944)

$$\Delta G° = -17,530 - 60.4T - 0.0141T^2 + 18.19T \ln T$$

Values for K_γ as a function of T and P are shown in Fig. 11-2, and the effect of temperature and pressure on the enthalpy change due to the synthesis reaction is shown in Fig. 11-3.

Natta (1955) made a thorough study of the kinetics of the methanol synthesis reaction. The equation he developed for the rate of reaction, which assumes a trimolecular reaction in the absorbed phase, is

$$r = \frac{f_{CO} f_{H_2}^2 - f_{CH_3OH}/K_f}{(A + Bf_{CO} + Cf_{H_2} + Df_{CH_3OH})^3}$$

where r is the gram moles of methanol produced per gram of catalyst per hour.

Table 11-2 Standard free energy of reaction $\Delta G°$, kcal/mol

Reaction	T, K				
	300	400	500	600	700
$CO + 2H_2 \rightleftharpoons CH_3OH$	-6.30	-0.80	5.00	10.80	16.70
$2CO \rightleftharpoons CO_2 + C$	-28.57	-24.31	-20.02	-15.73	-11.44
$CO + 3H_2 \rightleftharpoons CH_4 + H_2O$	-33.87	-28.56	-23.01	-17.29	-11.44
$2CO + 2H_2 \rightleftharpoons CH_4 + CO_2$	-40.67	-34.36	-27.87	-21.22	-14.49
$2CO + 5H_2 \rightleftharpoons C_2H_6 + 2H_2O$	-51.32	-40.46	-29.18	-17.62	-5.87

Source: From Natta (1955), by permission.

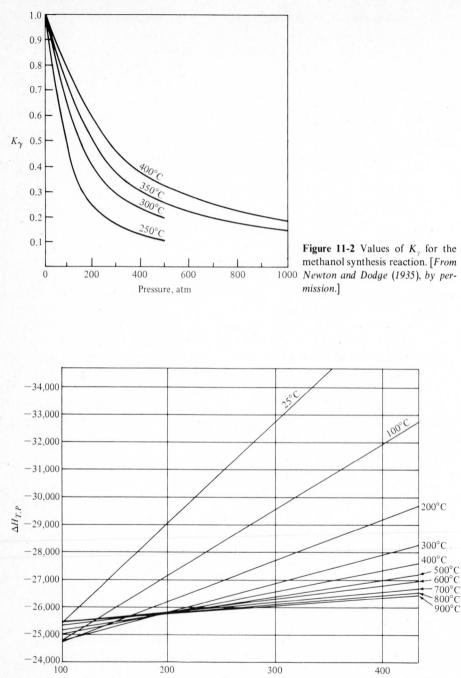

Figure 11-2 Values of K_γ for the methanol synthesis reaction. [*From Newton and Dodge (1935), by permission.*]

Figure 11-3 Effect of pressure on methanol heat of reaction. [*From W. J. Thomas and S. Portalski, Ind. Eng. Chem.,* **50**:967 *(1958), by permission. Copyright by American Chemical Society.*]

A, B, C, and D must be determined experimentally and are functions of temperature. For the ZnO–Cr_2O_3 catalyst, A increased with temperature whereas B, C, and D decreased. The heats of adsorption vary as follows: $CH_3OH > CO > H_2$. Activation energies are approximately 28,000 cal/g for ZnO–Cr_2O_3 catalysts and 14,000 for copper-based catalysts. Copper-containing catalysts are very active but have a short life at temperatures above 300°C.

Questions for Discussion

Some of the questions that will be discussed are as follows:

1. Why is the reaction carried out at elevated temperature and pressure?
2. What is the rationale that resulted in two competitive processes, a high-pressure process and a low-pressure process?
3. Why is CO_2 usually added to the gas fed to the reformer?
4. Why is the $H_2/(2CO + 3CO_2)$ ratio in the makeup gas greater than the stoichiometric ratio?
5. Why is the $H_2/(2CO + 3CO_2)$ ratio in the synthesis-loop gas much higher than the stoichiometric ratio?
6. A large number of reactions involving CO and H_2 are possible. If only thermodynamic considerations are invoked, what can be said about the likelihood of methanol synthesis occurring?
7. The methanol synthesis reaction is exothermic. What ramifications does this have with respect to reactor operation and design?
8. Although synthesis gas does not appear to be corrosive, it used to be customary to line the converter and process lines with copper or bronze. In present practice, the converter and process lines are fabricated of stainless-steel alloys. Analyze.
9. What effect can cooling-water temperature have on methanol production costs?

Most of these questions can be answered by appropriate consideration of the thermodynamic and kinetic aspects of the methanol synthesis process.

Discussion

Question 1 The reaction rate becomes appreciable only at elevated temperatures, hence the requirement for high temperature even though the equilibrium characteristics would favor a low temperature. Two optimum temperatures can be distinguished depending upon the objective function: (1) the optimum temperature for maximum production rate per unit of catalyst volume and (2) the optimum temperature for maximum production rate per unit volume of circulating gas. For the first, the production of methanol at a given temperature varies directly with the space velocity, while for the second it varies inversely with the space velocity provided that the temperature of operation is such as to ensure good selectivity of the catalyst. Above 400°C, for example, with a $ZnO-Cr_2O_3$ catalyst secondary reactions occur unless extremely high velocities of gas circulation are used, while at temperatures lower than 390°C raw methanol of high purity (98 percent) is obtained.

The synthesis of methanol proceeds with a considerable volume contraction; hence a high pressure has a marked favorable effect on equilibrium conversion. In addition, the reaction rate is a function of the fugacities of CO, H_2, and CH_3OH, which are, of course, also functions of pressure. Since the copper-based catalysts are very active, lower temperatures and pressure can be used than those required with $ZnO-Cr_2O_3$ catalysts.

Question 2 The production of methanol over a zinc-chromium catalyst requires, as stated above, elevated temperatures (325 to 400°C) to obtain satisfactory reaction rates. This, in turn, requires elevated pressures of the order of 300 atm in order to obtain reasonable percentage conversion of the synthesis gas to methanol in the converter. An important contributor to the production cost is the energy required to compress the synthesis gas to synthesis pressure. A reduction in pressure would result in a reduction in the cost of producing methanol. This was the reason for seeking a low-pressure methanol synthesis process.

With conventional zinc-chromium catalysts a reduction in pressure from 340 to 200 atm would result in a drop of percentage conversion from 5.5 to 2.4 percent, and if operation was at 50 atm, the maximum methanol concentration would be so low that condensation would not occur. Therefore, the key to a low-pressure process lay with a catalyst that would be active at temperatures lower than those required for zinc-chromium catalysts. It had been known for many years that copper-based catalysts promoted by ZnO and Cr_2O_3 have high selectivity and high initial activity, but they were sensitive to overheating and were short-lived.

The ICI low-pressure process uses a copper-based catalyst developed by ICI which is sensitive to temperatures above 300°C and is very sensitive to sulfur. The ICI process is therefore operated at 275°C, and the synthesis gas is maintained virtually sulfur-free. The operating pressure is 50 atm. The comparison between the high and low pressure processes presented by Strelzoff (1970) is reproduced in Table 11-3. The assumptions made are 25 percent inert gases, feed $H_2/(CO + CO_2)$ and effluent CH_3OH/H_2O molar ratios both equal to 2, and a fixed-percentage approach to equilibrium for both processes.

The low methanol content in the converter effluent (about 2.5 percent compared with 5.5 percent) is a disadvantage of the low-pressure process but also has some advantage. The lower rate of production results in a lower rate of heat generation compared with the high-pressure process and means that some of the excess hydrogen serving as coolant in the loop gas can be eliminated. This point will be discussed further below. An additional advantage of the low-pressure process results from the temperature used and the high selectivity of the catalyst. Fewer side reactions take place at the lower temperature and fewer

Table 11-3 Comparison between high- and low-pressure methanol synthesis processes†

Pressure, atm	CH₃OH in converter effluent, mol %	
	Z_nO–Cr_2O_3 catalyst, exit temp. 375°C	Cu-based catalyst, exit temp. 270°C
330	5.5	18.2†
200	2.4	12.4†
100	0.6†	5.8†
50	0.15†	2.5

† Hypothetical cases.
Source: From Strelzoff (1970), by permission.

impurities appear in the crude methanol. The crude methanol from the low-pressure process does not require pretreatment with potassium permanganate followed by neutralization to eliminate the aldehydes before distillation, as required in the high-pressure process.

The chief advantages of the low-pressure over high-pressure process are lower investment costs for gas compressors and synthesis loop equipment and lower energy costs. Catalyst costs per ton of methanol produced are higher, however.

Question 3 The logical starting point for the inlet-gas composition is the stoichiometric ratio of $2H_2$ to CO. On this basis the ideal hydrocarbon feedstock would have the empirical formula $(CH_2)_n$, and the steam reforming reaction would be

$$(CH_2)_n + nH_2O \rightleftharpoons nCO + 2nH_2$$

With the majority of the feedstocks, particularly with natural gas, which is the preferred feedstock for synthesis-gas production, the H_2/C ratio is too high. A gas stream containing CO_2 is added to the system to adjust the makeup gas to the desired composition. Additional CO would then be produced according to

$$CO_2 + H_2 \rightleftharpoons CO + H_2O$$

Any unreacted CO_2 can be converted into methanol in the converter

$$CO_2 + 3H_2 \rightleftharpoons CH_3OH + H_2O$$

so in plant operation the $H_2/(2CO + 3CO_2)$ ratio is used in place of H_2/CO and percentage conversion is based on $CO + CO_2$ rather than on CO only.

If the CO_2 content in the synthesis gas is not too high in relation to the hydrogen content, no adjustment in the synthesis gas need be made, as CO_2 will also be converted into methanol. The presence of small quantities of CO_2 actually has a favorable effect on the conversion to methanol at high space velocities, due probably to a number of factors: the CO_2 probably causes a decrease in the formation of dimethyl ether, and it allows better temperature regulation because the heat involved during methanol synthesis from CO_2 is about 10 kcal/mol less than for the synthesis from CO. CO_2 also retards the conversion of CO to CO_2. In addition, the reaction

$$CO_2 + H_2 \; \rightleftharpoons \; CO + H_2O$$

is endothermic and is catalyzed by zinc oxide. This reaction would act to regulate the temperature of the system and prevent overheating the catalyst.

Question 4 The reactants that are actually consumed and the compounds lost by purge must be supplied to the synthesis loop by the makeup gas. If the only reaction that took place were the methanol synthesis reaction from H_2 and $CO + CO_2$, the makeup gas would have to be supplied at a $H_2/(2CO + 3CO_2)$ ratio of 1.

In practice the $H_2/(2CO + 3CO_2)$ ratio in the makeup gas is in the range of 1.1 to 1.3 rather than 1. These deviations stem from two major factors: (1) extra hydrogen is consumed by side reactions, and (2) excess hydrogen is desirable in the converter gas because it suppresses side reactions and serves as an effective coolant. This extra hydrogen in the loop gas, in turn, means that extra amounts of hydrogen are lost from the loop gases by purging although the fuel value of the purge gas is recovered. This extra hydrogen must be supplied by the makeup gas.

Question 5 As pointed out earlier, the heats of adsorption of the methanol synthesis reactants and product rank as follows: $CH_3OH > CO > H_2$. This means that to obtain a H_2/CO ratio of 2:1 in the adsorbed phase it would be necessary to operate with the H_2/CO ratio much greater than 2 in the gaseous phase. In practice the gas entering the converter will have a $H_2/(2CO + 3CO_2)$ ratio as high as 2.2. These adsorption characteristics also explain why methanol acts as a depressant of the synthesis reaction rate. Its strong adsorption energy reduces the number of active sites available for the adsorption of CO and H_2.

It has already been pointed out that excess hydrogen is desirable in the synthesis loop for two major reasons, the suppression of undesirable side reactions and the use of hydrogen as a reactor coolant.

Question 6 Standard-free-energy changes for the methanol synthesis reaction and several other reactions between CO and H_2 are presented in Table 11-2. The equilibrium constants for several of these reactions will be calculated at 600 and 700 K from

$$\Delta G° = -RT \ln K_f$$

Equation	$K_{f_{600}}$	$K_{f_{700}}$	No.
$CO + 2H_2 \rightleftharpoons CH_3OH$	116×10^{-6}	6.1×10^{-6}	(1)
$2CO \rightleftharpoons CO_2 + C$	537×10^3	3.73×10^3	(2)
$CO + 3H_2 \rightleftharpoons CH_4 + H_2O$	1.99×10^6	3.73×10^3	(3)
$2CO + 2H_2 \rightleftharpoons CH_4 + CO_2$	53.7×10^6	33.4×10^3	(4)

In order to obtain the composition that would be expected if equilibrium were attained with respect to these reactions, it would be necessary to set up the appropriate mass-balance equations, to incorporate them into the equilibrium-constant expressions, and then to solve the resulting set of simultaneous equations. Alternatively, the composition could be calculated by the free-energy-minimization method, described in Chap. 4. The answer to the question posed for discussion can be obtained by inspection of the values shown above for the equilibrium constants. It is obvious that the percentage of methanol to be expected at thermodynamic equilibrium would be negligible.

This emphasizes that the success of the methanol synthesis is due to the activity and specificity of the catalysts used in the commercial process. The catalysts are so specific that the undesirable side reactions occur to a negligible extent in spite of their favorable equilibrium conversions.

Question 7 This situation will first be examined by comparing equilibrium conversions to be expected for isothermal and adiabatic operation. To obtain the full impact of the comparison we shall assume a feed gas to the converter that consists only of CO and H_2 in the proper stoichiometric ratio with no inert gasses present. We shall therefore assume a mixture of 2 mol of H_2 and 1 mol of CO fed to a converter, from which methanol, CO, and H_2 leave at equilibrium; operating conditions are similar to those of the low-pressure process, 270°C and 50 atm.

The thermodynamic data used are those developed by Dodge (1944), summarized below:

$$\log K_f = \frac{3835}{T} - 9.15 \log T + 0.00308T + 13.20$$

$$\Delta H_T = -17{,}530 - 18.19T + 0.0141T^2$$

Heat capacities (assumed independent of P):

$$C_p = \begin{cases} 6.65 + 0.00070T & \text{for } H_2 \\ 6.89 + 0.00038T & \text{for } CO \\ 2.0 + 0.03T & \text{for } CH_3OH \end{cases}$$

Units are calories, gram moles, and kelvins.

We shall assume that the fractional conversion of CO and H_2 at equilibrium to CH_3OH is X. The equilibrium mixture will therefore contain $1 - X$ mol of CO, $2(1 - X)$ mol of H_2, and X mol of CH_3OH, and

$$K_f = \frac{X(3 - 2X)^2}{4(1 - X)^3} P^{-2} K_\gamma$$

Isothermal operation At 270°C (543 K) $K_f = 8.159 \times 10^{-4}$, and at 543 K and 50 atm $K_\gamma = 0.85$. Substituting these values into the expression for K_f above and solving for X, we obtain

$$X = 0.412$$

Adiabatic operation Two simultaneous equations will be obtained for this case, an energy-balance equation and the equilibrium equation, both of which express the relationship between X and T.

The energy-balance equation can be written

$$X \, \Delta H_{543} = (1 - X) \int_{543}^{T} (6.89 + 0.00038T) \, dT$$

$$+ 2(1 - X) \int_{543}^{T} (6.65 + 0.00070T) \, dT$$

$$+ X \int_{543}^{T} (2.0 + 0.03T) \, dT$$

Solving for X gives

$$X = \frac{20.2T + 0.00089T^2 - 11,225}{18.2T - 0.0141T^2 + 17,534}$$

The energy-balance equation as written above assumes for the purposes of the calculation that all the enthalpy of reaction is produced at the inlet temperature and that this energy is used to heat up product and unreacted reactants to temperature T. In the equilibrium relationship the equilibrium conversion will be a function of temperature through the effect of temperature on the values of K_f and K_r. The two simultaneous equations between X and T can be solved by a trial-and-error procedure. The results are

$$T = 617 \text{ K } (344°C) \qquad X = 0.0683$$

The difference in equilibrium conversions between the two extremes of isothermal and adiabatic operation is impressive. The effect would be somewhat attenuated by the presence of inert gases, but it still would be striking. It is obvious that cooling the catalyst bed is desirable. If cooling by heat-transfer surfaces cannot be performed conveniently, an excess of one of the reactants should be present to minimize the adiabatic temperature rise. The heat to be dissipated from the reactor is substantial. A 1000 ton/day methanol reactor produces an amount of heat roughly equivalent to the combustion of fuel oil at a rate of more than 100 tons/day.

In order to deal with the thermal energy produced by the methanol synthesis reaction, the converter in large plants is designed on the quench principle.

No cooling tubes are embedded in the catalyst bed, and the entire amount of catalyst is placed in four or five baskets or beds in series. Although each bed is operated adiabatically, interbed cooling is performed by injecting cool synthesis gas between beds and by auxiliary jets of cool gas directed into the top of each bed. Cooling coils using water may also be installed between beds. It has already been mentioned that a large excess of hydrogen is used to serve as a coolant and as an inhibitor of side reactions.

The reactor design takes into account the fact that the reaction rate decreases as the reaction proceeds toward equilibrium. The thickness of the beds in the converter is varied, the thickness increasing progressively in the direction of gas flow. The thickness in each bed would be such as to result in equal conversion per bed. An example of the temperature and conversion profile is shown in Fig. 11-4.

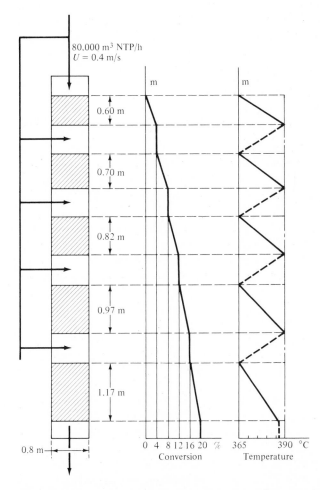

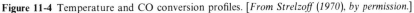

Figure 11-4 Temperature and CO conversion profiles. [*From Strelzoff (1970), by permission.*]

Question 8 Iron carbonyl is readily formed by reaction between CO and carbon steels and low-alloy steels. Not only is the steel attacked by the CO, but additional undesirable effects can occur. The synthesis of CH_4 is catalyzed by the presence of iron, and the most active form of iron is obtained by the thermal decomposition of iron carbonyl. In order to avoid the presence of iron carbonyl, the converter and process lines must be fabricated of materials that will not permit carbonyl formation.

Question 9 The cooling-water temperature has two effects: (1) the condenser surface required in the separator increases as cooling-water temperature increases, and (2) an increase in condensation temperature will result in an increase in the methanol content in the gas recirculating to the converter. The latter factor is of negligible importance in the high-pressure process but is of relevance in the low-pressure process. An increase in condensation temperature from 30 to 35°C will result in an increase in the concentration of methanol in the loop gas recycling to the converter from 0.42 to 0.53 mol %. This is significant when it is remembered that the concentration of methanol leaving the converter is about 2.5 percent. The result of an increase in cooling-water temperature would be an increase in production costs.

11-2 STYRENE PRODUCTION

Styrene is a large-volume, basic chemical commodity with production greater than 3 million tons per year. Its first large-scale use was during World War II in the production of synthetic rubber. Although synthetic rubber is still an important consumer, the major use is now in the production of polystyrene and various copolymers. Polystyrene is a clear, thermoplastic resin that can easily be molded, cast, or extruded. Foams of expanded polystyrene have good insulating and cushioning properties.

In this case study, instead of the procedure of questions and answers as used in Sec. 11-1, the reasons and considerations that led to the choice and selection of specific process and operating conditions will be discussed as part of the process description.

Production Methods

Styrene can be produced by a number of methods:

1. *Decarboxylation.* A very pure styrene can be produced by the dry distillation of cinnamic acid and its salts.
2. *Dehydration.* Phenylethyl alcohol can be dehydrated to give styrene in good yields.
3. *Pyrolysis.* Styrene is produced as one of the products of the pyrolysis of xylene, coal, oil, and propane at temperatures between 600 and 1000°C.

4. *Via ethylbenzene.* Several routes are available to styrene via ethylbenzene:
   a. Chlorination followed by removal of HCl

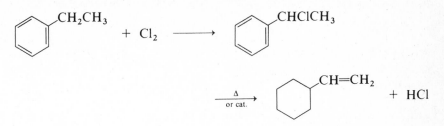

+ HCl

   b. Bromination followed by treatment with magnesium

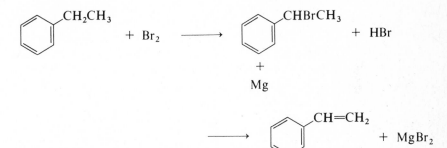

   c. Via acetophenone, yielding phenylethylalcohol, which is dehydrated as in item 2

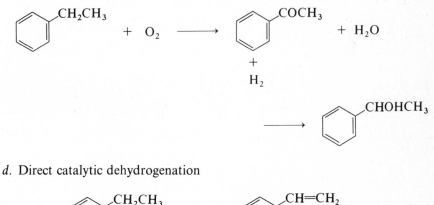

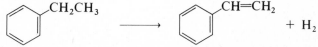

   d. Direct catalytic dehydrogenation

Commercially, routes 4a and 4d have been the most important although the chlorination route has now been completely superseded by the direct catalytic dehydration of ethylbenzene.

It is left as an exercise for the reader to study the other routes and to analyze and explain why none of the other routes have had commercial success or are likely to.

Three major steps are therefore involved in the production of styrene:

1. Production of ethylbenzene
2. Dehydrogenation of ethylbenzene
3. Purification of styrene

Each of these steps will be discussed and analyzed separately. The analysis will lead us to the reasons for the choice of the process and equipment and to the appropriate operating conditions.

Ethylbenzene Production

Although some ethylbenzene is recovered by fractionation of a petroleum fraction, most of it is produced by the Friedel-Crafts alkylation of benzene with ethylene

Two types of processes are used commercially, a vapor-phase and a liquid-phase process. The vapor-phase reaction is carried out at about 300°C and 40 atm pressure over a catalyst of phosphoric acid supported on kieselguhr, alumina, or silica gel. Most of the ethylbenzene produced, however, is made by the liquid-phase process with aluminum chloride catalyst.

The desired reaction is exothermic ($\Delta H^\circ_{298} = -27{,}203$ cal/g mol) and the standard free energy of reaction is highly negative ($\Delta G^\circ_{298} = -16{,}063$); thus the equilibrium conversion would be essentially 100 percent. In both processes, however, di-, tri-, and higher ethylbenzenes up to hexaethylbenzene are produced, in addition to a small amount of tars. The reasons and the process implications will now be discussed, primarily from the thermodynamic viewpoint.

Listed in Table 11-4 are the free energies of formation of the compounds that participate in the benzene alkylation reactions. The desired alkylation reaction is

$$C_6H_6 + C_2H_4 \rightleftharpoons C_6H_5R \qquad (11\text{-}1)$$

where R represents the ethyl group. The reactions leading to polyalkylated benzenes can be written as

$$C_6H_6 + 2C_2H_4 \rightleftharpoons C_6H_4R_2 \qquad (11\text{-}2)$$

$$C_6H_6 + 3C_2H_4 \rightleftharpoons C_6H_3R_3 \qquad (11\text{-}3)$$

$$C_6H_6 + 6C_2H_4 \rightleftharpoons C_6R_6 \qquad (11\text{-}4)$$

Table 11-4 Some free energies of formation

Compound	Free energy of formation, kcal/g mol	
	$\Delta G^\circ_{f,\,400}$	$\Delta G^\circ_{f,\,600}$
Ethylene†	17.7	20.9
Benzene†	35.0	43.7
Ethylbenzene†	39.7	57.6
Diethylbenzene‡	45.6	73.5
Triethylbenzene‡	52.0	89.9
Hexaethylbenzene‡	70.9	139.1

† Values from Reid and Sherwood (1966).
‡ Estimated by the method of Franklin.

The standard free energies of reaction in kilocalories per gram mole for the alkylation reactions are:

Reaction	ΔG°_{400}	ΔG°_{600}
Ethylbenzene	−13.0	−7.0
Diethylbenzene	−24.8	−12.0
Triethylbenzene	−36.1	−16.5
Hexathylbenzene	−70.3	−30.0

It is evident that thermodynamic equilibrium would favor the production of the higher ethylbenzenes rather than the production of monoethylbenzene. As would be expected for the exothermic reaction, ΔG° increases with temperature but the equilibrium at high temperatures is still in favor of the production of the polyethylbenzenes.

Fortunately, dealkylation reactions can also take place if no or little ethylene is present:

$$C_6H_4R_2 + C_6H_6 \; \rightleftharpoons \; 2C_6H_5R \qquad (11\text{-}5)$$

$$C_6H_3R_3 + C_6H_6 \; \rightleftharpoons \; C_6H_5R + C_6H_4R_2 \qquad (11\text{-}6)$$

$$\cdots\cdots\cdots\cdots\cdots\cdots\cdots\cdots\cdots\cdots\cdots\cdots\cdots$$

$$C_6R_6 + C_6H_6 \; \rightleftharpoons \; C_6HR_5 + C_6H_5R \qquad (11\text{-}7)$$

The standard free energies of reaction for these reactions are −1.2, −1.7, and −1.7 kcal/g mol, respectively, at 400 K, and these values are practically independent of temperature. Thus, in the presence of ethylene, the equilibrium condition would favor the production of the hexaethylbenzene, whereas in the absence of ethylene the dealkylation reactions would favor the production of ethylbenzene from the polyethylbenzenes.

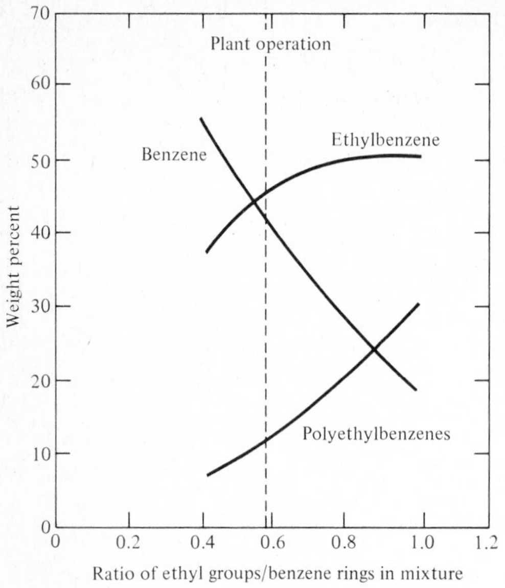

Plant operation
Benzene
Ethylbenzene
Polyethylbenzenes
Weight percent
70
60
50
40
30
20
10
0
0
0.2
0.4
0.6
0.8
1.0
1.2
Ratio of ethyl groups/benzene rings in mixture

Figure 11-5 Equilibrium conditions for desired and competing reactions. [*From Boundy and Boyer (1952), chap. 2, by permission.*]

The production of polyethylenes is therefore minimized in practice by operating at a high ratio of benzene to ethylene in the feed to the alkylation reactor. The ratio of benzene to ethylene commonly used in the vapor-phase process is 5 : 1 rather than the stoichiometric ratio of 1 : 1. Typical equilibrium conditions and plant operating conditions for the liquid-phase process are shown in Fig. 11-5. Although production alkylators do not operate at equilibrium conditions, production experience substantiates the optimum operating point as predicted by theory. Although operating at a high benzene level increases the yield of ethylbenzene, it also increases the cost for the fractionation of the excess benzene for recycle.

The liquid-phase process for the alkylation of benzene is shown in Fig. 11-6. Since water in the feed would increase the consumption of $AlCl_3$, increase sludge formation, and cause corrosion problems because of the wet HCl produced, the benzene feed to the reactor must be dry. Makeup benzene is dried azeotropically in a fractionating column to less than 30 ppm H_2O. Ethylene purity is not critical so long as unsaturated materials such as acetylene or higher olefins are not present. The feed material is usually 90 to 95 percent ethylene, the balance being inerts such as hydrogen and saturated paraffins. The anhydrous $AlCl_3$ catalyst must be activated by anhydrous HCl. This is usually done by continuously vaporizing a small amount of ethyl chloride into the ethylene reactor feed. The ethyl chloride decomposes into ethylene and HCl. Although anhydrous HCl can be used directly, it is more convenient from the handling viewpoint to use the ethyl chloride.

The alkylation reactor is jacketed, glass-lined, or made of Hastelloy, and is equipped with side-stream coolers to remove the heat of reaction. The reactor is operated at 85 to 95°C and at a pressure slightly above atmospheric to ensure

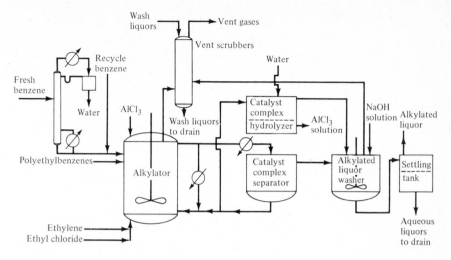

Figure 11-6 Flow diagram for liquid-phase benzene alkylation.

that air does not enter the system. The ethylene is sparged into the bottom of the reactor, where it meets fresh and recycle benzene, recycled polyethylbenzene, and the aluminum catalyst complex. Simultaneous alkylation and dialkylation take place. At a ratio in the total feed to the reactor of 0.5 to 0.6 ethylene groups per benzene ring (Fig. 11-5), the product stream from the alkylator would consist of about 40% ethylbenzene, 43% benzene, 15% diethylbenzene, and 2% higher-molecular-weight products. At higher ratios higher conversions would be expected, but more polyalkylated benzene and tar would be produced. Lower ratios of ethylene to benzene in the feed would produce lower conversions but less polyalkylation. The optimum ratio to use in any given plant depends upon economic considerations, in which investment in fractionation and recycle equipment and operating costs would have to be balanced against product yield and raw-material cost.

The reaction products include not only alkylated and unreacted benzene but also the hydrocarbon-insoluble catalyst complex formed between the $AlCl_3$ and the hydrocarbon fed to the alkylator. The reaction products are cooled, and the catalyst-complex layer is allowed to separate from the hydrocarbons. A small stream of this layer is bled off to keep the aluminum chloride active. It is hydrolyzed with water and allowed to settle. The oil layer is either recycled back to the reactor or joins the methylated liquor product for purification. The aqueous layer contains aluminum chloride and must be disposed of. The hydrocarbon layer from the catalyst-complex settling tank is washed with caustic soda solution and water and then purified.

The purification is done by fractionation, and the alkylated liquor can be regarded as a four-component mixture for this purpose. Three continuous fractionating columns are therefore required to separate the four components, benzene, ethylbenzene, polyethylbenzenes (di- to hexaethylbenzenes), and alkylation tars. The first column operates at atmospheric pressure, and pure benzene is taken overhead. Column 2 operates at about 200 mmHg condenser pressure, and monoethylbenzene is taken over head. The third column, in which the polyethylbenzenes are taken overhead, operates at an even lower pressure. The polyalkylated benzene in most plants is returned to the alkylator, where it dealkylates and suppresses the formation of additional polyethylbenzenes. In some plants the polyalkylate is sent to a separate dealkylation unit operating at 200°C.

The vapor-phase process operates with an acidic catalyst, usually phosphoric acid supported on a kieselguhr carrier. The fixed-bed reactor is operated at about 40 atm and 300°C. The major disadvantage of the vapor-phase process is its inability to dealkylate conveniently the polyethylbenzenes produced. The process is therefore operated at a high ratio of benzene to ethylene to minimize the production of di-, tri-, and higher ethylbenzenes. The major advantage is its ability to accept a relatively impure ethylene feedstock.

Dehydrogenation of Ethylbenzene

The desired dehydrogenation reaction is

$$
\text{C}_6\text{H}_5\text{—CH}_2\text{CH}_3 \rightleftharpoons \text{C}_6\text{H}_5\text{—CH}=\text{CH}_2 + \text{H}_2 \qquad (11\text{-}8)
$$

A number of additional reactions have been identified as being of importance. The desired and the competing reactions and consequences must be considered when the reactor operating conditions are being selected. Stull (1952) compiled a number of calculated and observed equilibrium constants for reaction (11-8) and calculated the thermodynamic equilibrium as a function of temperature at 1 atm pressure. The equilibrium compositions are presented in Fig. 11-7. If we now

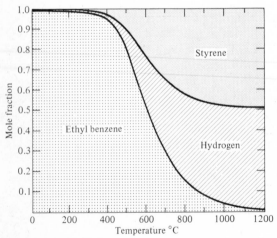

Figure 11-7 Equilibrium compositions for reaction (11-8). [*From Boundy and Boyer (1952), chap. 3, by permission.*]

consider only this reaction, it is obvious that it would be desirable to operate the reactor at a temperature in excess of 500°C. In view of the fact that the reaction involves an increase in volume, the conversion of ethylbenzene into styrene would be increased by operating the reactor under a vacuum or in the presence of an inert gas. Operating under a vacuum is undesirable in practice primarily for safety reasons but also because it requires compressing the by-product hydrogen. The use of an inert gas would therefore appear to be desirable to increase the styrene conversion. This inert gas could also be considered as a medium for heating the feed ethylbenzene to reactor conditions and as a possible means for supplying the endothermic heat of reaction.

At this stage of our analysis we can assume that the reactor would operate at about atmospheric pressure and a temperature above 500°C. The dehydrogenation reaction would take place in the presence of an inert gas to increase the conversion to styrene. If the reaction could be carried out at 750°C or higher, almost complete conversion of ethylbenzene into styrene could be expected at equilibrium conditions.

We shall now look at some of the possible competing or parallel reactions and examine their effects on the picture that has thus far emerged. The first reaction to be considered is the formation of ethynylbenzene by the dehydrogenation of styrene

$$\text{(11-9)}$$

Stull (1952) calculated the thermodynamic equilibria for the formation of this undesirable by-product, and his results are presented graphically in Fig. 11-8. Inspection of this figure shows that the formation of ethynylbenzene can be effectively controlled by operating the reactor at a temperature below 700°C.

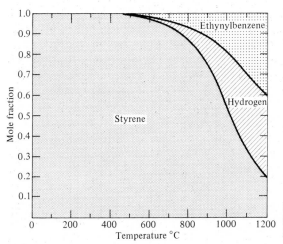

Figure 11-8 Equilibrium compositions for reaction (11-9). [*From Boundy and Boyer (1952), chap. 3, by permission.*]

Two additional reactions that result in the formation of benzene and toluene from ethylbenzene are

$$(11\text{-}10)$$

and

$$(11\text{-}11)$$

Reaction (11-10) is, of course, the reverse of the benzene alkylation reaction. Wenner and Dybdal (1948) calculated the equilibrium constants for these reactions at temperatures above 300°C, and their results are shown graphically in Fig. 11-9. The equilibrium constants for the competing reactions (11-10) and (11-11) are larger than the equilibrium constants for the desired styrene-formation reaction (11-8). Thus, if only the thermodynamic factors were considered, we could conclude that the possibility of preparing styrene by the dehydrogenation of ethylbenzene was bleak indeed. Once again, however, and as in the case of methanol synthesis, the development of highly selective catalysts has resulted in dehydrogenation catalysts that promote the rate of the desired reaction relative to the competing reactions. We shall not go into any additional kinetic analysis in this study, however.

An additional set of reactions is possible if steam is used as the inert-gas medium. The reactions are of the water-gas type. The important one from the standpoint of catalyst life is

$$H_2O + C \rightleftharpoons CO + H_2 \qquad (11\text{-}12)$$

At the temperature conditions to be expected in the reactor $(T > 500°C)$ this reaction would prevent carbonization of the catalyst and thereby help extend catalyst life.

The conclusions that can be reached from the thermodynamic considerations can now be summarized. In order to ensure a reasonable conversion of ethylbenzene into styrene, the reactor should be operated at a temperature greater than 500°C. From the equilibrium-conversion viewpoint, temperatures greater than 750°C would be desirable, but a number of competing undesirable reactions must also be considered. The cracking of ethylbenzene to benzene and the formation of toluene from ethylbenzene and hydrogen are reactions favored by the equilibrium conversion in preference to styrene formation at all temperatures. A dehydrogenation catalyst will therefore be necessary for styrene formation to occur. But since a dehydrogenation catalyst would also result in the further dehydrogenation of styrene to ethynylbenzene, an undesired reaction whose equilibrium is negligible at temperatures lower than 700°C, we can expect a styrene reactor to operate at $500 < T < 700°C$ with a dehydrogenation catalyst and probably with an inert gas.

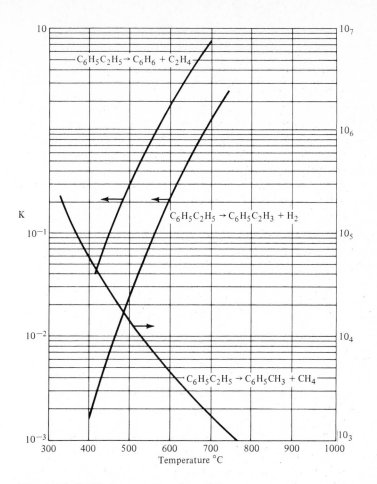

Figure 11-9 Equilibrium constants for ethylbenzene dehydrogenation. [*From Wenner and Dybdal (1948), by permission.*]

Commercial Ethylbenzene Dehydrogenation

We shall now describe the commercial styrene process and compare commercial practice with the conclusions reached by the a priori analysis. The two types of processes in use, adiabatic and isothermal, differ in how the endothermic heat of reaction is supplied. Both processes use a pelleted dehydrogeneration catalyst comprising Fe_2O_3, Cr_2O_3, and potassium salts or zinc oxide activated with Al_2O_3, CaO, K_2CrO_4, and K_2CO_3.

Adiabatic process In the adiabatic process the heat of reaction is supplied by mixing the ethylbenzene feed with superheated steam, whereas in the isothermal process heat is supplied to the reactor through the walls of the tubes which contain the catalyst. The adiabatic process is the more common. The adiabatic-process reactor is typically a large-diameter chrome-steel vessel in which the catalyst is contained in a catalyst bucket. The ethylbenzene, mixed with part of the steam, is preheated to about 500 to 550°C by heat interchange with the reaction products leaving the reactor. The preheated ethylbenzene is mixed with the remaining steam that has been superheated to about 720°C. A total of about 3 kg of steam is used per kilogram of ethylbenzene. The resulting mixture enters the reactor at about 600 to 620°C when a fresh catalyst is being used. As the catalyst ages, the temperature needs to be increased so conditions would have to be adjusted to raise the temperature up to about 650°C to maintain conversion to styrene.

The reaction products leave the reactor at about 560°C and are cooled and condensed in a spray-type desuperheater, followed by tubular heat exchangers, after having exchanged heat with the incoming feed. The vent gas contains H_2, CO, and CO_2 plus some methane and benzene. It may be further refrigerated to recover the aromatics, and the noncondensibles are used as fuel. The crude styrene product typically analyzes about 57% ethylbenzene, 38.5% styrene, 1 to 2% toluene, 0.5 to 2% benzene, and 0.5% or less tars.

Isothermal process In the typical isothermal process an ethylbenzene-steam mixture is superheated to 585°C by interchange with the reactor effluent stream and hot flue gas. The reactor is similar, in principle, to a shell-and-tube heat exchanger with the pelleted catalyst contained in the tubes. Reactor temperature is maintained at 585 to 610°C by hot flue gas circulating around the outside of the tubes.

By serving as an inert medium the steam reduces the partial pressure of the products and reactants, thus enhancing the equilibrium conversion of ethylbenzene into styrene. The steam also prevents carbonization of the catalyst by promoting the reaction between steam and coke. In the adiabatic process the steam also provides the endothermic heat of reaction. It is therefore added in ratios up to 3 : 1 steam to ethylbenzene, compared with 1 : 1 for the isothermal process, where the heat is supplied through the reactor walls.

The major advantage of the adiabatic process over the isothermal process is the lower capital cost for the steel-vessel reactor compared with the multitubed isothermal reactor. The advantages of the isothermal process over the adiabatic are (1) lower steam requirement because it is used only as an inert gas and to maintain catalyst activity, (2) fewer cracking reactions to benzene and toluene because of lower temperatures, and (3) slightly higher yields (92 to 94 percent compared with 88 to 91 percent) due to advantage 2 plus the thermodynamic advantage of isothermal over adiabatic operation.

We can summarize this section by pointing out that the commercial production of styrene by the dehydrogenation of ethylbenzene takes place within the range of process conditions that we were able to deduce from thermodynamic considerations.

Purification of Styrene

The problems arising in the purification of styrene are the result of two factors: (1) the relatively close proximity of the normal boiling points of styrene and ethylbenzene, 145 and 136°C, respectively, and (2) the danger of polymerization of styrene. Polymerization rates of styrene without inhibitors at 130, 110, 90, and 80°C are about 40, 10, 1.7, and 0.7 percent per hour, respectively. Either one of these factors taken by itself would pose no great problem. The close proximity of the boiling points and the vapor-liquid equilibrium relationships between styrene and ethylbenzene would merely mean that a large number of plates would be needed for the separation. For the required separation a minimum of about 70 actual plates are necessary. The danger of polymerization can be avoided by operating at such a temperature that thermally induced polymerization rates are low and tolerable. Rates of less than 1 percent per hour can be expected at temperatures lower than 90°C.

These two requirements appear, as we shall see, to be irreconcilable. An efficient bubble-cap plate would operate with a pressure drop of 3 mmHg. Thus, the pressure drop through the column would be 210 mmHg. To ensure condensation with cooling water a minimum head pressure of about 20 mmHg would be required. The pressure in the reboiler would therefore be 230 mmHg, and the minimum boiling temperature of the styrene in the reboiler would be 107°C, corresponding to a polymerization rate of 7 percent per hour.

The problem of limiting the boiling temperature in the reboiler was solved by splitting the required 70 plates into two columns operated in series with complete condensation of the overhead vapor of each column. By this means it was possible to operate so that the reboiler temperature remained below 90°C, thereby minimizing the possibility of thermally induced polymerization.

In spite of this relatively low temperature in the styrene-purification column the danger of polymerization still remains because the rate of polymerization can either be inhibited or catalyzed by the presence of even a few parts per million of impurities. For this reason a polymerization inhibitor is added to the crude styrene. Sulfur is an efficient inhibitor, and it is added by percolating the crude styrene through a bed of elemental sulfur.

A typical styrene purification train is shown in Fig. 11-10. Benzene and toluene are first taken overhead and then fractionated in separate columns. The mixture of ethylbenzene, styrene, and tars is fed to two or more columns operating in series, and up to 100 actual plates are used to make the ethylbenzene-styrene separation. The reboilers are designed to operate with a small temperature difference across the heating tubes and with low liquid heads to prevent high skin temperatures. *tert*-Butyl catechol is added to the reflux condenser of the pure styrene column to ensure polymerization inhibition.

The technological achievement involved in the developments described above made styrene production by ethylbenzene dehydrogenation a commercial proposition. The disadvantage of the two-column distillation train, however, is

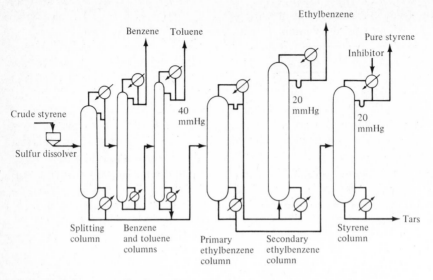

Figure 11-10 Styrene purification train.

the high energy requirement because of the necessity for complete condensation and revaporization of overhead. This doubles the steam and cooling-water duty compared with requirements of a single tower. A single tower would be a feasible proposition only if trays with a pressure drop less than 3 mmHg each were available. In the 1960s advances made in the design of distillation trays and notably the sieve tray made a single column system for styrene-ethylbenzene separation possible. The pressure drop per tray is 2.5 mmHg, and the 70 trays in the column have a total pressure drop of 175 mmHg. If the head pressure is 20 mmHg, the pressure in the reboiler would correspond to a boiling temperature of 100°C. Although this temperature would result in an uninhibited polymerization rate of over 4 percent per hour, the use of efficient polymerization inhibitors permits the distillation operation to proceed satisfactorily. The steam requirement for the single-column operation is 1.3 kg of steam per kilogram of styrene, compared with twice that for the two-column operation.

11-3 ETHYLENE GLYCOL

Ethylene glycol is one of the more important organic chemicals. Its production rate in the United States exceeds 2.0 million tons per year, and its selling price is in the range of 30 to 50 cents per kilogram. The major use of ethylene glycol is as an automotive antifreeze and coolant, and it finds additional uses as a chemical intermediate. In the 1930s practically all ethylene glycol was produced via ethylene chlorhydrin, but this process has now been completely displaced. From Table 11-5 it is apparent that by 1975 all synthetic ethylene glycol was produced

Table 11-5 Ethylene glycol process technology (percent distribution)

Year	Technology	
	Chlorhydrin	Ethylene oxide
1930	100	0
1940	90	10
1945	70	30
1950	68	32
1960	36	64
1965	18	82
1975	0	100

Source: From Brownstein (1975), by permission.

via ethylene oxide from ethylene oxidation. This process itself is now under attack by newer technologies.

In this study we shall first examine the chlorhydrin process and analyze why it is no longer in commercial use. The present commercial process based on ethylene oxide hydration will then be discussed and compared with newer technologies.

Ethylene Chlorhydrin Process

Ethylene glycol can be obtained from the hydrolysis of ethylene chlorhydrin or ethylene oxide. The direct hydrolysis of ethylene chlorhydrin with sodium bicarbonate is inconvenient because of the difficulty of separating the product from the salts in the mixture. Ethylene chlorhydrin, however, can be converted into ethylene oxide, which can then be separated easily from the mixture and further hydrolyzed to the glycol.

The addition of hypochlorous acid to ethylene gives ethylene chlorhydrin. The reactions are

$$Cl_2 + H_2O \; \rightleftharpoons \; HOCl + HCl \qquad (11\text{-}13)$$

$$C_2H_4 + HOCl \; \rightleftharpoons \; HOCH_2CH_2Cl \qquad (11\text{-}14)$$

The reaction is carried out in a contacting tower by passing ethylene and chlorine simultaneously into water. The competing reaction is the direct addition of chlorine to the ethylene double bond, giving ethylene dichloride. The equilibrium concentration of hypochlorous acid in water is small, but its rate of addition to ethylene is much faster than that of the addition of chlorine to ethylene. If a slight excess of ethylene is maintained, a concentration of 6 to 8 percent of chlorhydrin is reached in the solution with little simultaneous production of the dichloride. The reaction is carried out at 40 to 50°C, the temperature being maintained by the heat of reaction. The major byproduct obtained is ethylene dichloride, and small amounts of $\beta\beta'$-dichlorodiethyl ether are also formed. Since the HCl produced makes the solution extremely corrosive, construction material must be ceramic, plastic, or rubber-lined.

The next stage in the proress is the hydrolysis of ethylene chlorhydrin to ethylene oxide. Lime is used as the alkali, and the reaction is

$$2HOCH_2CH_2Cl + Ca(OH)_2 \longrightarrow 2\overset{\displaystyle O}{\overset{\displaystyle \diagup \diagdown}{CH_2-CH_3}} + CaCl_2 + 2H_2O \quad (11\text{-}15)$$

It is performed in a baffled vessel heated with live steam and maintained at a pressure of 60 to 80 mmHg. The vapor product is condensed and then fractionated to yield ethylene oxide as the overhead product and a bottom product containing ethylene dichloride, unreacted ethylene chlorhydrin, and water.

The ethylene oxide is now hydrated to produce the glycol product

$$\overset{\displaystyle O}{\overset{\displaystyle \diagup \diagdown}{CH_2-CH_2}} + H_2O \rightleftharpoons HOCH_2CH_2OH \quad (11\text{-}16)$$

The hydration is carried out in the liquid phase at about 200°C and 20 atm. A large excess of water is used to minimize the formation of higher glycols according to

$$HOCH_2CH_2OH + \overset{\displaystyle O}{\overset{\displaystyle \diagup \diagdown}{CH_2-CH_2}} \rightleftharpoons HOCH_2CH_2OCH_2CH_2OH \quad (11\text{-}17)$$

The ratio of mono- to diethylene glycols produced is about 10 : 1. The yield of the ethylene oxide is from 75 to 85 percent of the ethylene consumed, and the ethylene glycol yield is about 65 to 75 percent of the ethylene.

Demise of the Chlorhydrin Process

We shall now look at the weak points in this technology for the production of ethylene glycol. In reviewing the process it is apparent that two chemical products, ethylene oxide and ethylene glycol, are being made. The first part of the overall process is the production of ethylene oxide from ethylene, chlorine, water, and lime. In the second part, ethylene oxide is the raw material, along with water, for the production of ethylene glycol. Any process that can produce ethylene oxide can become a candidate for the first step in the process for producing ethylene glycol. This provides the key to the demise of the commercial ethylene chlorhydrin process.

If we summarize reactions (11-13) to (11-15), the overall reaction for the production of ethylene oxide by this route is

$$C_2H_4 + Cl_2 + \tfrac{1}{2}Ca(OH)_2 \rightleftharpoons \overset{\displaystyle O}{\overset{\displaystyle \diagup \diagdown}{CH_2-CH_2}} + HCl + \tfrac{1}{2}CaCl_2 \quad (11\text{-}18)$$

From reaction (11-18) we see that we are using chlorine as an oxidizing agent and converting it into HCl plus unsalable calcium chloride. In practice, the HCl would be neutralized by lime before disposal, so in this process, all the chlorine is converted into $CaCl_2$. Total consumptions per kilogram of ethylene oxide produced are approximately 2 to 2.1 kg of chlorine, 0.9 to 1 kg of ethylene, and 2 kg of lime (Salt, 1961). Chlorine is a relatively expensive oxidizing agent, and attempts were made quite long ago to develop a process that would not require chlorine or lime or result in the coproduction of calcium chloride. Oxygen is an inexpensive oxidizing agent, and in 1930 direct oxidation of ethylene to ethylene oxide was discovered in France.

Ethylene Oxide Process

The process based on direct oxidation will now be described. The desired reaction is

$$C_2H_4 + \tfrac{1}{2}O_2 \rightleftharpoons \overset{\displaystyle O}{\overset{\displaystyle \triangle}{CH_2-CH_2}} \qquad \Delta G^\circ_{298} = -19,410 \text{ cal/g mol} \quad (11\text{-}19)$$

and the obvious competing reaction would be

$$C_2H_4 + 3O_2 \rightleftharpoons 2CO_2 + 2H_2O \qquad \Delta G^\circ_{298} = -314,080 \text{ cal/g mol} \tag{11-20}$$

It is clear from the values shown for ΔG° that the thermodynamic equilibrium is strongly in favor of complete oxidation of ethylene to carbon dioxide and water. The key to the technology is therefore a selective catalyst on which the partial oxidation to ethylene oxide takes place while the competing combusion reaction is inhibited. The selective catalyst used in practice is supported silver, which is rugged and long-lived. A flowsheet for ethylene oxide and ethylene glycol production is shown in Fig. 11-11.

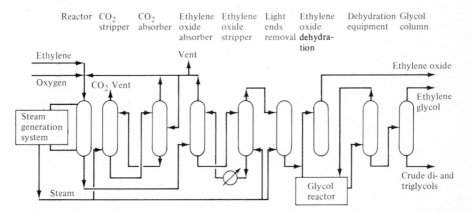

Figure 11-11 Flow diagram for ethylene oxide and ethylene glycol manufacture. [*From Hydrocarbon Process.*, *52(11):130 (1973), by permission.*]

The general method is to pass a mixture of oxygen (or air) and ethylene over the catalyst, which is contained in the tubes of a shell-and-tube reactor. The reaction is highly exothermic, and the reaction heat is removed by an organic coolant circulating through the shell side. The reaction heat can be used to generate steam. The reactor is operated at between 260 and 290°C with ethylene-air mixture and 220 to 230°C with ethylene-oxygen mixtures and usually at atmospheric pressure. The ethylene concentration in the gas is kept low, less than 3 percent, to ensure that the mixture will be outside the explosive limits.

The conversion of ethylene per pass is 40 to 50 percent, and the overall yield of ethylene oxide on ethylene can be about 70 percent although yields as high as 73 percent have been claimed.

The gas leaving the reactor is cooled by heat exchange with the recycle gas and compressed. The gas then passes to a scrubber, where the ethylene oxide is recovered by absorption in water. The gases from the scrubber are recycled except for a small amount that is vented. The product ethylene oxide is stripped by heating the "fat" absorbent and then distilled to remove light ends.

There are some important differences in the oxygen oxidant process compared with the air-oxidant process. The flowsheet in Fig. 11-11 is for the production of ethylene oxide by direct oxidation of ethylene with oxygen. It can be seen that part of the recycle gas remaining after ethylene oxide absorption is scrubbed for the removal of excess CO_2. Ethanolamine is the usual solvent. Only a small amount of gas is vented for inerts control. In the air-oxidant process there is no necessity for CO_2 absorption, but it is necessary to purge substantial amounts of nitrogen from the system. Because of the relatively low conversion of ethylene per pass through the reactor this purge stream would contain significant quantities of ethylene. To recover and utilize this ethylene the purge stream from an air-oxidant plant is passed through a secondary reactor that is operated at a high conversion. The exit gases from this reactor are cooled, compressed, and scrubbed with water to recover the ethylene oxide, and the residual gases are then vented from the system. The oxygen-oxidant process does not require such a secondary reactor and recovery system.

Ascendancy of Ethylene Oxide Process

We now have enough information to analyze why the ethylene chlorhydrin process has been completely displaced by the direct oxidation route. We shall examine the raw-material price situation that existed in the 1950s when, as is apparent from Table 11-5, the shift to direct-oxidation plants was in full swing. Price and chemical consumption data are presented in Table 11-6 for the two technologies. The table was prepared on the assumption that ethylene oxide molar yields from ethylene were 80 and 65 percent for the chlorhydrin and direct oxidation routes, respectively. The hydration step produces di- and triethylene glycols ethers, and it is assumed that 80 percent of the ethylene is converted into the monoglycol. No credit was taken for by-product sales.

Table 11-6 Raw material consumption, prices, and costs per kilogram for ethylene glycol production

Material	1955 price per kilogram†	Technology			
		Chlorhydrin		Direct oxidation	
		Use, kg	Cost	Use, kg	Cost
Ethylene glycol	$0.28	1	$0.28	1	$0.28
Ethylene	0.065	0.7	0.0455	0.87	0.0566
Chlorine	0.065	1.8	0.117		
Lime	0.01	1.7	0.017		

† Price data from Aries and Newton (1955).

We see from Table 11-6 that the costs in 1955 of the major raw materials for ethylene glycol via the chlorhydrin route totaled 18 cents per kilogram of product, whose selling price was about 28 cents, compared with 5.66 cents for raw-material cost via direct oxidation of ethylene oxide. The raw-materials cost for the chlorhydrin route represented more than 50 percent of the product selling price, compared with about 20 percent for the direct oxidation route. The ethylene chlorhydrin process could have survived in the face of this raw-material cost situation only if its capital costs had been extremely small compared with those for the ethylene oxidation route. This was not the case. Actually, capital costs for both processes are quite similar, so that the commercial production of ethylene oxide and ethylene glycol via the ethylene chlorhydrin route is no more.

Bypassing Ethylene Oxide

Let us now examine the present situation with respect to the direct oxidation route. We recall that it is a two-step process, in which ethylene is converted into ethylene oxide in step 1 and ethylene oxide is hydrated to ethylene glycol in step 2. The first step is a profligate user of ethylene, and yields of ethylene oxide are about 70 percent of the ethylene consumed. In other words, about 30 percent of the ethylene that enters the ethylene oxide plant leaves it as CO_2 and water. This situation was tolerable in the 1950s and 1960s, when ethylene prices were about 6 cents per kilogram, but the situation becomes more difficult when, as in the 1970s, ethylene prices range from 15 to 25 cents per kilogram.

One obvious approach to solving this problem would be to increase the ethylene oxide selectivity of the catalyst. This approach has been tried, but, at least at the present, there appears to be no hope for a substantial increase in selectivity that would bring it above the presently attained 70 to 75 percent.

From observation of the process it is apparent that the problem of low yield on ethylene stems from step 1 of the process, oxidation of ethylene. The question that can then be asked is: Can ethylene glycol be prepared without having to go through the ethylene oxide step? After all, ethylene glycol is built up from carbon, oxygen, and hydrogen, and there should be alternative routes to this chemical. Alternative routes have been under investigation for a number of years, and two routes have emerged, oxychlorination and acetoxylation.

The oxychlorination route actually proceeds through a chlorhydrin intermediate. Since this is hydrolyzed in situ, no lime or chlorine is required in the process. The reaction is carried out at 140°C in concentrated aqueous HCl with a thallium catalyst. The overall chemical equation is

$$C_2H_4 + \tfrac{1}{2}O_2 + H_2O \rightleftharpoons HOCH_2CH_2OH \tag{11-21}$$

Since reported yields (Brownstein, 1975) are about 75 percent based on ethylene, the process offers no yield advantages over the direct oxidation route. It has not yet been commercialized.

The acetoxylation route is being commercialized (Brownstein, 1975) and offers substantial raw-material savings over the direct oxidation route. HOAc represents acetic acid in the reactions

$$O_2 + 2C_2H_4 + 3HOAc \rightleftharpoons AcOCH_2CH_2OAc + HOCH_2CH_2OAc$$
$$+ H_2O \tag{11-22}$$

$$HOCH_2CH_2OAc + AcOCH_2CH_2OAc + 3H_2O \rightleftharpoons 2HOCH_2CH_2OH$$
$$+ 3HOAc \tag{11-23}$$

Reaction (11-22) is carried out at 160°C with a Pd/Cu, Pd/NO$_3$, Te/Br, or Ta catalyst. The hydration reaction (11-23) is carried out at 110°C, and the acetic acid is recycled to the reactor. The overall reaction is

$$2C_2H_4 + O_2 + 2H_2O \rightleftharpoons 2HOCH_2CH_2OH \tag{11-24}$$

which is the same as for the direct oxidation route.

The advantage of the acetoxylation route lies in its high selectivity, which has been reported to be 95 percent. This can provide the acetoxylation process with a significant raw-material cost advantage over the direct oxidation route. This advantage is not as dramatic as that of the direct oxidation route over the chlorhydrin route to ethylene glycol, but is good enough to provide a significant threat to the direct oxidation route.

Bypassing Ethylene

The next question that can be asked is: Why go through ethylene? Can ethylene glycol be prepared from more elementary building blocks? The answer, of course, is yes, and a possible reaction starting from synthesis gas is

$$2CO + 3H_2 \rightleftharpoons HOCH_2CH_2OH \tag{11-25}$$

Such a process has been patented, and Brownstein (1965) shows that its prospects appear quite attractive if it can be developed to a commercial practicability. A number of reactions would compete with the desired reaction, and additional chemicals of commerce with by-product value could be obtained. Some of the possibilities include

$$
3CO + 5H_2 \; \rightleftharpoons \; \overset{\displaystyle \overset{OH}{|}}{CH_3CHCH_2OH} + H_2O \tag{11-26}
$$

$$
3CO + 4H_2 \; \rightleftharpoons \; \overset{\displaystyle \overset{OH}{|}}{HOCH_2CH_2OH} \tag{11-27}
$$

and, of course,

$$
CO + 2H_2 \; \rightleftharpoons \; CH_3OH \tag{11-28}
$$

Thus, the ethylene chlorhydrin process for ethylene glycol production was displaced by a newer technology based on the direct oxidation of ethylene. This technology itself is now under attack by the acetoxylation process, and in the wings are developments strongly indicating that the future technology for ethylene glycol production may reside in its direct synthesis from CO and hydrogen.

11-4 TAMING THE RUNAWAY REACTION

A real industrial problem in the scale-up of a commercial reaction was discussed and described by Kladko (1971). This study is based on Kladko's paper. The chemicals involved are of a proprietary nature and are referred to only by letter.

Background

In the process to be studied raw material A undergoes isomerization to B at an elevated temperature in a solvent medium C. Solvent C is a relatively high-boiling and inert aromatic, and both A and B are liquids at room temperature and have high boiling points. At the early development stages of the process there was no reliable analytical procedure available for A and B. The procedure was to use the product of the isomerization in the next reaction in the sequence of reactions and to measure the final product yield. If the yield was satisfactory, it was then assumed that A had isomerized to B. Laboratory runs were made with a 1 : 1 ratio of A to C at a temperature of 163°C for 12 h. The unavailability of quantitative analytical procedures made it impossible to make kinetic measurements.

The next step in the development program required larger quantities of B than could be prepared in the laboratory, and a 50-gal steam-heated reactor was used to prepare pilot quantities. Insufficient steam pressure caused difficulties in heating up to 163°C, the reaction temperature used in the laboratory tests. The reaction was therefore carried on for 18 h at a slightly lower temperature, and because the next step in the synthesis sequence was satisfactory, the reaction was judged to have been complete.

The Sting

Manufacturing requirements then demanded an even larger batch, and it was necessary to prepare 4000 lb of B. A 750-gal glass-lined jacketed kettle with a hot-oil heating system was available. The hot-oil system would ensure that 163°C could be reached easily and would also be able to supply the small amount of endothermic heat believed to be required by the reaction. It was decided to use this reactor but with 4000 lb of A and 2000 lb of solvent C to prepare the desired quantity in one batch. The reactant-to-solvent ratio would therefore be 2 : 1 rather than 1 : 1, as used in the previous runs. It was believed that the increase in concentration would have no effect on the reaction since the reaction was a uni-molecular isomerization.

The solvent and A were charged to the reactor and heating started. When the temperature recorder showed 150°C, the reaction ran out of control; the contents heated up rapidly and finally erupted through the safety valve and vented out over the building area.

What Happened?

The first conclusion that can be reached is that the reaction is probably exothermic and not endothermic, as had previously been assumed. Therefore, calculations based on the method of group contributions were made, and it became clear that the reaction was indeed exothermic to the extent of 83 cal/g. Although this is not a high heat of reaction, it is not insignificant.

Why then were no difficulties with heat generation noted in the lab work or in the 50-gal reactor? Though quite simple, the answer is two-pronged: (1) since a reactant-to-solvent ratio of 1 : 1 was used, an efficient heat sink of solvent was present and (2) the laboratory flask is small and has a high ratio of surface area to volume. The exothermic heat would easily be transferred to the "heating" medium. With the 50-gal reactor the area-to-volume ratio is smaller but still high; in addition, the steam pressure was not high, and it was not possible to reach reaction temperature with the steam available. The exothermic heat was therefore effectively dissipated.

With the 750-gal reactor, however, the situation was different. Since the reactant-to-solvent ratio was 2 : 1, the moderating effect of the solvent as a heat sink was lessened. In addition, the hot-oil system, which was still bringing the system up to temperature, would have been at a temperature above 163°C. Therefore, when the reaction took off, there was no way to remove heat through the agency of the hot-oil system. When the reactor contents exceeded the hot-oil temperature, the area-to-volume ratio was too small to permit the heat of reaction to be removed by the oil. The result was a continuing temperature rise of the reactor contents and the accompanying increase in the rate of reaction until the solvent boiling point was reached. The continuing temperature rise resulted in a pressure buildup until the safety valve opened and vented the reactor contents with violence.

Back to the Lab Bench

As a result of the runaway incident, the reaction was now treated with a great deal of respect. It was decided that some quantitative information about reaction rates was necessary if the reaction was to be scaled up with confidence. The first step was to develop a convenient method of analyzing the reaction mixture quantitatively for species A and B. A quantitative gas-chromatographic procedure was developed, and it opened the door to kinetic measurements.

The next step was to examine the effects of elevated temperatures on the reaction more carefully. It was found that undesired by-products began to be formed at temperatures in excess of 170°C. At lower temperatures the only reaction that occurred at appreciable rates was the A isomerization. The previously determined operating temperature of 163°C was thereby confirmed as being satisfactory, and reaction-rate measurements were made at this temperature. Measurements of the conversion of A into B were made in laboratory flasks at initial A concentrations of 33, 50, 67, and 100 percent. Samples were withdrawn at timed intervals and analyzed for A by the gas-chromatography procedure. The results are presented in Fig. 11-12. The finding that the concentration of A does have an effect on the isomerization reaction rate was unexpected.

The isomerization reaction is

$$A \; \rightleftharpoons \; B$$

which suggests that first-order kinetics might apply. If they do apply, then the data at different dilutions presented as percentage A remaining vs. time on semilogarithmic coordinates should plot as straight lines. A plot of the data in this form is presented in Fig. 11-13; for each concentration the data plot as two straight lines. Thus, we are in a position to make a fairly good prediction by using the rate constants calculable from one of the two straight lines obtained for each initial concentration. Since most of the conversion was obtained before the break point was reached, the apparent rate constants were calculated from the slope of the upper portions of the line.

The apparent rate constant for a batch first-order irreversible reaction at constant volume can be calculated from

$$k = \frac{1}{t_2 - t_1} \ln \frac{1 - X_1}{1 - X_2} \tag{11-29}$$

where X is the fraction of A converted. The results are presented in Fig. 11-14. The results, interestingly enough, plot up as a straight line which extrapolates to the origin. This suggests that the apparent reaction rate constant can be represented empirically as a function of solvent concentration by

$$k = k_i(1 - b) \tag{11-30}$$

where b is the weight fraction of solvent C and k_i is the slope of the apparent-reaction-rate-constant line. At $b = 0$ (100% A) the value of the apparent reaction

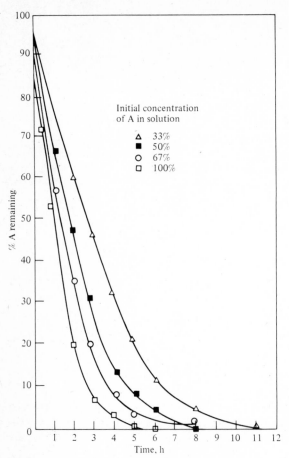

Figure 11-12 Rate of isomerization of A at various dilutions at 163°C. [*From M. Kladko, CHEMTECH, 1(3):141(1971), by permission. Copyright by the American Chemical Society.*]

rate constant was 0.8 h^{-1}; hence the slope is 0.8. Thus we can write

$$t = \frac{1}{0.8(1-b)} \ln \frac{1}{1-X} \qquad (11\text{-}31)$$

where t is the time required to reach X fractional conversion of A and no B is present initially.

The effect of temperature on reaction rate when no solvent was present was studied by measuring the apparent reaction rate constant at 150 and 183°C. The results are presented in Fig. 11-15 as an Arrhenius plot. The equation for the reaction rate constant can be extracted from the plot, and

$$k = 274 \times 10^{12} \exp \frac{-1.45 \times 10^4}{T} \qquad (11\text{-}32)$$

Analysis of the findings to this point showed quite clearly that the presence of solvent was not necessary. By eliminating the solvent from the process the

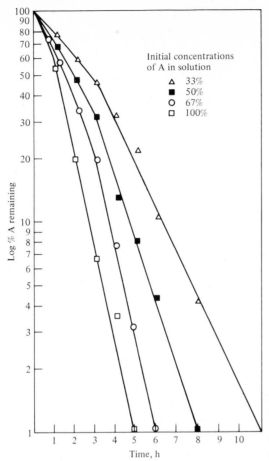

Figure 11-13 Isomerization kinetics of A at 163°C. [*From M. Kladko, CHEM-TECH,* **1**(*3*):*141 (1971), by permission. Copyright by the American Chemical Society.*]

apparent reaction rate is doubled compared with the 1 : 1 solvent-to-A case and the volume of the vessel is halved. Thus, the required reactor volume for a given conversion would be effectively reduced by a factor of 4. Also, it had been assumed that only a batch reactor would be feasible for this slow reaction. It was now apparent that continuous operation could not be rejected a priori as a feasible proposition.

Continuous Operation

Sufficient data are now in hand to calculate reactor volume requirements and the heat-removal load. Batch operation, however, would be difficult if no solvent were used. The reader is referred to Kladko (1971) for a summary and analysis of this batch problem and for the solution. We pursue the route of continuous operation instead.

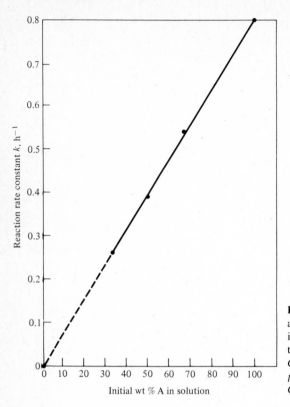

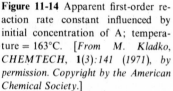

Figure 11-14 Apparent first-order reaction rate constant influenced by initial concentration of A; temperature = 163°C. [*From M. Kladko, CHEMTECH,* **1**(*3*):*141* (*1971*), *by permission. Copyright by the American Chemical Society.*]

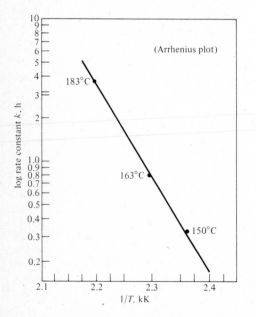

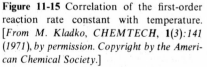

Figure 11-15 Correlation of the first-order reaction rate constant with temperature. [*From M. Kladko, CHEMTECH,* **1**(*3*):*141* (*1971*), *by permission. Copyright by the American Chemical Society.*]

For the purpose of this analysis we shall assume a temperature of 163°C. Obviously, higher-temperature operation would result in increased isomerization rates. The question of what the higher temperature would do to the kinetics of formation of undesirable by-products would have to be resolved. We shall assume a series of CSTRs. Equations for a series of CSTRs can easily be derived from the expressions developed in Chap. 6. MacMullin and Weber (1935) derived the equations for a first-order constant-volume reaction in a series of equal volume CSTR's

$$t = \frac{nX}{k[(1 - X) + (1 - X)^2 + \cdots + (1 - X)^n]} \tag{11-33}$$

where t is the total residence time and n is the number of reactors in series. We shall calculate the residence-time requirements for $n = 1$, 2, and 3. The equation assumes $X = 0$ at $t = 0$. For reaction at 163°C with no solvent, $k = 0.8$ h^{-1} and $X = 0.97$.

The results are tabulated in Table 11-7, in which the residence times per reactor for each case are presented along with the predicted values of the concentration of A leaving each reactor and the actual values obtained in a bench-scale setup of one and two stirred reactors in series. The experimental results for the three reactors are actually for two reactors in series where the residence times used were those predicted for the three-reactor sequence. From these results it is apparent that the kinetic equation predicts conservative results. We now have in hand experimental verification of the kinetic model and could proceed with confidence to design a CSTR sequence.

Our case study of a real industrial problem ends at this point. The process evidently did not undergo large-scale commercialization, and so we cannot report or compare commercial practice with the analysis made here. We can infer from Kladko (1971) that a market for B did not develop and that the

Table 11-7 Predicted residence times and predicted and observed conversion in CSTR system

Number of reactors	Residence time per reactor, h	Exit concentration of A, %			
		Predicted		Experimental†	
1	40.4		3		1.1–2.7
2	6.9	First	16	First	12.3–14.5
		Second	3	Second	0.7–1.8
3	3.9	First	24	First	22.4–23.7
		Second	6	Second	2.9–3.7
		Third	3		

† Results are for replicate samples.

Source: From M. Kladko, *CHEMTECH*, March 1971, p. 141, by permission. Copyright by the American Chemical Society.

quantities required could be made by occasional operation of a 4-kL glass-lined kettle. For this operation a computer simulation of the reaction was used to guide and direct the operators in the heating, feeding, and cooling sequences of this kettle. The interested reader is referred to Kladko for further details.

11-5 CAPACITY-EXPANSION PROJECT

We consider a hypothetical but realistic study of a plan to expand production capacity for an existing product in the face of an increase in market. The case study is based on one presented in a series of articles by Leibson and Trischman (1971) covering the plans by the XYZ Company ("our" company) to expand the production of its product B. We present only part of the study here. The interested reader is referred to the original series of articles.

Background Information

The XYZ Co. manufactures and markets, among other products, a liquid resin. In our company jargon the resin is referred to as product B. Product B is used as an additive in paints, a drying agent in inks, and a tackifying agent in adhesives. We are not the only manufacturer of this product, and we have five competitors who produce a similar resin. Our company's current sales of product B are 2.9×10^6 gal/yr from our present plant, which has a capacity of 3×10^6 gal/yr. Our sales represent about 10 percent of the market.

Marketing has made a study and projection of the potential total market growth for product B for the next 8 years. The projection was based on information obtained in field calls on product B consumers, on exchange of nonconfidential marketing information with our competitors, and on a simulation of the market growth of the three markets served by product B. An average annual growth rate for product B in the ink market of 19 percent is anticipated, about 10 percent in the adhesives market, and 3 percent in the paint market. The forecast is presented in Table 11-8.

Table 11-8 Forecast of all-industry market demand and growth for product B, millions of gallons per year

Application area	Project year†							
	1	2	3	4	5	6	7	8
Paint additive	14.4	14.8	15.2	15.7	16.2	16.7	17.2	17.7
Ink-drying agent	13.0	15.5	18.5	21.9	25.7	29.8	34.7	39.8
Adhesive	4.3	4.9	5.4	5.9	6.5	7.2	7.9	8.7
Total	31.7	35.2	39.1	43.5	48.4	53.7	59.8	66.2

† Present time is project year 0.

Source: From Leibson and Trischman (1971), by permission.

In view of the anticipated growth in the markets for product B our marketing department proposed that additional production capacity be installed. Marketing's forecast for our product is shown in Table 11-9. This forecast was based on an aggressive marketing strategy that would increase our company's share of market from 10 to 15 percent by project year 3. The specific proposal was that an additional 4×10^6 gal/yr capacity be installed no later than the middle of year 2.

Present State of Study

In the first stage of the complete study that resulted from the marketing department proposal the engineering and financial departments produced cost figures and comparisons for the base case (the present facility) and base case (BC) plus purchase of product B material from a competitor for resale under our label. They also considered three expansion possibilities: BC plus 2×10^6 gal/yr expansion, BC plus 4×10^6 gal/yr expansion, and BC plus 8×10^6 gal per/yr expansion.

A thorough study of these cases, their earnings, and expected profitability and a heart-searching discussion of the likelihood of marketing achieving its forecast market penetration resulted in the following assignments for the three departments involved in this study.

1. Marketing department to propose a new sales forecast based on retention of our current share of market for each grade of product.
2. Engineering department to develop investment and cost estimates for a 2×10^6 gal per/yr plant expandable to 4×10^6 gal per/yr; the initial installation of a 4×10^6 gal per year plant and, finally, the successive installation of two 2×10^6 gal/yr minimum-investment plants. Engineering is also to develop the project construction schedule. The 8×10^6 gal/yr expansion was eliminated from consideration.
3. Financial department to develop, along with engineering department, project profitability for the following cases:
   a. Base case (BC) plus installation of a minimum-investment 2×10^6 gal/yr plant with subsequent addition of 2×10^6 gal/yr in accordance with revised sales forecast.
   b. BC plus initial installation of 2×10^6 gal/yr expandable plant with subsequent expansion to 4×10^6 gal/yr.
   c. BC plus initial installation of 4×10^6 gal/yr plant.

It was also concluded that for the first two project years before additional capacity could be added and in order to meet market demand, product B would be purchased from a competitor for resale. Not only would the profitability with purchase and resale be higher than operations with only BC production but we also would not lose sales momentum or part of our market by inability to meet our customers' orders. In the subsequent discussion, therefore, we shall move our zero time (case year 0) forward 3 years into the future to correspond to the time when production in an expanded facility would begin.

Table 11-9 Market forecast for XYZ Company product B†

	Project year																	
	0		1		2		3		4		5		6		7		8	
Product B grade	V	P	V	P	V	P	V	P	V	P	V	P	V	P	V	P	V	P
Paint additive	1.20	69.3	1.2	62.5	1.2	66.0	1.7	61.5	1.8	65.0	1.8	64.0	1.8	63.0	1.9	62.0	1.9	61.0
Ink-drying agent	1.14	100.0	1.5	98.5	2.2	98.0	2.9	96.5	3.5	97.0	4.2	96.5	5.0	96.0	5.7	95.5	6.5	95.0
Adhesive	0.56	95.0	0.8	94.0	1.1	94.2	1.4	94.0	1.7	93.5	2.0	93.0	2.2	92.5	2.4	92.0	2.6	91.5
Total	2.9	86.3	3.5	85.1	4.5	88.6	6.0	86.0	7.0	87.9	8.0	88.3	9.0	88.5	10.0	88.3	11.0	88.3

† V = volume, 10^6 gal/yr; P = sales price, cents per gallon.
Source: Leibson and Trischman (1971), by permission.

Market Forecasts, Investment, and Earnings

The revised market forecast based on retention rather than expansion of our market share that was assigned to marketing is presented in Table 11-10. The forecasts are, of course, smaller than the original projection presented in Table 11-9. The engineering and financial departments used the revised forecasts to determine the most profitable product mix that could be attained for the three alternatives to be studied and the annual gross sales income to be expected for each case. Their results are shown in Table 11-11.

The capital-cost estimates made by engineering are presented in Table 11-12. These estimates are of budget accuracy, that is, ± 10 percent, and include a 10 percent contingency. As would be expected, the estimate for a 4×10^6 gal/yr plant is less than the estimate for two separate 2×10^6 gal/yr plants. The latter is slightly more expensive than an expandable 2×10^6 gal/yr facility followed by expansion to 4×10^6 gal/yr. The second step of the two-step expansion, if undertaken, would be scheduled for the middle of case year 3.

DCF Rate-of-Return Evaluation

An economic analysis will be necessary for us to be able to arrive at a well-founded decision as to which, if any, of the three alternatives should be recommended to our board of directors. To do this we evaluate the DCF rate of return and the present worth for each of the three cases. Such an evaluation must also consider earnings beyond the 7-case-year detailed forecast made by the marketing department. The project life was assumed to be 15 years. To generate cash-flow information for the final 8 years of the project life it was assumed that the cash flow decreased by 1 percent per year due to price deterioration during that period. The calculations, as actually made, assumed a uniform annual cash flow during the final 8 years that was 4 percent less than that attained in the seventh case year.

Table 11-10 Revised market forecast for XYZ Company's product B

Product grade	Case year† 0		1		2		3		4		5		6	
	V	P	V	P	V	P	V	P	V	P	V	P	V	P
Paint	1.3	61.5	1.4	65.0	1.4	64.0	1.5	63.0	1.5	62.0	1.6	61.0	1.7	60.0
Ink	2.0	96.5	2.3	97.0	2.7	96.5	3.2	96.0	3.8	95.5	4.5	95.0	5.3	94.5
Adhesive	0.8	94.0	0.9	93.5	1.0	93.0	1.1	92.5	1.2	92.0	1.3	91.5	1.4	91.0
Total	4.1	84.9	4.6	86.6	5.1	86.9	5.8	86.8	6.5	87.1	7.5	87.0	8.4	86.9

† Revised time scale: case year 0 corresponds to project year 3 in Tables 11-8 and 11-9.
Source: From Leibson and Trischman (1971), by permission.

Table 11-11 Revised product mix and gross income from sales

Case year	Product grade	Base case + 2 × 10⁶ gal/yr		Base case + 4 × 10⁶ gal/yr†	
		10^6 gal	GSI, millions	10^6 gal	GSI, millions
0	Paint	1.3		1.3	
	Ink	2.0		2.0	
	Adhesive	0.8		0.8	
			$3.482		$3.482
1	Paint	1.4		1.4	
	Ink	2.3		2.3	
	Adhesive	0.9		0.9	
			3.983		3.983
2	Paint	1.3		1.4	
	Ink	2.7		2.7	
	Adhesive	1.0		1.0	
			4.368		4.432
3	Paint	0.7		1.5	
	Ink	3.2		3.2	
	Adhesive	1.1		1.1	
			4.531		5.035
4	Paint	0		1.5	
	Ink	3.8		3.8	
	Adhesive	1.2		1.2	
			4.733		5.663
5	Paint	0		1.2	
	Ink	3.7		4.5	
	Adhesive	1.3		1.3	
			4.705		6.197
6	Paint	0		0.3	
	Ink	3.6		5.3	
	Adhesive	1.4		1.4	
			4.676		6.463

† This corresponds to BC plus either (1) installation of 4 × 10⁶ gal/yr, or (2) two separate 2 × 10⁶ gal/yr, or (3) installation of 2 × 10⁶ gal/yr expandable to 4 × 10⁶ gal/yr.

Source: From Leibson and Trischman (1971), by permission.

Although Leibson and Trischman (1971) make a complete analysis for the three alternatives, we present a detailed analysis for only one alternative, BC plus the expandable 2 × 10⁶ gal/yr plant followed by its subsequent expansion to 4 × 10⁶ gal/yr. The earnings picture for this alternative is shown in Table 11-13. In the development of this table the profit after taxes for BC plus expansion was first calculated. The after-tax profit for BC only (calculations not presented here) was then deducted from the after-tax profit for BC plus expansion. The additional profit and additional cash flow due to the additional investment for expansion is thereby calculated, making a valid comparison between BC and BC plus expansion possible.

Table 11-12 Investment estimates for alternative expansion schemes

Expansion scheme	Capacity, 10^6 gal/yr	Investments, millions
One-step expansion	2	$1.8
	4	3.1
Two separate expansions:		
a. First 2 × 10^6 gal/yr	2	1.8
With addition of second plant	4	3.8
b. Expandable 2 × 10^6 gal/yr plant	2	2.5
With expansion to 4 × 10^6 gal/yr	4	3.3

Source: From Leibson and Trischman (1971), by permission.

Table 11-13 Earnings and cash flow for base case plus expandable 2 × 10^6 gal/yr plant, dollar figures in millions

	Case Year						
	0	1	2	3†	4	5	6
Gross sales, 10^6 gal/yr	4.1	4.6	5.0	5.8	6.5	7.0	7.0
Gross sales income	$3482	$3983	$4368	$5035	$5663	$6197	$6463
Returns, allowances, etc.	26	30	33	38	42	46	48
Sales income	$3456	$3953	$4335	$4997	$5621	$6151	$6415
Freight	324	337	354	406	454	489	491
Net sales income	$3132	$3616	$3981	$4591	$5167	$5662	$5924
Cost of sales:							
Variable cost	$1137	$1286	$1442	$1674	$1914	$2148	$2336
Fixed cost‡	1272	1272	1272	1415	1615	1615	1615
Other operating expenses	127	127	127	145	156	156	156
Total cost of sales	$2536	$2685	$2841	$3234	$3685	$3919	$4107
Profit before taxes	596	931	1140	1357	1482	1743	1817
Profit after taxes	298	466	570	679	741	872	909
Base case profit	372	392	390	385	379	373	366
Profit due to expansion	−$74	$74	$180	$294	$362	$499	$543
Additional depreciation due to expansion	$167	$167	$167	$189	$220	$220	$220
Cash flow	93	241	347	483	582	719	763

† Expansion to 4 × 10^6 gal/yr for last 5 months of case year 3.

‡ Includes depreciation.

  Source: From Leibson and Trischman (1971), by permission.

Table 11-14 Calculation of DCF rate of return, dollar values in millions

2×10^6 gal/yr plant expanded to 4×10^6 gal/yr

Case year	Item	Cash flow	10% Discount factor	10% Present value	13% Discount factor	13% Present value	12.5% DCF Discount factor	12.5% DCF Present value
0	Investment	$-2800	1.0000	$-2800	1.0000	$-2800	1.0000	$-2800
0-1	Earnings	93	0.9516	88	0.9377	87	0.9400	87
1-2	Earnings	241	0.8611	208	0.8234	198	0.8296	200
2-3	Earnings	347	0.7791	270	0.7230	251	0.7322	254
$3\frac{1}{2}$	Investment	-1050	0.7047	-740	0.6345	-666	0.6456	-678
3-4	Earnings	483	0.7050	341	0.6349	307	0.6462	312
4-5	Earnings	582	0.6379	371	0.5575	324	0.5703	332
5-6	Earnings	719	0.5772	415	0.4895	352	0.5034	362
6-7	Earnings	763	0.5223	399	0.4299	328	0.4444	339
7-15	Earnings†	5860	0.3418	2003	0.2502	1466	0.2622	1536
15	Working-capital recovery	550	0.2231	123	0.1423	78	0.1533	84
15	Net salvage	0						
	Total			677.3		-74.7		29.4

† (8)(763)(0.96) = 5860. By interpolation, DCF rate of return = 12.64%.
Source: From Leibson and Trischman (1971), by permission.

The calculation of the DCF rate of return is shown in Table 11-14. The investments shown at case years 0 and $3\frac{1}{2}$ include the provision of working capital, which is recovered at project termination by sale of inventories, collection of outstanding accounts, etc. The net salvage value was taken as zero on the assumption that dismantling costs would approximate the sum received by sale of the used plant equipment.

The discount factors for the instantaneous cash flows shown in Table 11-14 were taken from Table 7-1, and those for the uniform annual flows were taken from Table 7-3. The discount factor for the case years 7 to 15 is the product of the discount factor over an 8-year period from Table 7-2 and the factor for an instantaneous flow after 7 years from Table 7-1. By this procedure the cash flow from case years 7 to 15 was first discounted to its present value at case year 7, which was, in turn, discounted to its present value at case year 0.

The DCF rate of return calculation was by trial and error. The first trial was made at an assumed interest rate of 10 percent, yielding a positive present worth of $677,300. For the next trial a higher interest rate was assumed. With $i = 13$ percent the present worth dropped to $-\$74,700$. The DCF rate of return will be between these two interest rates. The next trial was made at 12.5 percent with the result being a small positive value for the present worth. The DCF rate of return of 12.64 percent was obtained by linear interpolation between the final two trial values.

Present-Worth Calculation

The cost of capital for the XYZ Company is 10 percent; thus any investment with a positive worth when discounted at 10 percent interest rate is a profitable investment. The present-worth calculation for 10 percent interest rate is shown in Table 11-14 as the first trial-and-error calculation for the DCF rate of return. The present worth of the 2×10^6 gal/yr plant expanded to 4×10^6 gal/yr is $677,300.

A summary of the DCF rates of return and present worths for the three expansion alternatives is presented in Table 11-15.

Recommendations

After reviewing the earnings statements and the profitability summary the marketing, financial, and engineering departments decided to recommend to the board of directors that production capacity for product B be increased by the construction of a 2×10^6 gal/yr facility that would be designed for easy expansion to 4×10^6 gal/yr capacity after $3\frac{1}{2}$ years. The present worth of this expansion at 10 percent interest rate would be $677,300, and the expected DCF rate of return is 12.64 percent. The positive present worth and the rate of return of more than 10 percent meet management guidelines. The cumulative positive cash flow during the first seven case years will be $3,228,000.

The other two alternatives are not as attractive; the present worths, rates of return, and positive cash flows are smaller than those expected for the recommended alternative. The board of directors will decide on this proposal on the basis of overall company plans, goals, and constraints.

Comment

It should be apparent from this study that a great deal of time and effort was invested in developing the market and price forecasts, investment figures, and earnings projections for the cases and alternatives considered. The additional

Table 11-15 Summary of profitability study for product B expansion

Expansion scheme	Profitability	
	DCF rate of return, %	Present worth at 10%, millions
One-step expansion	10.82	$244.7
Two separate 2×10^6 gal/yr expansions	12.25	518.4
Expandable 2×10^6 gal/yr expansion to 4×10^6 gal/yr	12.64	677.3

effort involved in converting these figures into present worths and DCF rates of return to permit a well-founded and valid economic comparison between the competing alternatives is minimal. Although uncertainties are not considered in the case study as presented here, some are considered by Leibson and Trischman (1971) in their complete study.

PROBLEMS

Problems appropriate to this chapter are, of course, additional case studies. Some case studies and case study sources are listed below:

Bodman, S. W.: "The Industrial Practice of Chemical Engineering," MIT Press, Cambridge, Mass., 1968.

"Chemical Engineering Case Problems," Education Projects Committee, American Institute of Chemical Engineers, New York, 1967.

Clausen, C. A., and G. C. Mattsen: "Case Studies in Industrial Chemistry," Florida Technological University, Orlando, Fla., 1974.

Engineering Case Library, Engineering Case Program, Stanford University, Stanford, California.

ICI Limited: "Case Studies in Chemical Engineering," Millbank, London, 1974.

Rase, H. F.: "Chemical Reactor Design for Process Plants," vol. 2, "Case Studies and Design Data," Wiley, New York, 1977.

Some case studies dealing primarily with business, management, and economics are available from:

Case Research Programme, School of Management, Cranfield Institute of Technology, Cranfield, Bedford, England.

Intercollegiate Case Clearing House, Soldiers Field, Boston, Mass.

Many case-study topics suggest themselves as one teaches a course, and although the preparation of a case study is not a trivial matter, it can be done by the instructor, often with the help of the students. *CHEMTECH* is a fertile source of material that can be converted into case studies; see, for example:

Slack, A. V. and G. M. Blouin: Urea Technology: A Critical Review, January 1971, p. 32.

Neier, W., and J. Woellner: Isopropyl Alcohol by Direct Hydration, February 1973, p. 32.

Mocearov, V., et al.: Benzene Alkylation with Dilute Olefins, March 1973, p. 182.

Chemical Engineering Progress also can provide material that can be easily converted to case studies; see, for example,

Casten, J. W.: Mechanical Recompression Evaporators, **74** (7): 61 (1978).
Bennett, R. C.: Recompression Evaporation, **27** (7) 67 (1978).
Malloy, J. B.: Risk Analysis of Chemical Plants, **67** (10): 68 (1971).

The AIChE. Student Contest Problems should not be overlooked as a source for case studies.

A case study recommended by J. Wei for student study followed by class discussion is C. R. Scott and T. L. Harrell, Living with Change, *CHEMTECH*, April 1979, p. 234. This case evolves around an environmental problem and the resulting economic and social issues.

The two case-study-type problems presented below are to be discussed in class after several hours of student preparation.

11-1 Our fertilizer production operations require a 52 to 56°Bé sulfuric acid, but the sulfuric acid contact process produces acid of a much higher concentration. Your supervisor recalls that the obsolete chamber process† produces the desired concentration directly. Recalling also that dilution is a thermodynamically inefficient operation, your supervisor suggests that you examine the chamber process to determine whether by use of present technology, presently available materials of construction, appropriate research, etc., it may be possible to revive this process. You should, of course, consider a number of aspects such as investment requirements, operating costs, space and energy requirements, and environmental considerations.

11-2 Many refinery and natural gases contain substantial concentrations of H_2S. One conventional treatment is to absorb the H_2S in a liquid absorbent such as monoethanolamine, which upon heating releases the absorbed gas so that a gas containing more than 90% H_2S, the remainder being primarily CO_2, is obtained. This gas can then be treated by the Claus process to produce sulfur according to

$$H_2S + 1\tfrac{1}{2}O_2 \longrightarrow SO_2 + H_2O$$

$$SO_2 + 2H_2S \longrightarrow 3S + 2H_2O$$

You are to explain, primarily from the kinetic and thermodynamic viewpoints, why the typical Claus process plant is designed the way it is designed and why it is operated at the conditions at which it is operated.‡

---

† See, for example, R. N. Shreve, "Chemical Process Industries," 1st or 2d ed., McGraw-Hill, New York; A. M. Fairlie, "Sulfuric Acid Manufacture," Reinhold, New York, 1936.

‡ See, for example, R. N. Shreve, "Chemical Process Industries," p. 234, McGraw-Hill, New York, 1967; F. A. Lowenheim and M. K. Moran, "Industrial Chemicals," 4th ed., p. 787, Wiley, New York, 1975; P. Grancher, *Hydrocarbon Process.*, July 1978, pp. 155–160. Several processes for cleaning Claus process tail gas are described in *Hydrocarbon Process.*, April 1979.

REFERENCES

Allen, D. H. (1967): *Chem. Eng.,* **74**(14): 75.

Am. Assoc. Cost Eng. Bull. (1958): **1:** 12, November.

Anderson, J. W., G. H. Beyer, and K. M. Watson (1944): *Natl. Petrol. News Tech. Sec.,* **36:** R476 (July 5).

Aries, R. S., and R. D. Newton (1955): "Chemical Engineering Cost Estimation," McGraw-Hill, New York.

Aris, R. (1964): "Discrete Dynamic Programming," Blaisdell, New York.

—— and N. R. Amundson (1958): *Chem. Eng. Sci.,* **7:** 121, 132.

Avriel, M., M. J. Rijckaert, and D. J. Wilde (1973): "Optimization and Design," Prentice-Hall, Englewood Cliffs, N.J.

Bauman, H. C. (1964): "Fundamentals of Cost Engineering in the Chemical Industry," Reinhold, New York.

Beightler, C. S., and D. T. Phillips (1976): "Applied Geometric Programming," Wiley, New York.

Bellman, R. (1957): "Dynamic Programming," Princeton University Press, Princeton, N.J.

Benedict, M., G. B. Webb, and L. C. Rubin (1940): *J. Chem. Phys.,* **8:** 334.

——, ——, and —— (1942): *J. Chem. Phys.,* **10:** 747.

Bojnowski, J. H., J. W. Crandall, and R. M. Hoffman (1975): *Chem. Eng. Prog.,* **71**(10): 50.

Boundy, R. H., and R. F. Boyer (eds.) (1952): "Styrene: Its Polymers, Copolymers and Derivatives," Reinhold, New York.

Bridgewater, A. V., (1968): *The Chem. Eng.,* no. 217, p. CE75.

Bross, I. D. F. (1953): "Design for Decision," Macmillan, New York.

Brownstein, A. M. (1975): *Chem. Eng. Prog.,* **71**(9): 72.

Burford, C. L., H. D. Liles, and R. D. Dryden (1977): *Chem. Tech.,* **7**(2): 129.

Calderbank, P. H. (1953): *Chem. Eng. Prog.,* **49:** 585.

Chenery, H. B., and T. Watanabe (1958): *Econometrica* (Amsterdam), **26:** 487 (October).

Chinn, J. S., and W. A. Cuddy (1971): *Chem. Eng. Prog.,* **67**(6): 17.

Clausen, C. A., and G. C. Mattsen (1973): Case Studies in Industrial Chemistry, Styrene; Dept. of Chemistry, Florida Technological University, Orlando, Fla.

Coleman, J. R., and R. York (1964): *Ind. Eng. Chem.,* **56**(1): 28.

Cox, N. D. (1976): *Chem. Eng. Prog.*, **72**(6): 77.

Crowe, C. M., A. E. Hamielic, T. W. Hoffman, A. I. Johnson, D. R. Woods, and P. T. Shannon (1971): "Chemical Plant Simulation," Prentice-Hall, Englewood Cliffs, N.J.

Dawson, S. H. (1961): Styrene and Polystyrene, in H. Steiner (ed.), "Introduction to Petroleum Chemicals," Pergamon, Oxford.

Denbigh, K. G. (1944): *Trans. Faraday Soc.*, **40**: 352.

Dluzniewski, J. H., and S. B. Adler (1972): *Inst. Chem. Eng. Symp. Ser.*, **35**(4): 21.

Dodge, B. F. (1944): "Chemical Engineering Thermodynamics," McGraw-Hill, New York.

Forman, J. C., and G. Thodos (1958): *AIChE J.*, **4**: 356.

Forney, R. C., and J. M. Smith (1951): *Ind. Eng. Chem.*, **43**: 1841.

Fredenslund, A., J. Gmehling, and P. Rasmussen (1977): "Vapor-Liquid Equilibrium Using UNIFAC," Elsevier, Amsterdam.

Frith, K. M. (1972): *Chem. Eng.*, **79**(4): 72.

Generoso, E. I., and L. B. Hitchcock (1968): *Ind. Eng. Chem.*, **60**: 15.

Goldstein, R. F. (1958): "Petroleum Chemicals Industry," 2d ed., Spon, London.

Gregg, J. V., C. H. Hossel, and J. T. Richardson (1964): "Mathematical Trend Curves: An Aid to Forecasting," Oliver & Boyd, Edinburgh.

Guthrie, K. M. (1970): *Chem. Eng.* **77**(12): 140.

—— (1974): "Process Plant Estimating, Evaluation and Control," Craftsman, Solana Beach, Calif.

Hayden, J. G., and J. P. O'Connell (1975): *Ind. Eng. Chem. Process Des. Dev. Q.*, **14**: 209.

Hendry, J. E., D. F. Rudd, and J. D. Seader (1973): *AIChE J.*, **19**: 1.

Hirsch, J. H., and E. M. Glazier (1960): *Chem. Eng. Prog.*, **56**(12): 37.

Hirschmann, W. B. (1964): *Harvard Bus. Rev.*, **42**(1): 25.

Hougen, O. A., K. M. Watson, and R. A. Ragatz (1959): "Chemical Process Principles," pt. 2, 2d ed., Wiley, New York.

Jelen, E. C. (ed.) (1970): "Cost and Optimization Engineering," McGraw-Hill, New York.

Johnson, A. I., and C. J. Huang (1955): *Can. J. Technol.*, **33**: 421.

Kehat, E., and M. Shacham (1973): *Process Technol.*, **18**(3): 115.

King, C. J. (1971): "Separation Processes," McGraw-Hill, New York.

—— (1974): *AIChE Monogr. Ser.*, vol. 70, no. 8.

——, D. W. Gantz, and F. J. Barnes (1972): *Ind. Eng. Chem. Process Des. Dev. Qt.*, **11**: 271.

Kirk, R. E., and D. F. Othmer (1970): "Encyclopedia of Chemical Technology," 2d ed., vol. 21, Wiley-Interscience, New York.

Kladko, M. (1971): *CHEMTECH*, **1**(3): 141.

Kordbachen, R., and Chi Tien (1959): *Can. J. Chem. Eng.*, **37**: 162.

Lang, H. J. (1947): *Chem. Eng.*, **54**(10): 117.

—— (1948): *Chem. Eng.*, **55**(6): 112.

Leibson, I., and C. A. Trischman, Jr. (1971): *Chem. Eng.*, Sept. 6, p. 86; Oct. 4, p. 85; Nov. 1, p. 78; Dec. 13, p. 97.

Liebskind, D. (1973): *CHEMTECH*, **3**: 543.

Luce, R. D., and A. Raiffa (1957): "Games and Decisions," Wiley, New York.

Lyderson, A. L., R. A. Greenkorn, and O. A. Hougen (1955): *Univ. Wis. Eng. Exp. Stn. Rep.* 4 (October).

MacMullin, R. B. (1948): *Chem. Eng. Prog.*, **44**: 183.

—— and M. Weber, Jr. (1935): *Trans. AIChE*, **31**: 409.

Malloy, J. B. (1974): *Chem. Eng. Prog.*, **70**(9): 77.

Masso, A. H., and D. F. Rudd (1969): *AIChE J.*, **15**: 10.

May, D., and D. F. Rudd (1976): *Chem. Eng. Sci.*, **31**: 59.

Melnechuk, T. (1963): *Intl. Sci. Technol.*, February, p. 26.

Motard, R. L., M. Shacham, and E. M. Rosen (1975): *AIChE J.*, **21**: 47.

Nathanson, D. M. (1972): *Chem. Eng. Prog.*, **68**(11): 89.

Natta, G. (1955): in P. H. Emmett, (ed.), "Catalysis" chapter 8, vol. III, Reinhold, New York.

Newton, R. H., and B. F. Dodge (1935): *Ind. Eng. Chem.*, **27**: 577.

Norman, R. L. (1965): *AIChE J.*, **11**: 450.

Obert, E. F. (1960): "Concepts of Thermodynamics," McGraw-Hill, New York.

O'Connell, F. P. (1962): *Chem. Eng.*, **69**(4): 150.

Oliver, R. C., S. E. Stephanou, and R. W. Baier (1962): *Chem. Eng.* **69**(4): 121.

Perry, R. H., and C. H. Chilton (eds.) (1973) "Chemical Engineers' Handbook," 5th ed., McGraw-Hill, New York.

Peters, M. S., and K. D. Timmerhaus (1968): "Plant Design and Economics for Chemical Engineers," 2d ed., McGraw-Hill, New York.

Pings, C. J. (1965): *Ind. Eng. Chem. Fundam. Q.*, **4**: 260.

Polya, G. (1945): "How to Solve It: A New Aspect of Mathematical Method," Princeton University Press, Princeton, N.J.

Quale, O. R. (1953): *Chem. Rev.*, **53**: 439.

Reid, R. C., and T. K. Sherwood (1958): "The Properties of Gases and Liquids," McGraw-Hill, New York.

—— and —— (1966): "The Properties of Gas and Liquids," 2d ed., McGraw-Hill, New York.

——, J. M. Prausnitz, and T. K. Sherwood (1977): "The Properties of Gases and Liquids," 3d ed., McGraw-Hill, New York.

Reidel, L. (1954): *Chem. Ing. Tech.*, **26**: 83.

Rihani, D. N., and L. K. Doraiswamy (1965): *Ind. Eng. Chem. Fundam.*, **4**: 17.

Robbins, L. A., and C. L. Kingrea (1962): *Am. Petrol. Inst. Div. Refining*, **42**: 111.

Rose, L. M. (1976): "Engineering Investment Decisions: Planning under Uncertainty," Elsevier, Amsterdam.

—— (1977): *Eng. Process Econ.*, **2**: 17.

—— (1978): *1st Mediterr. Cong. Chem. Eng.*, Nov. 21–24.

——, J. Myhre, and O. H. D. Walter (1974): *CHEMTECH*, **4**: 494.

Royal, M. J., and N. M. Nimmo (1969): *Hydrocarbon Process.*, March, p. 147.

Rudd, D. F. (1962): *Ind. Eng. Chem. Fundam.*, **1**: 138.

—— (1968): *AIChE J.*, **14**: 343.

—— and C. C. Watson (1968): "Strategy of Process Engineering," Wiley, New York.

Salt, F. E. (1961): Products from Ethylene and Propylene in H. Steiner (ed.), "Introduction to Petroleum Chemicals," Pergamon, London.

Siddall, J. N. (1972): "Analytical Decision-Making in Engineering Design," Prentice-Hall, Englewood Cliffs, N.J.

Siirola, J. J., and D. F. Rudd (1971): *Ind. Eng. Chem. Fundam.*, **10**: 353.

Stephenson, R. M. (1966): "Introduction to the Chemical Process Industries," Van Nostrand Reinhold, New York.

Stiel, L. I., and G. Thodos (1964): *AIChE J.*, **10**: 275.

Stobaugh, R. B. (1964): *Chem. Eng. Prog.*, **60**(12): 13.

Strelzoff, S. (1970): *Chem. Eng. Prog. Symp. Ser.*, **66**(98), 54.

Stull, D. A. (1952): "Styrene: Its Polymers, Copolymers and Derivatives," chap. 3, Reinhold, New York.

Stull, J. K. (1968): "Industrial Organic Chemistry," chap. 2, Prentice-Hall, Englewood Cliffs, N.J.

Twaddle, W. W., and J. B. Malloy (1966): *Chem. Eng. Prog.*, **62**(7): 90.

Van Arnum, K. J. (1964): *Chem. Eng. Prog.*, **60**(12): 18.

Van Krevelin, D. W., and H. A. G. Chermin (1951): *Chem. Eng. Sci.*, **1**: 66.

Van Ness, H. C. (1964): "Classical Thermodynamics of Non-Electrolyte Solutions," Pergamon, London.

Verma, K. K., and L. K. Doraiswamy (1965): *Ind. Eng. Chem. Fundam.*, **4**: 389.

Watson, F. A., and F. A. Holland (1977): *Eng. Process Econ.*, **2**: 207.

Wei, J., T. W. F. Russell, and M. W. Swartzlander (1979): "Structure of the Chemical Processing Industries," McGraw-Hill, New York.

Wells, G. L. (1973): "Process Engineering with Economic Objective," Leonard Hill, Aylesbury.

Wenner, R. R., and E. C. Dybdal (1948): *Chem. Eng. Prog.*, **44**: 275.

Wessel, H. E. (1952): *Chem. Eng.*, **59**(7): 209.

White, W. B., S. M. Johnson, and G. P. Dantzig (1958): *J. Chem. Phys.*, **28**: 751.

Wilde, D. J. (1964): "Optimum Seeking Methods," Prentice-Hall, Englewood Cliffs, N.J.
——— and C. S. Beightler (1967): "Foundations of Optimization," Prentice-Hall, Englewood Cliffs, N.J.
Williams, R. (1947): *Chem. Eng.*, **54**(6): 124.
Wilsher, P. (1970): "The Pound in Your Pocket, 1870–1970," Cassell, London.
Woods, M. F., and G. B. Davies (1972): *Inst. Chem. Eng. Symp. Ser. 35*, vol. I; p. 3.
Wroth, W. F. (1960): *Chem. Eng.*, **67**(10): 204.
Yen, L. C., and S. S. Woods (1966): *AIChE J.*, **12**(7): 95.

INDEX